D0488135

STATISTICS
With a View Toward Applications

Leo Breiman
Private Consultant
Los Angeles

STATISTICS

With a View Toward Applications

Houghton Mifflin Company · Boston
Atlanta
Dallas
Geneva, Illinois
Hopewell, New Jersey
Palo Alto

Printed in the United States of America

Library of Congress Catalog Card Number: 72-3131

ISBN: 0-395-04232-1

To my brothers and sisters and our children

Preface

Think of a final examination like this: A student is placed in an unknown, unformulated, statistical situation. For example, he is given some data, told how they were derived, and, vaguely, what he is to determine. From then on, he is on his own. He must go through a number of steps in an attempt to understand what questions are to be answered. Then he must construct a model and use a statistical technique, say, an estimate, test, or confidence region, to arrive at some answers. Now he has to confront further questions: Why did you use the model you selected? What assumptions were used? How do you know they are reasonable? Can you check your model for fit? What happens if the assumptions are violated? Are there other, better models? How good is your statistical method compared to other methods? How accurate is it? How good are your answers?

This text is aimed at preparing students for such a hypothetical final examination. A student should emerge from a first course in applied statistics with a basic knowledge of

(1) the construction and tailoring of statistical models to experimental situations,

(2) the use and performance of a few simple, widely applicable statistical methods.

This book is intended as a text for a junior, senior, or early graduate course in applied statistics for students in quantitative fields, such as engineering and physical sciences. The prerequisites are a knowledge of calculus and a first course in probability theory. The first five chapters of *Probability and Stochastic Processes: With a View Toward Applications* (Houghton Mifflin Company, 1969), the companion volume to this book, give a sufficient background.

The approach taken in both of these books is based on the belief that students learn most profoundly from examples and problems. No theory is introduced unless the way has been paved for it by examples which illustrate both the point and the difficulties. Through the examples both models and statistical techniques are explored in a variety of backgrounds. The numerous examples impart a leisurely character. It is important not to rush through the examples to get to the theory. The examples give the student valuable training in the use of statistics.

The theory is stated as precisely as possible in theorems, propositions, and corollaries. A proof is given only if it is simple and illuminating. The emphasis is on understanding the theory and knowing how to apply it. The growth of the student as a user of statistics is developed by working problems. Many good, solid problems are included in the text. The class time taken in discussing them is time well spent.

Following is a brief outline of the contents of the book, which shows how its purpose has shaped its content:

In Chapter 1 the concept of a general statistical model is formulated.

In Chapter 2 a variety of parametric families of distributions are defined and contrasted. The chapter gives the student a "dictionary" of models to practice on.

In Chapters 3, 4, and 5 estimation, confidence regions, and hypothesis testing for parametric models are studied, starting with general definitions and working toward specific techniques.

The development of these first chapters differs from the usual one. Many concepts, such as uniformly most powerful tests and sufficient statistics, are not included because of their limited applicability. In estimation, squared error loss is used to compare estimates. Asymptotic efficiency is defined, and the large sample properties and distribution of maximum likelihood estimates are studied. The latter makes possible the derivation of the simple, large sample confidence intervals based on maximum likelihood estimates. Exact confidence intervals are computed in examples and compared to the large sample results.

In hypothesis testing the Newman–Pearson Lemma forms the gateway for the introduction of likelihood ratio tests. The detection capability of a test is defined to give a measure of how far the state of nature has to be from the null hypothesis in order for the test to reject with high probability. Finally, the large sample properties of the likelihood ratio tests are developed.

Familiar estimates and tests, such as the sample mean and variance, the t test, and the F test, do not appear as isolated instances. Rather, they appear as special examples of maximum likelihood estimates or likelihood ratio tests.

In Chapter 6 a detailed study is made of goodness of fit tests, primarily the chi square and Kolmogorov–Smirnov tests. The measure of detection capability is used to point up the limited ability of these tests to discriminate between the different underlying distributions. The chapter sets the stage for an examination of what happens and what can be done if a parametric model is suspect.

In Chapter 7 the trade-off between robustness and efficiency of estimates is examined in examples. A number of familiar nonparametric estimates and confidence intervals are defined and studied.

In Chapters 8 and 9 the most commonly used nonparametric tests are explored. Specifically, these are the sign and median tests, the Wilcoxon rank tests, the Kolmogorov–Smirnov two sample tests, the Spearman and Kendall rank correlation tests, the Kruskal–Wallace rank test, and the many faces of the chi square test. The concept of detection capability allows a simple definition of asymptotic relative efficiency, which we use to compare a number of the above nonparametric tests to their parametric competitors.

In Chapter 10 a fairly theoretical treatment is made of analysis of variance and regression. Vector space and matrix notation are used to build a simple, unified approach from which the distributional properties of the various statistics follow. Regression and the various analyses of variance layouts are treated as special cases. One eye is always kept on the problems of robustness and goodness of fit.

The book contains more material than can be covered in one quarter; perhaps more than can be covered in one semester or even in two quarters. If time must be saved, cover less ground instead of rushing through with less depth and detail. For whatever material is covered, treat the relevant examples and problems with care. I know of no better way of preparing the students for the hypothetical examination posed above.

The sections starred in the table of contents deal with more specialized topics and could be omitted. If more drastic cutting is necessary, cut one or more of the last four chapters. Doing Chapter 10 properly is time consuming. I would omit Chapters 7 and 10.

The probabilty notation here (see pp. 393–394) is mostly standard and is defined in *Probability and Stochastic Processes* (Houghton, 1969).

I am indebted to the Literary Executor of the late Sir Ronald A. Fisher, F.R.S., to Dr. Frank Yates, F.R.S., and to Oliver and Boyd, Edinburgh, for permission to reprint Tables . . . from their book *Statistical Tables for Biological, Agricultural and Medical Research*.

The final version of the book bears a strong imprint of Herman Chernoff's hand. I am deeply indebted to him for his conscientious and creative editorship. Mrs. Ruth Goldstein typed this manuscript, as she has my previous ones, with efficiency, dispatch, and very few errors.

Leo Breiman

Contents

1 The Background of Statistics

Introduction

The main thing to learn about statistics is what is sensible and honest and possible. If you are a statistician, people may come to you with all sorts of vaguely formulated questions and hopes. A good part of your work seemingly has nothing to do with mathematics. For example, you must find out exactly what it is the client wants to know, what data is available, and what additional data are necessary.

The construction of a statistical model forms the bridge from the physical context into statistical theory. It needs to be emphasized that the construction of an adequate model is nine-tenths of the statistician's battle. All the background information he can get in his questioning of how the data were gathered and what reasonable assumptions can be made concerning the physical characteristics of the situation are summarized in his model.

The best way to illustrate this is to work from examples.

The First Example

You have "total operating time until failure" data for 1000 A-type transistors. Denote these times by $t_1, \ldots, t_{1000}$ and call $\mathbf{t} = (t_1, \ldots, t_{1000})$ the *data vector*. You have been asked to use $\mathbf{t}$ to find out something about the failure time distribution of A-type transistors. But what does failure time distribution mean? Where does randomness come in? Where does probability come in?

We look at the data vector $\mathbf{t}$ as the outcome of some random experiment. If we took another 1000 transistors and measured failure times, we would get a different data vector. So we formulate the problem: Construct a reasonable probability model.

Think of putting all A-type transistors, past, present, and future, into an enormous pot. Now imagine an equally enormous experiment in which every transistor in this pot is tested until failure. To every time interval I assign the number $P(I)$, defined as the proportion of failure times that fall into the interval I. Call this

the underlying failure time distribution.

We can conceptualize the way we got our data vector as follows: We draw the first transistor "at random" from the enormous pot (that is, all transistors are equally likely to be drawn) and measure its failure time. The probability that this failure time is in the interval I is $P(I)$. In more familiar words, the first failure time is the outcome of a random variable T_1 with distribution equal to the underlying failure time distribution:

$$P(\mathsf{T}_1 \in I) = P(I).$$

Now we assume that the second transistor is drawn independently of the first but also "at random," and continue this way until 1000 transistors have been selected and tested.

We have obtained a model:

> *The data vector* **t** *is the outcome of a random vector* $\mathbf{T} = (\mathsf{T}_1, \ldots, \mathsf{T}_{1000})$, *where* $\mathsf{T}_1, \ldots, \mathsf{T}_{1000}$ *are independent with common distribution.*

The failure time distribution completely specifies the distribution of **T**.

Obviously, you need to do some questioning to see if this model is reasonable. Were all transistors tested under the same conditions? Were the selected transistors typical of A-type transistors? If, for example, they were all selected from a single day's run, do the parameters of the transistor vary significantly from Monday's production to Tuesday's production? The independence of the tests seems like a reasonable assumption. There is no reason to believe that knowledge of when transistor No. 1 fails has any effect on the probability distribution of the failure times of the other transistors. Their distributions remain $P(dt)$ no matter when No. 1 goes out. However, in many experimental situations, the assumption of independence has to be closely questioned. Assuming all of our questions have been satisfactorily resolved, we accept the model and refer to the distribution of **T** as the *underlying distribution.*

Now let us look at the problem in a different way. We do not know the underlying failure time distribution, since we want to sell the A-type transistors, not destroy them all. Thus the problem is to make some deductions concerning this distribution from the given data. So actually the model we have consists of 1000 independent identically distributed random variables with an unknown distribution $P(dt)$.

> *This is the difference between a probabilistic model and a statistical model. In a probabilistic model, the probability is completely specified. In a statistical model, the exact probability distribution is unknown. What are specified are some general characteristics of the probability distribution.*

In our current example, the probability distribution on the outcome space, or equivalently, on the random vector **T**, is specified to be such that $\mathsf{T}_1, \ldots, \mathsf{T}_{1000}$ are independent and identically distributed. The unknown is the one-dimensional distribution $P(dt)$.

The common feature of all statistical problems is the use of sample data to make inferences concerning the unknown underlying distribution from which the data vector was drawn. In this example, the problem is to use $\mathbf{t} = (t_1, \ldots, t_{1000})$ to get some idea of the underlying failure time distribution $P(dt)$. This too reverses the order of things in probability theory. There the problem was:

> *Given the probability distribution of a group of random variables, what can be said about the properties of their outcomes?*

For example, if $\mathsf{Y}_1, \mathsf{Y}_2, \ldots$ are independent and identically distributed, then the law of large numbers applied to the variables

$$\mathsf{X}_k = \begin{cases} 1, & \text{if } \mathsf{Y}_k \in I \\ 0, & \text{otherwise} \end{cases}$$

leads to the conclusion that for large n the proportion of Y_k, $k = 1, \ldots, n$, falling in the interval I is usually close to the probability $P(\mathsf{Y}_1 \in I)$. In probability theory we assumed $P(I)$ was known and used this to deduce, for instance, that in tossing a fair die, the proportion of the times that Face 1 comes up converges to one-sixth as the sample size increases. But in our statistical problem, what is known is a group of 1000 outcomes and what is unknown is the distribution from which they were selected. The result above, based on the law of large numbers, still holds. But now we reverse direction and instead of thinking: I can use my knowledge of $P(dt)$ to deduce how the outcomes $t_1, \ldots, t_{1000}$ will be distributed, we think statistically! That is, we think: I can use any knowledge of how my data $t_1, \ldots, t_{1000}$ are distributed to form some conclusions about the unknown distribution from which they were drawn. Specifically, if I compute the proportion of my sample falling in the interval I, then this should be a reasonably good estimate of the probability $P(I)$.

This last remark suggests ways of using our data to get an idea of what the underlying distribution looks like. For instance, define the sample cumulative distribution function $\hat{F}(t)$ to be the proportion of the total measured failure times that are less than t. Graph $\hat{F}(t)$ (see Figure 1.1). Since $\hat{F}(t)$ should be a good estimate of $P([0, t))$, this graph should look something like the graph of the cumulative distribution function for the underlying failure time distribution. Another approach is to divide the t-axis into subintervals $\Delta t_1, \Delta t_2, \ldots$ of equal length and graph the function whose value in the ith subinterval Δt_i is the proportion of the

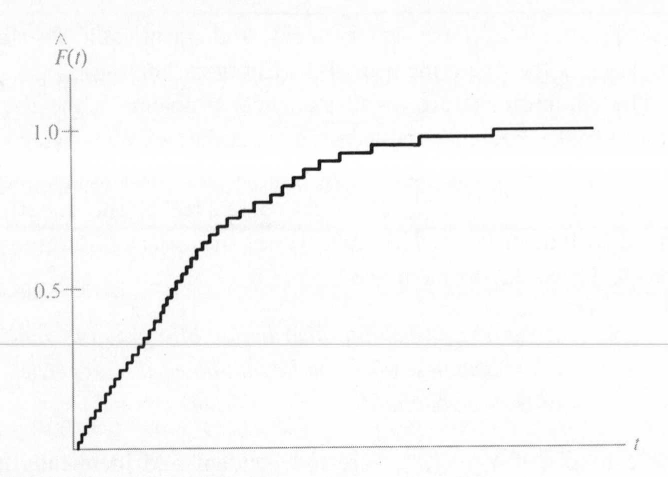

Figure 1.1 $\hat{F}(t)$ *with 40 outcomes*

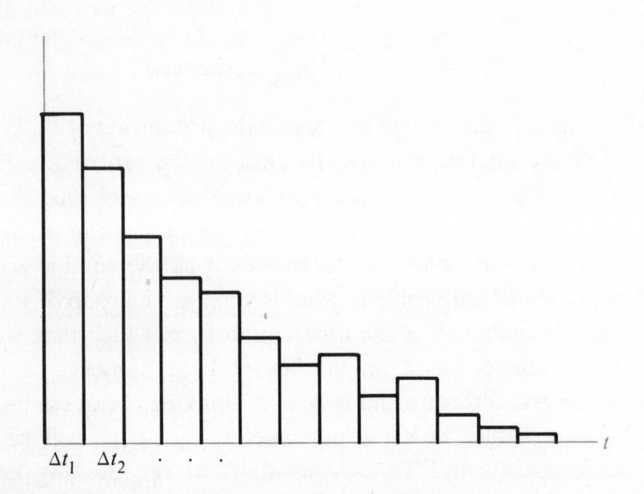

Figure 1.2 Normalized histogram. Equal width intervals Δt, $\Delta t_2, \ldots$

recorded failure times with values in Δt_i (see Figure 1.2). This graph is called the *normalized histogram* of the data.

The proportion of outcomes in Δt_i is usually close to $P(\Delta t_i)$. If the underlying distribution has a density $f(t)$, then

$$P(\Delta t_i) \simeq f(t_i)\, \Delta t,$$

where t_i is the endpoint of the ith interval and Δt is the interval length. Putting these two together, we conclude that the proportion of failure times in Δt nearly equals $f(t)\, \Delta t$. Therefore, the histogram, with the appropriate rescaling, should resemble the graph of the density $f(t)$.

So far, so good. We have managed to make a start toward using our data **t** to get a rough picture of the underlying distribution. Obviously, we are trying to get a lot out of a little. From the finite group of outcomes **t**, we are trying to estimate the entire distribution of the failure time variable T. Our results cannot be precise—it is nonsensical to claim that the sample cumulative distribution function is the cumulative distribution function of the underlying failure time distribution. At best, we can say that it is close. Then the question becomes: How close? This will depend on the sample size. A thousand observations ought to give fairly good accuracy. But with ten observations, $\hat{F}(t)$ can give only an extremely crude approximation to the cumulative distribution function of the underlying distribution.

In the preceding discussion we have tacitly assumed that we have no prior knowledge about the distribution of T. We reflected this in our modeling by taking the set of possible underlying distributions for T to be all distributions with nonnegative outcomes. Any information regarding the characteristics of the distribution of T should be used. Generally, the smaller, more restricted the class of possible underlying distributions, the more accurate the estimates. For example, suppose we know from physical considerations that the transistors deteriorate very little with age. That is, a transistor that has been running for time t will have very nearly the same failure probability during the following hour as it had during its first hour of operation. Then the failure times have an exponential distribution: $P(\mathsf{T} > t)$ is of the form $e^{-\alpha t}$. Our model now is that the failure data $t_1, \ldots, t_{1000}$ are the outcomes of independent identically distributed random variables having some exponential distribution. The only unknown is the value of the parameter α. If we want to pinpoint the failure time distribution, all we need to do is to estimate from the data the real number α. The nature of the problem has been entirely changed by the specification that the underlying distribution is exponential. How do we estimate α? The basic idea is very simple. Here is one way: If the underlying distribution has parameter α, then $\hat{F}(t)$ should approximate one of the cumulative distribution functions

$$F(t) = 1 - e^{-\alpha t}.$$

Estimate α by finding which one of the functions $\{1 - e^{-\alpha' t}\}, 0 < \alpha' < \infty$, "most closely" approximates $\hat{F}(t)$.

Here is another way: If T has an exponential distribution with parameter α, then $E\mathsf{T} = 1/\alpha$, where $E\mathsf{T}$ is the expectation of T. If $\mathsf{T}_1, \ldots, \mathsf{T}_n$ are independent identically distributed random variables, then

$$\frac{\mathsf{T}_1 + \cdots + \mathsf{T}_n}{n}$$

is usually close to $E\mathrm{T}_1$. Therefore, unless some unhappy fluke has occurred, if $t_1, \ldots, t_{1000}$ are the observed failure times, then

$$\bar{t} = \frac{t_1 + \cdots + t_{1000}}{1000} \simeq \frac{1}{\alpha},$$

so we estimate α as $1/\bar{t}$.

These two examples illustrate that we can have many different and sensible ways of estimating α, all based on selecting the exponential distribution that in some way gives the "best match" to the observed data. Of these many ways which one shall we use?

A closely associated question is: How accurate is our estimate of α? What is meant by *accurate*; what measure of accuracy shall we use? If we could answer this last question, we could get an idea of how to select an estimate of α. Some methods of estimation might turn out to be quite inaccurate by our measure—these we would discard. In fact, what we would like to find is the "most accurate" estimation method, if one exists.

The accuracy of our estimate will depend on the size of the sample. For 1000 observed failure times we would expect fairly accurate estimates of α. But if we have observed only ten failure times, then the accuracy will be low—we do not have a large enough sample to "average out" the randomness.

Now let's back off for a minute. Suppose the underlying failure time distribution is not really exponential, but that its density looks like Figure 1.3.

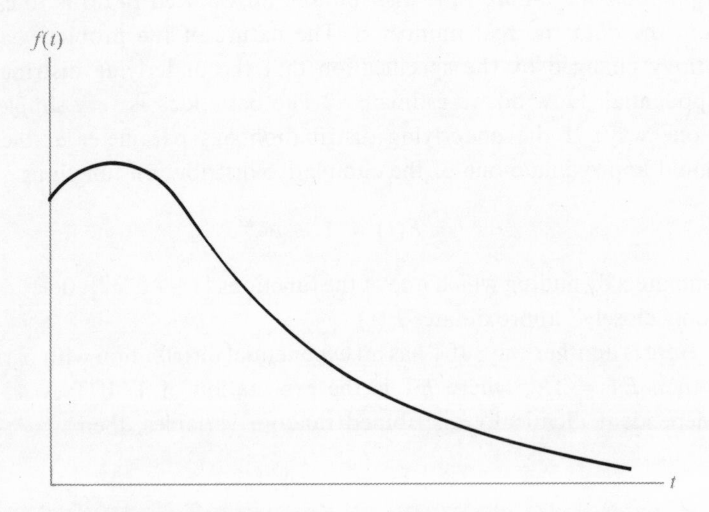

***Figure 1.3** Failure time distribution density*

If we estimate things based on the assumption that this distribution is exponential, what will happen? Suppose we know only that the distribution is "nearly" exponential in shape—can we use this knowledge and still avoid the extremes of either erroneously restricting the underlying distribution to be exponential or throwing away the information by taking the set of possible underlying distributions to be all distributions with nonnegative outcomes?

To a great extent, answers to these questions depend on what characteristics of the underlying distribution you want to estimate. If you are most interested in estimating the half-lifetime $t_{1/2}$ defined by

$$P(\mathsf{T} \leq t_{1/2}) = P(\mathsf{T} \geq t_{1/2}) = \tfrac{1}{2},$$

then a common sense estimate for $t_{1/2}$ is the 500th largest measured failure time. This estimate works fairly well for a variety of differently shaped underlying distributions.

But if what we want to get is some closed form expression for the density of T, then it would be better to work with some two- or three-parameter family of distributions that includes the exponential family and many other similar distributions and try to estimate parameters.

This leads us to the important question of whether we can detect departures from an exponential distribution by looking at the data. With 1000 observations, if the histogram of the sample looked like Figure 1.4, we would be very suspicious of the assumption of exponential distribution. In other words, with large sample size, we can detect small departures from an exponential density. But if we had only ten observations, the

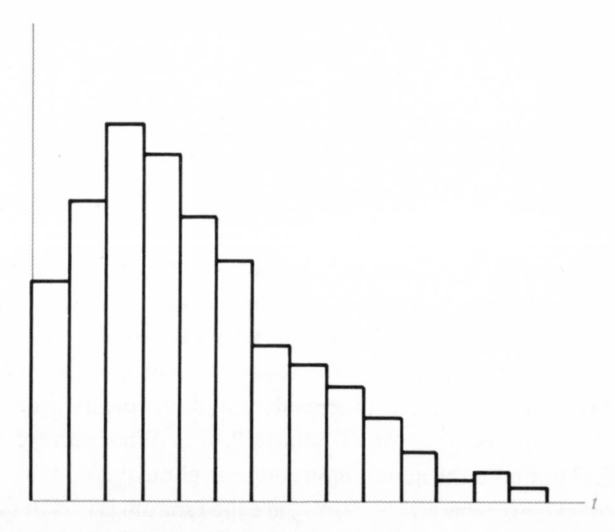

***Figure 1.4** Possible histogram plot*

underlying distribution would have to be drastically nonexponential for us to be able to detect the difference.

When we ask if the data could reasonably have come from an underlying exponential distribution, the data itself are addressed in search of an answer. This is, therefore, a statistical problem. We want to use the given data to form some conclusion regarding the unknown distribution.

Problem 1 It is desired to estimate the half-life of a given radioactive material by counting the numbers of disintegrations during each of 250 successive time intervals of equal length. Construct a reasonable statistical model.

Another Example

Ten repeated measurements $x_1, \ldots, x_{10}$ are made of the molecular weight of a chemical compound. What is the "true" molecular weight, and what is the error in our estimate?

There is a standard answer to the above questions that appears almost universally: The best way to get the "true" molecular weight is to take the average

$$\bar{x} = \frac{x_1 + \cdots + x_{10}}{10}$$

of all the measurements. To get an error estimate, compute the average squared error

$$\overline{e^2} = \tfrac{1}{10} \sum_{x=1}^{10} (x_k - \bar{x})^2.$$

Then state the results of your experiment as: The true molecular weight is

$$\bar{x} \pm 2\sqrt{\overline{e^2}}$$

(or $\bar{x} \pm \sqrt{\overline{e^2}}$ in some sources). But the question is: Why use this recipe? What assumptions underlie it? To understand any statistical recipe, you have to understand the model from which it was derived. So, let us try to construct a sensible model. We usually accept on faith that $x_1, \ldots, x_{10}$ are the outcomes of ten independent and identically distributed random variables with common distribution $P_X(dx)$. What can we assume about $P_X(dx)$? Suppose the actual molecular weight is m and that $X_i = m + Y_i$, $i = 1, \ldots, 10$, where $Y_1, \ldots, Y_{10}$ are the random errors in the experiment. Denote the common distribution of the $Y_1, \ldots, Y_{10}$ by $P_Y(dy)$. What can we specify as reasonable characteristics of the error distribution? A usual assumption is that there is no "systematic bias." It is a bit difficult to interpret this condition quantitatively. Roughly, it means that the errors are not consistently positive or consistently negative. If there is an appreciable undetected systematic bias in the errors, then there is no way of estimating m accurately—we will either overestimate or underestimate. The absence of systematic bias may be formulated by assuming that the errors are positive as frequently as they are negative. That is,

1.1 $$P(Y_i > 0) = P(Y_i < 0)$$

An alternate formulation is that the errors will tend to cancel each other out. More precisely, assume the average of the errors

$$\bar{y} = \frac{y_1 + \cdots + y_{10}}{10}$$

will be small compared to the size of the individual errors $y_1, \ldots, y_{10}$. By the law of large numbers, as $n \to \infty$,

$$\frac{Y_1 + \cdots + Y_n}{n} \to EY_1.$$

Therefore, imposing the condition that the average of the errors over many repetitions is small is equivalent to assuming that the underlying distribution has mean zero:

1.2 $$EY = \int y P_Y(dy) = 0.$$

Sometimes it is assumed that the distribution $P_Y(dy)$ is symmetric about 0. This means that the probability of a positive error in any range $[y_1, y_2]$, equals the probability of a negative error in the range $[-y_2, -y_1]$. If $P_Y(dy)$ has a density $f(y)$, an equivalent condition is that $f(y)$ is a function symmetric about 0. If the error distribution is symmetric about 0, then it satisfies conditions 1.1 and 1.2 proposed above as interpretations of lack of a systematic bias. The assumption of symmetry is a much more restrictive interpretation of an absence of bias. Still, it appears to be a not unreasonable model for many experiments.

However, the model in common use makes the strongest possible assumption: that the errors are normally distributed with mean 0 and unknown variance σ^2. Any normal distribution is symmetric about its mean. So this normal model forces on the possible underlying distributions a restriction even more severe than symmetry. One can give some sort of rationale for assuming normality. If the errors are the sums of many small errors coming from independent sources, then the central limit theorem can be called upon to proclaim the normality of the sum. But this is only a cut above a hopeful appeal to the ghost of Laplace. An equally reasonable argument is that any distribution with a smooth density, symmetric around zero and with a single maximum at 0 can be approximated in its central section by one of the $N(0, \sigma^2)$ distributions. But this does not eliminate the possibility that the larger errors may have a distribution that is totally unlike a normal distribution.

If we accept the validity of this last model, then the use of $\bar{x}$ to estimate m and $\sqrt{e^2}$ to gauge the accuracy of the estimate makes sense. If the Y_i are $N(0, \sigma^2)$, then the X_i are $N(m, \sigma^2)$, and the problem is reduced to estimating the two parameters of a normal distribution.

If the error characteristics of our measuring techniques are known to be closely normal by the analysis of many hundreds of past measurements, then we have an acceptable basis for assuming a normal model. But on the basis of ten measurements alone, there is no way of telling whether the distribution $P_Y(dy)$ is normal or departs drastically from normality.

If, after these warnings, you decide that the assumption of normality is dangerous, what then? A much safer model is letting the possible error distributions be the symmetric distributions. Now the problem has changed character—instead of estimating the two parameters of a normal distribution, all we know is that $x_1, \ldots, x_{10}$ are the outcomes of random variables symmetrically distributed around some point m. How can we use the data to estimate m and gauge the accuracy of the estimate?

One last disturbing but educational point is relevant to this example. Independence of successive measurements is usually casually accepted, without causing even a trace of the disturbance that assumptions such as that of normality cause. However, the famous statistician Student in 1927 published calculations on some industrial data obtained from 100 routine analyses made daily on samples from the same batch of thoroughly mixed material. He found a degree of dependence between successive measurements large enough to cause seriously misleading results if the methods derived from the model under the assumption of independence were uncritically applied. The moral is never to assume independence without some qualms and due consideration.

In the current state of statistics, the dangers caused by departures from normality can, to a great extent, be avoided by the use of more general models, leading to methods that are valid for a wide range of possible underlying distributions. But giving up the assumption of independence or of identical distributions may lead to models so general that nothing workable comes out of them. See Problem 3 for an exaggerated illustration.

Problem 2 If the errors are *median unbiased* in the sense that $P(\mathsf{Y} > 0) = P(\mathsf{Y} < 0)$, then what is a sensible way of estimating m?

Problem 3 As a model for 10 measurements $\mathbf{x} = (x_1, \ldots, x_{10})$, because of the possibilities of dependence, non-normality, etc., you decide to let $\mathbf{x}$ be the outcome of a random vector $\mathbf{X} = (\mathsf{X}_1, \ldots, \mathsf{X}_{10})$, where the joint distribution of $\mathsf{X}_1, \ldots, \mathsf{X}_{10}$ can be anything. What can be inferred from the data point $\mathbf{x}$ about the underlying distribution of $\mathbf{X}$? What distribution for $\mathbf{X}$ makes the appearance of the data point $\mathbf{x}$ most likely? Discuss why this latter distribution is not a realistic candidate for the actual distribution of $\mathbf{X}$.

The Third Example

To trace the effects of tribal structure on intelligence, two groups of 20 children of about the same age were selected by availability from two primitive African villages with widely disparate tribal cultures. Both groups were given a culture-free I.Q. test administered under carefully controlled conditions. The two groups of test scores are denoted by

$$\mathbf{x} = (x_1, \ldots, x_{20})$$
$$\mathbf{y} = (y_1, \ldots, y_{20}).$$

The problem is to use this data to decide if there are any real differences between the two groups. As before, the important problem is to find an appropriate model. Suppose we measure all test scores for all children of similar age of the first tribe. These scores would have some distribution P_1, where $P_1(I)$ is the proportion of scores in the interval I. We can think of $x_1, \ldots, x_{20}$ as twenty numbers we pulled at random out of a hatful of numbers whose distribution was given by $P_1(I)$. Thus $x_1, \ldots, x_{20}$ are the outcomes of 20 independent random variables $\mathsf{X}_1, \ldots, \mathsf{X}_{20}$ with common distribution $P_1(dx)$. Similarly, $y_1, \ldots, y_{20}$ are the outcomes of independent variables $\mathsf{Y}_1, \ldots, \mathsf{Y}_{20}$ with some common distribution $P_2(dy)$. To completely specify the model, we need to specify the joint distribution of the two groups of variables. But whether child No. 20 in the first group scores high or low should have no influence on whether child No. 7 in the second group scores high or low. We reflect this by assuming that $\mathbf{X} = (\mathsf{X}_1, \ldots, \mathsf{X}_{20})$ and $\mathbf{Y} = (\mathsf{Y}_1, \ldots, \mathsf{Y}_{20})$ are independent random vectors.

Why is it necessary to specify the joint distribution of $\mathbf{X}$ and $\mathbf{Y}$? In a statistical model we must specify the class of all possible distributions from which the data vector might be drawn. But here the data vector consists not of **x** by itself and **y** by itself, but of the couple (**x**, **y**). This means that it is the class of all possible distributions underlying the *forty* measurements (**x**, **y**) that must be characterized.

In our first try at building a model for this situation, the set of all possible underlying distributions is reduced to the set of all possible pairs of 1-dimensional distributions $(P_1(dx), P_2(dx))$. The problem may be rephrased: Use the data vector (**x**, **y**) to decide whether or not P_1 and P_2 are the same distribution. Actually this states the problem too generally. We are not terribly interested in whether the shape of P_1 is exactly the same as the shape of P_2. What interests us are the more specific questions of whether outcomes from P_1 tend to be larger, smaller, or nearly equal to outcomes from P_2.

One common sense approach is to compare the average scores of the two groups. Suppose Group 1 had average score $\bar{x} = 97.3$, and Group 2, average score $\bar{y} = 103.1$. What can be concluded? If the averages had been, say, 101.1 and 101.2, their difference would certainly be written off as insignificant. At what level does the difference become significant? For instance, by the variability of scores on the given test, we might find that the group average of another group of 20 children from the first tribe was 101.7. Now how significant would you judge the difference between 97.3 and 103.1 to be?

Furthermore, why should the average score be used as a measure of overall level of performance? For example, compare two groups of five each. Suppose the scores in the first group were (in order of magnitude)

$$93, 95, 98, 105, 140$$

and in the second group

$$101, 103, 105, 109, 111.$$

Except for the one very high score in Group 1, the second group scored higher. Yet, the average score in the first is larger than the average score in the second group.

Looking at averages makes good statistical sense if we assume a normally distributed model. Suppose on the basis of long experience with the test, we can comfortably assume that P_1 is $N(\mu_1, \sigma_1^2)$ and P_2 is $N(\mu_2, \sigma_2^2)$, where σ_1^2 and σ_2^2 may or may not be known. To simplify, assume further that $\sigma_1^2 = \sigma_2^2$. Then the two distributions P_1 and P_2 have exactly the same shape with P_1 centered around μ_1 and P_2 around μ_2. The two distributions are the same if $\mu_1 = \mu_2$. The first tribe has a generally higher distribution of scores if $\mu_1 > \mu_2$, and conversely. Now, to see if the data indicate any difference in the two distributions, it makes good sense to estimate μ_1 by $\bar{x}$, μ_2 by $\bar{y}$, and compare the two estimates. Furthermore, it turns out that for a normal $N(\mu, \sigma^2)$ distribution, the sample average $\bar{x}$ is a good estimate of μ. Thus, normality corroborates common sense.

But if the underlying distribution is not nearly normal, comparing the average scores may or may not be a sensible procedure. For some symmetric distributions, using the average to estimate the midpoint is an extremely inaccurate procedure. If P_1 and P_2 are distributions of this type, we would not want to compare group averages. So the question remains, if we do not want to make many assumptions about the shapes of P_1 and P_2, how can we sensibly use the data to arrive at a conclusion regarding the effect of the tribal cultures on the test scores?

The above example is in a different category from the first two. The data consist of two samples from two possibly different populations. We call it a *two-sample* problem. The first two examples are called *one-sample* problems.

Problem 4 The test scores for two groups are

Group 1

105, 101, 110, 109, 85, 93, 114, 95, 102, 86,
90, 100, 82, 99, 87, 118, 103, 82, 96, 89.

Group 2

104, 95, 103, 110, 87, 97, 102, 87, 113, 106,
109, 86, 108, 105, 123, 102, 96, 106, 119, 99.

Are the two groups significantly different? Answer this question by setting up several reasonable different criteria. Discuss what you think a sensible approach would be.

Problem 5 A certain fish powder is claimed to be growth producing. To test this claim, we separate two batches of 50 baby rats; one batch is given the usual diet, the other batch, the fish powder. At the end of a month all rats are weighed in. Set up one or more statistical models. In setting up the experiment what precautions would you insist upon?

Problem 6 In the situation of Problem 5, consider two different procedures for selecting the two groups of baby rats. In the first procedure, the two groups are picked at random from a large population of baby rats. In the second procedure, 50 pairs are selected so that the members of each pair are as alike as possible. Discuss the models for the two different procedures. In terms of the models, why would the second be preferable to the first?

Problem 7 To test the effects of adding various amounts of metal A to copper, a record is made of the breaking stress of a sample of 60 wires which differ only in that 20 have 10% of A, 20 have 5% of A, 20 contain no A. Discuss various statistical models for this experiment.

Problem 8 During a rainstorm, the number of inches of rainfall per minute was recorded simultaneously at two points 5 miles apart for one hour. Set up a statistical model for the observed data, using and criticizing various assumptions.

Types of Statistical Problems and Models

In a statistical model the data vector is considered to be the outcome of some random vector $\mathbf{X}$, which we call the *random data vector*. To specify the model, you have to specify *the set* $\mathscr{P}$ *of all possible underlying probability distributions for* $\mathbf{X}$. Recall that a probability distribution for $\mathbf{X}$ assigns to every observable event A a number $P(\mathbf{X} \in A)$ which is the probability that the outcome of $\mathbf{X}$ is in A. That is, let Ω be the space of all possible data vectors. Then a probability distribution for $\mathbf{X}$ assigns to every event A in Ω a probability that the observed data vector will be in A.

How is $\mathscr{P}$ selected? What heuristic is used? The entire point in constructing a statistical model and matching it to reality—in fact, the basic primitive concept of $\mathscr{P}$ is that it contains all distributions that we believe are possible underlying distributions for the experiment.

A translation between reality and the model that is awry will show up when attempts are made to predict future performance from the conclusions drawn by using the model; that is, when further repetitions of the experiment occur.

Since limiting $\mathscr{P}$ too severely leads to the possibility that the actual underlying distribution is not in $\mathscr{P}$ and opens the door to erroneous conclusions, why not take $\mathscr{P}$ so large that it must contain the underlying distribution? The situation is similar to looking for a needle in a haystack by pulling out that portion of it where you think the needle is located to take home and look through. If the portion you select is too small, the needle may not be in it. But if, to prevent this eventuality, you take home the entire haystack, then you have a most difficult search on your hands. Translated into statistical language this means that the larger you take $\mathscr{P}$, the less you can deduce from $\mathbf{x}$ about the character of the underlying distribution. For instance, let $\mathbf{X} = (X_1, \ldots, X_n)$, and take $\mathscr{P}$ to be *all* n-variate distributions. This implies that we are willing to assume that $X_1, \ldots, X_n$ could have any underlying distribution: dependent or independent, identically distributed or with widely varying individual 1-dimensional distributions. Now, what can we tell about the underlying distribution from the single data vector $\mathbf{x}$? Virtually nothing!

But if we take $\mathscr{P}$ to be defined by the restrictions that $X_1, \ldots, X_n$ are independent and have a common exponential distribution, then, at least for n large, we can use $\mathbf{x}$ to get an accurate estimate of the parameter α and feel confident that we are close to the actual underlying distribution (assuming, of course, that the underlying distribution is in $\mathscr{P}$).

Thus, we are in a squeeze play. Taking $\mathscr{P}$ too small may lead to erroneous conclusions. Taking $\mathscr{P}$ too large, and not using reasonable prior restrictions may lead to no meaningful conclusions. There is no easy way out of this dilemma. The model constructor must use caution, knowledge, and intelligence to strike a reasonable balance.

What one first needs in statistics is a dictionary of models: What are the most widely applied and useful models, what are their properties, and what techniques are useful for solving problems involving these models? As a first crude categorization of statistical models we can separate them, as in our examples, into one-sample models, two-sample models, and so on.

The second distinction is between *parametric* and *nonparametric* models. In a parametric model, we get all distributions in $\mathscr{P}$ by varying a finite number of numerical parameters. For instance, in the first example, if the failure time is taken to be exponential with parameter α, $0 < \alpha < \infty$, then all possible underlying distributions for the random data vector $\mathbf{T} = (T_1, \ldots, T_{1000})$ are obtained by varying α. Similarly, in the second example, if the measurements are taken to be the outcomes of independent $N(m, \sigma^2)$ variables, then we get all possible underlying distributions for the model by varying m and σ^2, $-\infty < m < \infty$, $0 < \sigma^2 < \infty$. But in nonparametric models the space of possible underlying distributions is so large that it cannot be characterized in any natural convenient way by a finite number of parameters. For instance, in our second example, assume that the measurements are drawn from a distribution restricted only by the condition that it is symmetric about its midpoint. This set of distributions cannot be nicely characterized by any finite set of parameters.

Methodology in parametric and nonparametric models differs greatly. In parametric models, the tools are sharp and specific—designed for the best performance within the very small finite-parameter family of possible underlying distributions. For nonparametric models, the tools have to be blunderbuss-like, capable of coping with a large variety of underlying distributions. Surprisingly enough, we will find that at times the blunderbuss is almost as sharp as the scalpel.

Given a model, we can ask different questions concerning the underlying distribution. For instance,

a In the first example, given that the underlying distribution is exponential, estimate the parameter α. In the second example, estimate the midpoint of the distribution.

b How accurate are the estimates of α or of the midpoint?

c In the first example, is it reasonable to assume that the distribution of T is exponential? In the third example, do the two groups of tribal children differ significantly in their test score distributions?

Questions of Type a. are *estimation* problems. What is required is to estimate one or more numerical characteristics of the underlying distribution? Questions of Type b. regarding accuracy require the formulation of *confidence intervals or regions*. The notable point about the questions of Type c. is that they have yes or no answers. In terms of the model, they ask: Given a subset $\mathscr{P}_1$ of the possible underlying distributions $\mathscr{P}$, is the actual underlying distribution in $\mathscr{P}_1$? The procedures of deciding yes or no from the data are called *hypothesis testing*.

As we have emphasized, what techniques are used depends essentially on the model. Statistical theory is by no means complete and polished. It is a recently born and rapidly growing field in which many important questions are unanswered or only very partially answered. As the complexity of the model increases, what is known becomes less complete and definitive.

The plan for this book is to start with the simplest and most well-known models and gradually develop more complex ones, up to the boundaries of what is known. We start with parametric models, covered in the next four chapters:

Chapter 2—general introduction to parametric models.
Chapter 3—estimation in parametric models.
Chapter 4—confidence intervals in parametric models.
Chapter 5—hypothesis testing in parametric models.

There are a number of reasons for spending a fair amount of time on parametric models. In early statistics they were the most frequently studied models. Statistical theory cut its teeth on techniques for these models, and the techniques that were developed formed the jumping off point for later developments. The techniques changed as the models got more complex, but the framework—the concept of a statistical model and the types of questions that could be addressed to the data—carried over.

In a natural sequence of use, the next problem is how to decide whether a given parametric model fits the data adequately. For example, can these 1000 failure times have come from some exponential distribution? Could a given set of observations have come from an $N(\mu, \sigma^2)$ distribution? And equally important, to what extent can we discriminate between distributions? For example, if we decide that the failure times could have come from an exponential distribution, this does not rule out the possibility that they could also have been drawn from a nonexponential distribution which lies "sufficiently close" to some exponential distribution. This general area, called *goodness of fit*, is discussed in Chapter 6.

What can be done if the possible underlying distributions are not a small finite-parameter family? Typically, suppose there are departures from normality; then what happens? What techniques are useful? Or, suppose that we take the underlying distribution of failure times to be restricted only by the condition that failure time is a nonnegative variable. Now what information about the underlying distribution can we get out of the data, and how accurate is it? These questions about one-sample nonparametric models are covered in Chapters 7 and 8.

Chapter 9 covers N-sample models. Here we have N samples drawn independently from N unknown underlying distributions. Do the underlying distributions differ in some significant respect? If so, how do we locate the differences? For instance, if there are three sets of observations from three unknown underlying distributions, are these three the same? If not, are any two similar, or do the three differ widely? This is an important question, and it appears constantly in statistical applications. To answer this question, famous techniques in statistics, both parametric and nonparametric, have been developed.

The final chapter is Chapter 10—multisample, multifactor techniques. Again we have N independent samples drawn independently from N unknown distributions. But the questions raised in this chapter are more complicated. The type of problem we study arises when a number of factors are varied simultaneously and one wishes to trace the effect of each factor separately. The main tools used are for models in which the distributions are assumed normal. The most important are regression and analysis of variance.

Throughout the text the emphasis is on the model, and on what is sensible and realistic. Always ask yourself: Is my translation of the physical context into a model unreasonable and the conclusions misleading? Doubt and suspicion, as well as technical knowledge, are indispensable tools in statistics.

2 Parametric Models

Introduction

To specify a statistical model you have to specify the class $\mathscr{P}$ of possible underlying distributions and the space Ω of possible outcomes for the data vector. The latter is simple and routine; the specification of $\mathscr{P}$ is the distinguishing feature of a statistical model. Again, in a model we have

a *a data vector* $\mathbf{x}$ *which consists of the set of numbers observed or measured in the experiment. The space of all possible data vectors is denoted by* Ω.
b *an outcome* $\mathbf{x}$ *of the random data vector* $\mathbf{X}$ *which takes values in* Ω.
c *a specified class* $\mathscr{P}$ *of probability distributions for* $\mathbf{X}$. *We call* $\mathscr{P}$ *the class of possible underlying distributions.*

There is always the tacit fundamental assumption that the actual distribution of $\mathbf{X}$ is in $\mathscr{P}$. Call this distribution the *actual underlying distribution*, or the *true underlying distribution*, or simply the *underlying distribution*.

In the single sample case $\mathbf{X} = (X_1, \ldots, X_n)$, where $X_1, \ldots, X_n$ are independent with a common unknown distribution $P(dx)$. Hence this 1-dimensional distribution $P(dx)$ completely defines the joint distribution for $\mathbf{X}$. Because of this the model is completely specified once the class of possible 1-dimensional distributions is given. We therefore use our terminology rather loosely and refer to the latter set as the class of all possible underlying distributions. If this class is a finite-parameter family of distributions, then we are in the parametric case.

In this chapter we focus on the construction of parametric statistical models and, in particular, for the single sample parametric case. Also, we give a short introduction to some of their properties. As usual, the best way to begin is with examples.

Examples of Model Construction

Example a To decide whether a die is unfair, it is tossed 600 times and a record is kept of the sequence of outcomes.

Obviously the space Ω of all possible data vectors consists of all sequences 600 long of digits ranging from 1 through 6. The random data vector $\mathbf{X}$ has the form $(X_1, \ldots, X_{600})$, where each of the random variables $X_1, \ldots, X_{600}$ has the possible outcomes 1, 2, 3, 4, 5, 6. What should we specify as the class $\mathscr{P}$ of all possible underlying probabilities for $\mathbf{X}$?

A reasonable assumption is that the successive throws are independent. Certainly, the statistician should try to design the experiment to ensure that complicating factors such as dependence between tosses are at a minimum. Another reasonable assumption is that the probabilities of the various faces do not change during the 600 throws. In other words, we assume that $\mathscr{P}$ consists of all probabilities such that $X_1, \ldots, X_{600}$ are independent and identically distributed. If we denote $P(X_k = 1)$ by $p_1, \ldots, P(X_k = 6)$ by p_6, then since $p_1, p_2, \ldots, p_6$ must add to one, the probabilities in $\mathscr{P}$ are completely specified by giving the values of $p_1, \ldots, p_5$. Hence $\mathscr{P}$ is a 5-dimensional space. For specified values of $p_1, \ldots, p_6$, the corresponding distribution for $\mathbf{X}$ is given by

$$P(\mathbf{X} = \mathbf{x}) = P(X_1 = x_1, \ldots, X_{600} = x_{600}) = p_1^{n_1} p_2^{n_2} \cdots p_6^{n_6},$$

where n_1 is the number of 1's in $x_1, \ldots, x_{600}$, n_2 the number of 2's, and so on.

Example b From a large batch of rods of identical manufacture, n are to be chosen and measured to see whether their lengths meet a given specification. This test is to be used to see whether the overall proportion of rods not meeting the given specifications is sufficiently small.

This example makes explicit what was hidden in the last example—that what you take $\mathscr{P}$ to be partly depends on what you want to find. Suppose that we record the measured lengths $x_1, \ldots, x_n$ of the n rods. The data vector has the form $\mathbf{x} = (x_1, \ldots, x_n)$. Denote the underlying distribution of lengths in the batch by $P(dx)$, and to model this situation, assume that $\mathbf{X} = (X_1, \ldots, X_n)$ has independent components with common (but unknown) distribution $P(dx)$. Since we are not primarily interested in $P(dx)$, but only in the proportion p of rods in the batch not meeting the specifications, we may decide to record, for each rod measured, only "Yes, it does meet specifications," or "No, it does not meet specifications." Give this recording a numerical translation: No = 1 and Yes = 0. Now the data vector $(x_1, \ldots, x_n)$ consists of a string of zeros and ones, and the underlying distributions for $\mathbf{X}$ are given by

$$P(\mathbf{X} = \mathbf{x}) = p^k(1 - p)^{n-k},$$

where k is the number of 1's in the data vector $\mathbf{x}$. To every value of p, $0 \leq p \leq 1$, corresponds a different distribution for $\mathbf{X}$. If we put no restrictions on the value of p, then the family $\mathscr{P}$ consists of all distributions for $\mathbf{x}$ corresponding to the various values of p between 0 and 1. Thus $\mathscr{P}$ is a 1-dimensional family.

But a small nagging doubt arises. For the sake of simplicity, we have thrown away information, namely, the exact measurements of the sample. Is it possible that we could get a better estimate of p, the overall proportion of defectives, by using all the measurement information on hand? We come back to this later, in Chapter 6.

A Formulation

In the first example each probability distribution in $\mathscr{P}$ was specified by the values of any five of $p_1, \ldots, p_6$. In Example b. each distribution in $\mathscr{P}$ was distinguished by the value of p, $0 \le p \le 1$. We generalize by introducing

2.1 ***Notation*** *If the probabilities in $\mathscr{P}$ are specified by the values of θ, where θ ranges over some space Θ, then write P_θ for the probability corresponding to θ, and $\{P_\theta\}$, $\theta \in \Theta$, for the family $\mathscr{P}$.*

Thus, in Example a. we write $P_{p_1,\ldots,p_6}$ for the probability distribution defined by

$$P_{p_1,\ldots,p_6}(\mathbf{X} = \mathbf{x}) = p_1^{n_1} \cdots p_6^{n_6}.$$

Here Θ is the space of points $(p_1, p_2, \ldots, p_6)$ such that $p_1 \ge 0, \ldots, p_6 \ge 0$, $p_1 + \cdots + p_6 = 1$.

In Example b. we write P_p for the probability distribution

$$P_p(\mathbf{X} = \mathbf{x}) = p^k(1 - p)^{n-k}.$$

Here Θ is the interval $0 \le p \le 1$. This gives us a neat, convenient notation. Now we can state

2.2 ***Definition*** *In parametric statistical models, $\mathscr{P}$ is a class of distributions $\{P_\theta\}$ where θ ranges over some finite dimensional space Θ; Θ is called the parameter space for the model, or the "possible states of nature."*

If the random data vector takes on only discrete values, then the distributions $\{P_\theta\}$ are specified by the values $P_\theta(\mathbf{X} = \mathbf{x})$ for all possible data vectors $\mathbf{x}$. At times we may abbreviate $P_\theta(\mathbf{X} = \mathbf{x})$ as $P_\theta(\mathbf{x})$. In the density case the distributions P_θ are specified by a density function $f_\theta(\mathbf{x})$ defined for all $\mathbf{x} \in \Omega$ such that the probabilities are given by single or multiple integrals

$$P_\theta(\mathbf{X} \in A) = \int_A f_\theta(\mathbf{x})\, d\mathbf{x}.$$

Let θ^* be the parameter value in Θ corresponding to the actual underlying distribution. We call this value "*the true state of nature*" or "*the underlying parameter value.*" Then the general statistical problem is: On the basis of a data vector $\mathbf{x}$, make some statement concerning the location of θ^*.

In a test of a hypothesis, a subset Θ_1 of the parameter space is chosen, you observe **x** and are then asked to decide whether or not θ^* is in Θ_1.

In estimation problems you are trying to answer the question: What is the location of θ^*? Then, on the basis of your observed data **x**, you have to decide on a single point in Θ. However, there are many other problems in which you only want to find the value of one coordinate of θ^* or of some function of θ^*. These problems are also called estimation problems. For instance, in the die-throwing experiment, Θ consists of all points $(p_1, \ldots, p_6)$ in the "plane" $p_1 + \cdots + p_6 = 1$ satisfying the restrictions $p_1 \geq 0, \ldots, p_6 \geq 0$. Then $\theta^* = (p_1^*, \ldots, p_6^*)$. To estimate θ^* we have to specify a single point in the 5-dimensional plane. But we may very well be interested in estimating the probability p_1^* of Face 1 only.

Obviously, the crucial question in estimation problems is: How good are our estimates? Actually, what we may sensibly want is to use **x** to find a region Θ_1 of Θ such that θ^* will be inside Θ_1 with high probability. This brings up the question of confidence regions.

Problem 1 An experiment to determine the molecular weight of a compound is repeated 50 times. Construct a parametric statistical model for this experiment. What is Θ and what is the set $\{P_\theta\}$, $\theta \in \Theta$?

Problem 2 A number of specimens of the same steel are stressed until they break, and then the angles that their surface microcracks make with the direction of stress are measured. The results are a total for all specimens of n_1 cracks with an angle between 0° and 5°, n_2, between 5° and 10°, and so on, up to the 85° to 90° range. Construct a statistical model. Define Θ and $\{P_\theta\}$, $\theta \in \Theta$.

Some Common Families of Distributions

In this section we list some of the most common and useful 1-dimensional families of distributions.

Discrete distributions:

a *Binomial* with parameters n, p; n a positive integer, $0 \leq p \leq 1$, $q = 1 - p$ (Figure 2.1):

2.3 $$p(k) = \frac{n!}{k!\,(n-k)!}\, p^k q^{n-k}, \qquad k = 0, 1, \ldots, n,$$

$$\text{mean} = np, \qquad \text{variance} = npq.$$

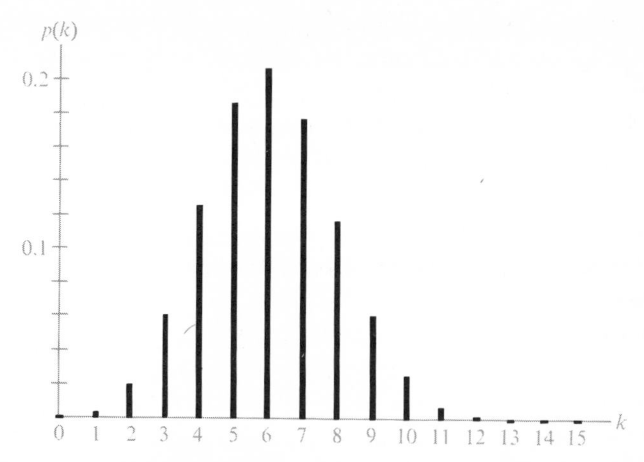

Figure 2.1 *Binomial probabilities with $n = 15$, $p = 0.4$*

If $p(n+1)$ is not an integer, the maximum of $p(k)$ occurs at the greatest integer $m \le p(n+1)$.

b *Poisson* with parameter λ, $0 \le \lambda < \infty$ (Figure 2.2):

2.4
$$p(k) = \frac{\lambda^k}{k!}e^{-\lambda}, \qquad k = 0, 1, 2, \ldots,$$

$$\text{mean} = \lambda, \qquad \text{variance} = \lambda.$$

If λ is not an integer, the maximum $p(k)$ occurs at the greatest integer $m \le \lambda$.

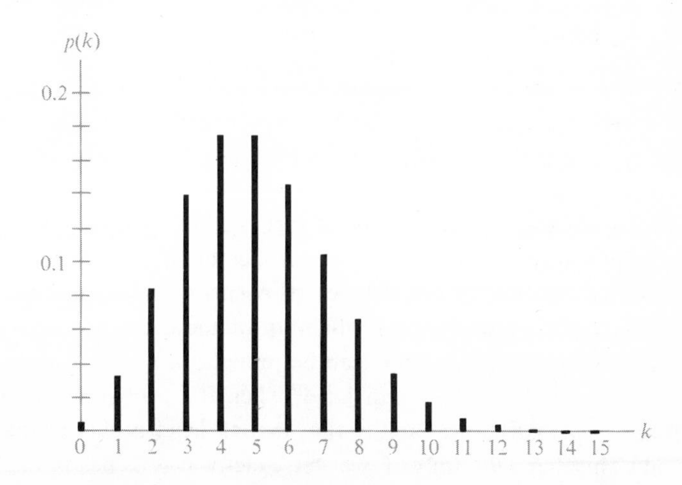

Figure 2.2 *Poisson probabilities with $\lambda = 5$*

c *Negative binomial* with parameters $r, p, r = 1, 2, \ldots, 0 \le p \le 1$, $q = 1 - p$ (Figure 2.3):

2.5 $$p(k) = \frac{(r+k-1)!}{k!(r-1)!} p^r q^k, \qquad k = 0, 1, 2, \ldots,$$

$$\text{mean} = r/p, \qquad \text{variance} = rq/p^2.$$

If $(q/p)(r - 1)$ is not an integer the maximum $p(k)$ occurs at the greatest integer $m \le (q/p)(r - 1)$.

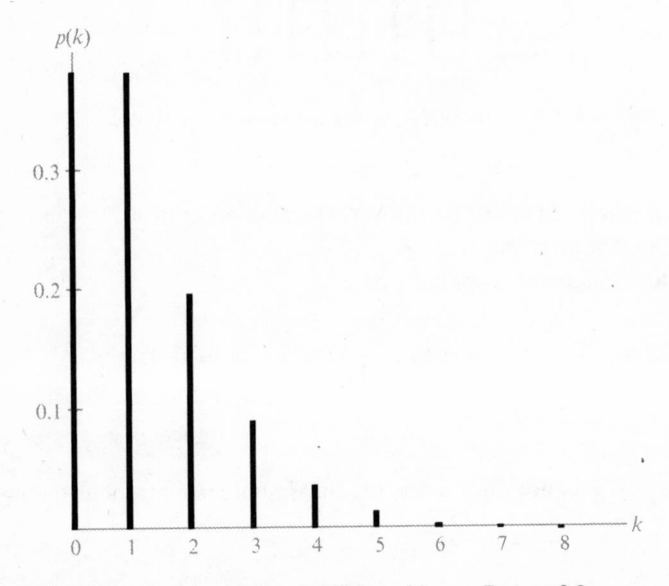

Figure 2.3 *Negative binomial probabilities with r = 5, p = 0.8*

These three discrete distributions can all be derived from the tossing of a biased coin. Think of the binomial distribution as giving the probability of exactly k heads in n independent tosses of a coin which has probability p of heads. Recall that the Poisson is the "rare event" distribution. It comes from the binomial distribution by taking $p \to 0$, $n \to \infty$ in such a way that the expected number of heads $np \to \lambda$. Roughly, it is the distribution of the number of occurrences of an event in a large number of trials when the probability of the event at each trial is very small. The negative binomial can be thought of as the distribution of the number of tails until the rth head, when the probability of heads is p. That is, $p(k)$ is the probability that the rth head occurs at the $(k + r)$th trial. This happens only if we get exactly $r - 1$ heads in $k + r - 1$ trials and a head on the $(k + r)$th trial. The probability of $r - 1$ heads in $k + r - 1$ trials comes from the binomial distribution. If we multiply by p, we get the given expression for $p(k)$.

Continuous distributions:

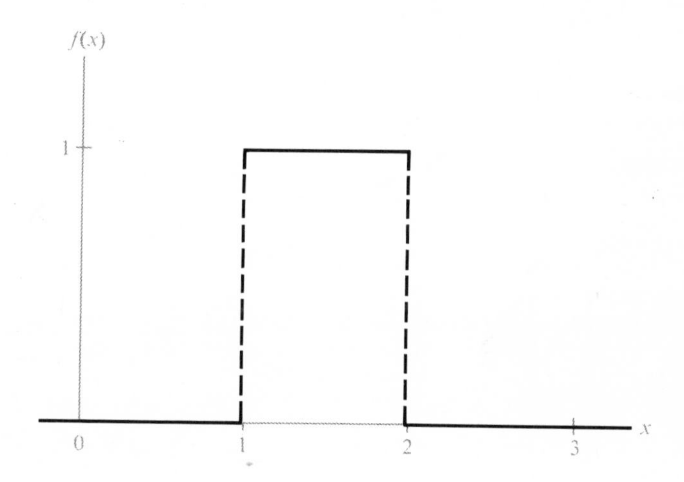

Figure 2.4 *Density of the uniform distribution on [1, 2]*

a *Uniform* on $[a, b]$, $a < b$ (Figure 2.4):

2.6
$$f(x) = \begin{cases} \dfrac{1}{b - a}, & a \le x \le b \\ 0, & x \notin [a, b] \end{cases}$$

$$\text{mean} = \frac{a + b}{2}, \qquad \text{variance} = \frac{(b - a)^2}{12}.$$

b *Exponential* with parameter $\beta > 0$ (Figure 2.5):

2.7
$$f(x) = \begin{cases} \dfrac{1}{\beta} e^{-x/\beta}, & x \ge 0 \\ 0, & x < 0 \end{cases}$$

$$\text{mean} = \beta, \qquad \text{variance} = \beta^2.$$

c *Gamma* with parameters α, β, $0 < \alpha < \infty$, $0 < \beta < \infty$ (Figure 2.6):

2.8
$$f(x) = \begin{cases} \dfrac{1}{\Gamma(\alpha)\beta^\alpha} x^{\alpha - 1} e^{-x/\beta}, & x \ge 0 \\ 0, & x < 0 \end{cases}$$

$$\text{mean} = \beta\alpha, \qquad \text{variance} = \beta^2\alpha, \qquad \text{maximum at } x = \beta(\alpha - 1), \alpha \ge 1.$$

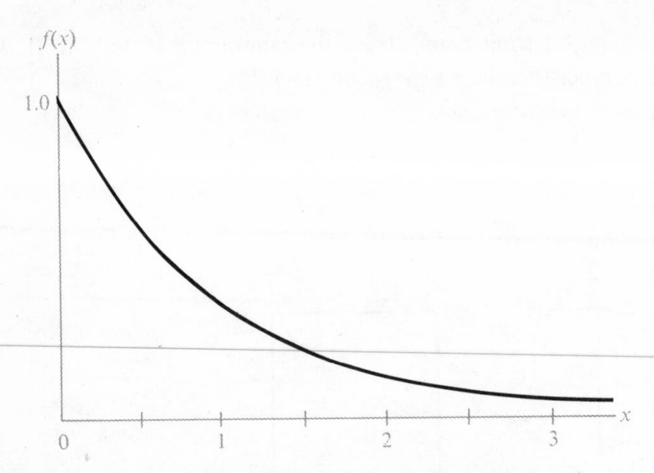

Figure 2.5 *Density of the exponential distribution with* $\beta = 1$

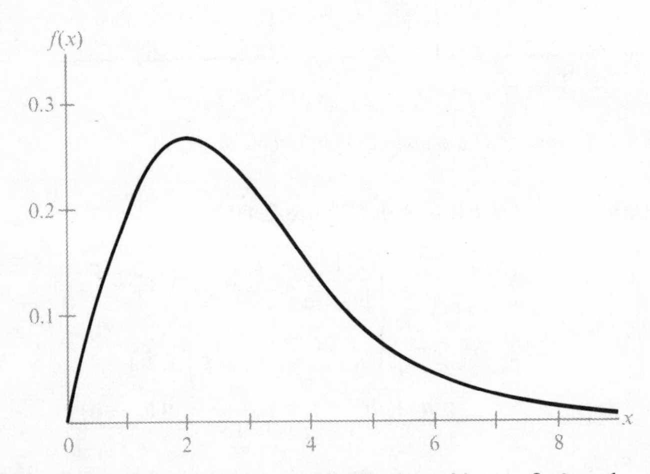

Figure 2.6 *Density of the gamma distribution with* $\alpha = 3,\ \beta = 1$

d *Weibull* with parameters $\alpha, \beta, 0 < \alpha < \infty, 0 < \beta < \infty$ (Figure 2.7):

2.9
$$f(x) = \begin{cases} \dfrac{\alpha}{\beta^{\alpha}} x^{\alpha-1} e^{-(x/\beta)^{\alpha}}, & x \geq 0 \\ 0, & x < 0 \end{cases}$$

$$\text{mean} = \beta\Gamma\left(\frac{1}{\alpha} + 1\right),$$

$$\text{variance} = \beta^2\left(\Gamma\left(\frac{2}{\alpha} + 1\right) - \Gamma^2\left(\frac{1}{\alpha} + 1\right)\right),$$

$$\text{maximum at } x = \beta\left(\frac{\alpha - 1}{\alpha}\right)^{1/2} \text{ for } \alpha \geq 1.$$

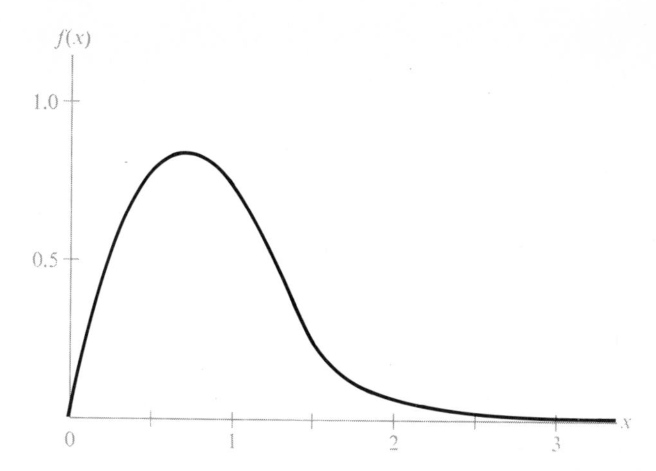

Figure 2.7 Density of the Weibull distribution with $\alpha = 2$, $\beta = 1$

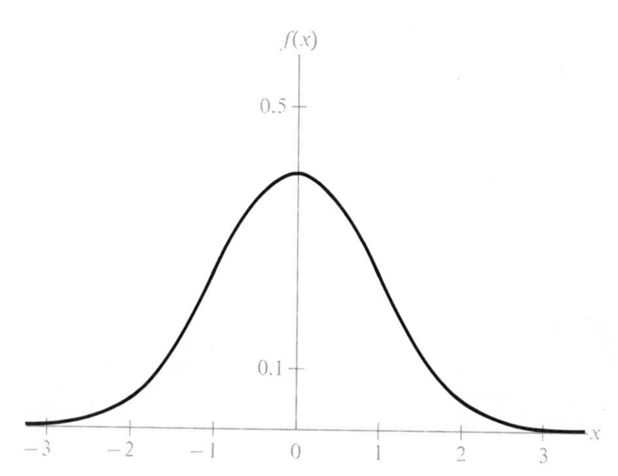

Figure 2.8 Density of the normal distribution with $\mu = 0$, $\sigma^2 = 1$

e *Normal* with parameters μ, σ^2, $-\infty < \mu < \infty$, $0 < \sigma^2 < \infty$ (Figure 2.8):

2.10
$$f(x) = \frac{1}{\sqrt{2\pi}\sigma} e^{-(x-\mu)^2/2\sigma^2}, \qquad -\infty < x < \infty,$$

$$\text{mean} = \mu, \qquad \text{variance} = \sigma^2, \qquad \text{maximum at } \mu.$$

f *Lognormal* with parameters μ, σ^2, $-\infty < \mu < \infty$, $0 < \sigma^2 < \infty$ (Figure 2.9):

2.11
$$f(x) = \begin{cases} \dfrac{1}{x} f_{\mu,\sigma^2}(\log x), & x \geq 0 \\ 0, & x < 0, \end{cases}$$

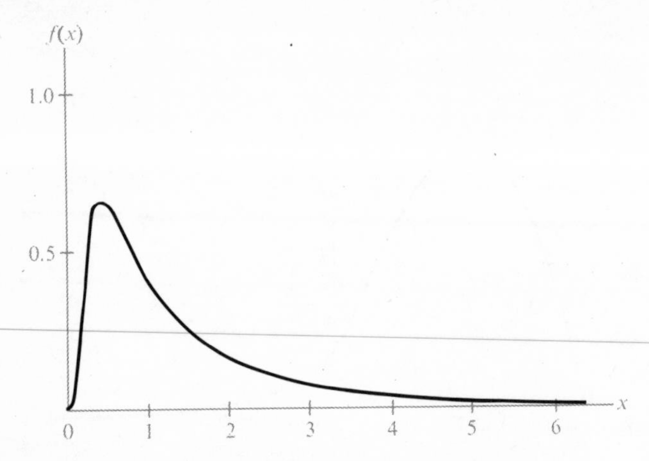

Figure 2.9 *Density of the lognormal distribution with* $\mu = 0, \sigma^2 = 1$

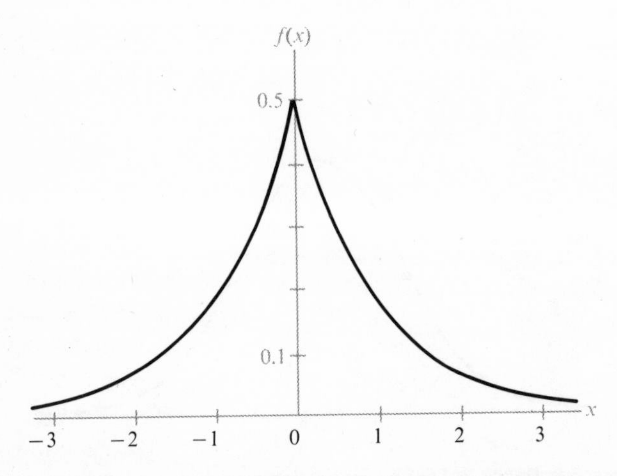

Figure 2.10 *Density of the double exponential distribution with* $\alpha = 0, \beta = 1$

where $f_{\mu,\sigma^2}(x)$ is the normal density given in e.

$$\text{mean} = e^{\mu}e^{\sigma^2/2}, \qquad \text{variance} = e^{2\mu}e^{\sigma^2}(e^{\sigma^2} - 1),$$

$$\text{maximum at } x = e^{\mu}e^{-\sigma^2}.$$

g *Double exponential* with parameters α, β, $-\infty < \alpha < \infty, 0 < \beta < \infty$ (Figure 2.10):

2.12
$$f(x) = \frac{1}{2\beta} e^{-|x-\alpha|/\beta}, \qquad -\infty < x < \infty,$$

$$\text{mean} = \alpha, \qquad \text{variance} = 2\beta^2, \qquad \text{maximum at } x = \alpha.$$

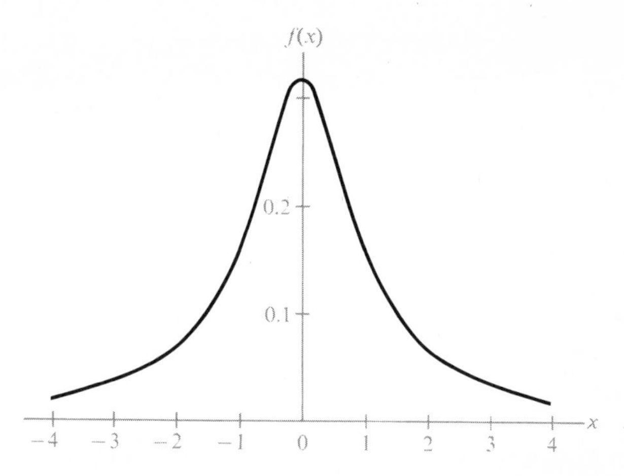

Figure 2.11 *Density of the Cauchy distribution with* $\alpha = 0$, $\beta = 1$

h *Cauchy* with parameters $\alpha, \beta, -\infty < \alpha < \infty, 0 < \beta < \infty$ (Figure 2.11):

2.13 $$f(x) = \frac{1}{\beta\pi} \frac{1}{1 + [(x - \alpha)/\beta]^2}, \qquad -\infty < x < \infty.$$

This distribution is symmetric around α and has a single maximum there. Its mean is undefined. The first absolute moment, $E|X|$, is infinite, and its second moment is likewise infinite ($EX^2 = \infty$).

Among these continuous distributions, the gamma and the Weibull are, in a way, extensions of the exponential. Both families include the exponential family. In both, $\alpha = 1$ gives an exponential distribution. Recall that the exponential is the distribution of the time until the first occurrence in a Poisson process. The gamma distribution, with $\alpha = n$, gives the distribution of time until the nth occurrence. Put another way, the gamma, for $\alpha = n$, is the distribution of the sum of n independent identically distributed exponentials. (See PSP, p. 95.)†

The Weibull distributions are failure time distributions when the failure rate is assumed to increase like some power of t. (See PSP, p. 133.) But a simpler way of looking at the Weibull distributions is this: Let Y have an exponential distribution with parameter 1. Then $X = \beta(Y)^{1/\alpha}$ has a Weibull distribution with parameters α, β.

The lognormal, like the exponential, Weibull, and gamma, is a family of distributions for random variables with positive outcomes. Let Y be an $N(\mu, \sigma^2)$ variable; then $X = e^Y$ is a lognormal variable with parameters μ and σ^2, and its density can be computed from this relationship.

† We shall use PSP as an abbreviation for Leo Breiman, *Probability and Stochastic Processes: With a View Toward Applications* (Boston: Houghton Mifflin Co., 1969).

The uniform, normal, double exponential, and Cauchy distributions have the property of being symmetric about their midpoints. The last three are *unimodal*, that is, their densities have a unique maximum.

The gamma ($\alpha \geq 1$), Weibull ($\alpha \geq 1$), and lognormal distributions are also *unimodal*, but the value of x at which the maximum occurs is not usually the mean value of the distribution.

The characteristic that most distinguishes the four symmetric families is their tail behavior. To compare these, take their midpoints to be at zero and, for each family, select the value(s) of the parameter(s) so that 50% of the outcomes drawn from the distribution will fall between -1 and $+1$. That is, select the parameter(s) such that $P(-1 \leq \mathsf{X} \leq 1) = .5$. The resulting distributions are: uniform on $[-2, +2]$, normal with $\sigma = 1.48$, double exponential with $\beta = 1.44$, and Cauchy with $\beta = 1.00$. In Figure 2.12 $P(|\mathsf{X}| > x)$ is graphed for each of these distributions. The uniform distribution puts none of its outcomes past ± 2, the normal puts 18% of its outcomes past ± 2, while the double exponential and Cauchy put 25% and 30% past ± 2, respectively. But as we go further out, the distinction becomes marked. The normal, double exponential, and Cauchy distributions put 4, 12, and 20% of their outcomes past ± 3, and 0.7, 6, and 16%, respectively, past ± 4. Beyond ± 6 these percentages drop to .00 (to two significant figures), 1.6, and 10%. Look at the contrast:

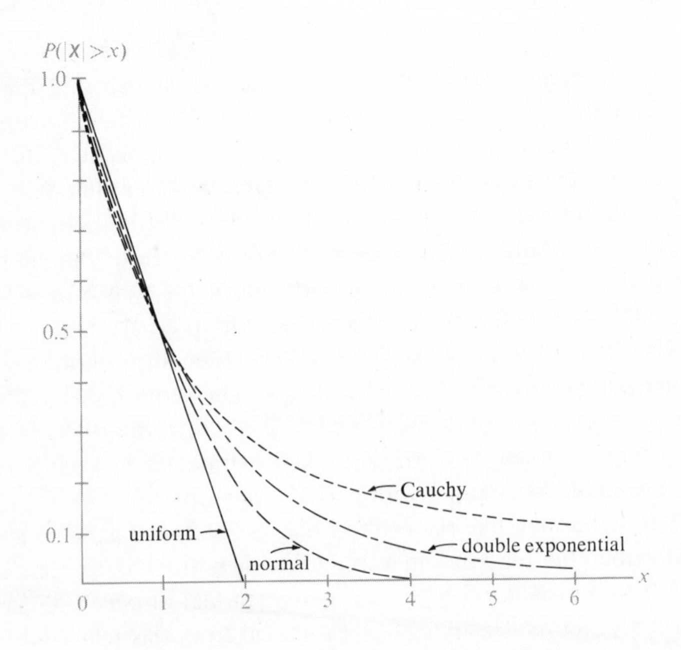

Figure 2.12 *$P(|\mathsf{X}| > x)$ for the uniform, normal, double exponential, and Cauchy distributions*

For a normal distribution you might expect one outcome in 100 outside of ± 4 and certainly none outside of ± 6. But for the Cauchy distribution 16 outcomes out of 100 should fall outside ± 4, and 10 outside ± 6.

This behavior reflects the fact that the tails of a normal density go down like e^{-cx^2}, the double exponential tails like $e^{-c|x|}$, and the Cauchy tails like c/x^2 (x large). However, the midsection behavior of the four symmetric families is not dissimilar. Figure 2.13 is a graph of the four cumulative distributions with the parameters given above ($P(|X| \leq 1) = .5$).

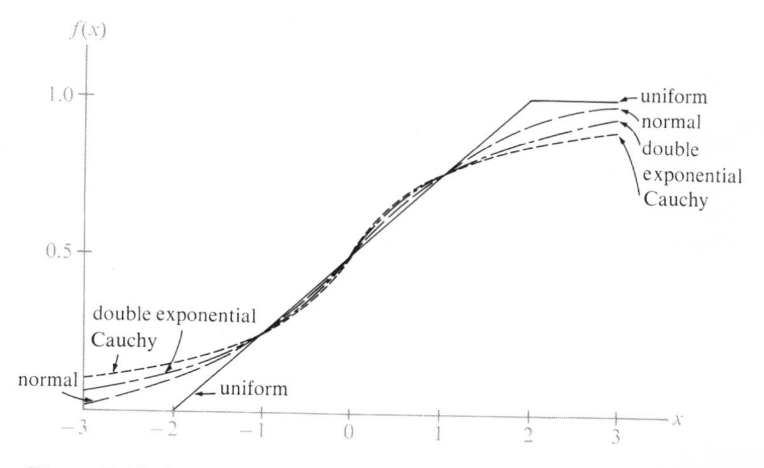

***Figure 2.13** Cumulative distribution functions of the uniform, normal, double exponential, and Cauchy distributions*

The tails of the exponential, gamma, Weibull, and lognormal distributions all decrease rapidly. For x large, the tail of the gamma is dominated by the term e^{-cx}, the Weibull by e^{-cx^2}, and the lognormal by $e^{-c(\log x)^2}$. All of these terms go to 0 faster than any power of x, the lognormal decreasing most slowly.

The small outcomes are more useful in distinguishing these families. The maximum density of the exponential is near $x = 0$. The gamma and Weibull ($\alpha > 1$) densities both go to 0 at the origin like the ($\alpha - 1$)st power of x. Hence small readings may not discriminate between the two families, but they will be useful in determining what power of x the density behaves like near the origin, and thus in estimating the parameter α. The lognormal goes to 0 near the origin faster than any power of x, and produces fewer small outcomes than any of the above distributions. To illustrate, we select parameters for the gamma, Weibull, and lognormal distributions such that the maximum of each occurs at $x = 1$. Now look at Figure 2.14.

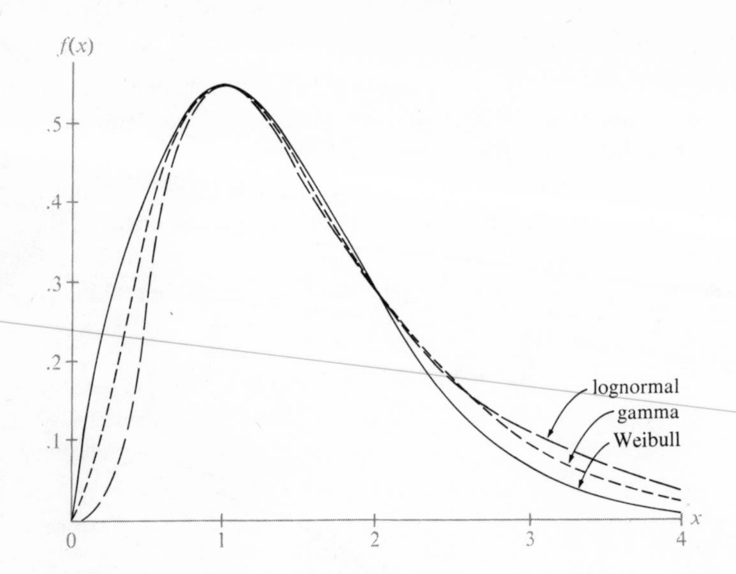

***Figure 2.14** Densities of the gamma, Weibull, and lognormal distributions with parameters selected to give the maximum value .54 at x = 1*

To compare cumulative distribution functions we select parameters so that $P(X \geq 1) = .5$.

It is important to realize that for parametric families there is *no unique parametrization.* If a family $\{P_\theta\}$ is parametrized by points in a space Θ, then any other space Φ such that there is a one-to-one correspondence between the points of Θ and Φ can also be used as a parameter space. For example, we parametrized the exponential family by a number β, $0 < \beta < \infty$, such that the densities are

$$f(x) = \frac{1}{\beta} e^{-x/\beta}, \qquad x \geq 0.$$

But by setting $\beta = 1/\alpha$, $0 < \alpha < \infty$, we could also parametrize this family by writing the densities as

$$f(x) = \alpha e^{-\alpha x}, \qquad x \geq 0.$$

Another parametrization of the normal family that is sometimes encountered uses μ and $\gamma = |\mu|/\sigma$. The new parameter space Φ is $-\infty < \mu < \infty$, $0 \leq \gamma < \infty$. In this new parametrization, the densities have the form

$$f(x) = \frac{1}{\sqrt{2\pi}} \frac{\gamma}{|\mu|} e^{-(\gamma^2/2)((x/\mu)-1)^2}$$

For the uniform family, a common parametrization uses the midpoint m of the interval and its length l. The original parameter space was

$\Theta = \{(a, b); a < b\}$. The new space is $\Phi = \{(m, l); -\infty < m < \infty, l > 0\}$, with

$$f(x) = \begin{cases} \dfrac{1}{l}, & m - \dfrac{l}{2} \leq x \leq m + \dfrac{l}{2} \\ 0, & \text{elsewhere.} \end{cases}$$

The choice of a parametrization is usually a matter of simplicity, convenience, and usefulness.

Problem 3 *Skewness* refers to the lack of symmetry of a distribution about its mean. If the distribution of X is symmetric about $\mu = E\mathsf{X}$, then the third moment of X about μ is zero: $E(\mathsf{X} - \mu)^3 = 0$. Denote $\mu_k = E(\mathsf{X} - \mu)^k$. Then a measure of skewness is

$$k_s = \frac{\mu_3}{\sigma^3},$$

where $\sigma^2 = E(\mathsf{X} - \mu)^2$.

a Find k_s for the gamma distributions. (Ans. $2/\sqrt{\alpha}$.)

b Why would you suspect (before the calculation was made in **a**.) that as $\alpha \to \infty$, $k_s \to 0$? (Hint: Recall the central limit theorem.)

Problem 4 (Difficult!) Let X and Y be independent, having gamma distributions with parameters α_1, β and α_2, β, respectively. Show that X + Y has a gamma distribution with parameters $\alpha_1 + \alpha_2$, β. Use the fact that

$$\int_0^1 (1 - v)^{\alpha_1 - 1} v^{\alpha_2 - 1}\, dv = \frac{\Gamma(\alpha_1)\Gamma(\alpha_2)}{\Gamma(\alpha_1 + \alpha_2)}.$$

Problem 5 The lognormal family is given the new parametrization $\beta = e^{\mu}$, $\gamma = e^{\sigma^2}$. What is the new parameter space? Write the densities in the new parametrization.

Problem 6 Two distributions $P_1(dx)$ are called close to each other if

$$\max |P_1(I) - P_2(I)| \ll 1,$$

where the maximum is over all intervals I. In other words, there is a small number δ such that for any interval I, $|P_1(I) - P_2(I)| \le \delta$. Let $F_1(x)$, $F_2(x)$ be the corresponding cumulative distribution functions. Since for any interval $I = [a, b)$, $P(I) = F(b) - F(a)$, another way of saying that two distributions are close is,

$$\max_x |F_1(x) - F_2(x)| \ll 1.$$

We say that $P(dx)$ is a nearly normal distribution if there is a normal distribution $P_{(\mu,\sigma)}(dx)$ such that $P(dx)$ is close to $P_{(\mu,\sigma)}(dx)$.

a Show that for any random variable X, if the distribution of $(\mathsf{X} - \mu)/\sigma$ is close to an $N(0, 1)$ distribution, then the distribution $P_{\mathsf{X}}(dx)$ of X is close to an $N(\mu, \sigma^2)$ distribution.

b If the random variable X in a. is integer-valued, how can it be close to a normal distribution?

Location and Scale Parameters

Add a fixed number θ to a random variable Y with distribution $P_0(dx)$ getting $\mathsf{X} = \mathsf{Y} + \theta$. This is called a change of location. Denote the distribution of X by $P_\theta(dx)$ and look at the family of all such distributions for $-\infty < \theta < \infty$. This family is called *the location family generated from the distribution* $P_0(dx)$, and θ is called *a location parameter*. More precisely, the distributions $P_\theta(dx)$ are a location family if for every θ, $P_\theta(dx)$ is the distribution of $\mathsf{Y} + \theta$, where Y has the standard distribution. There are equivalent definitions. For example, θ is a location parameter if each distribution $P_\theta(dx)$ in the family has the cumulative distribution function

$$F_\theta(x) = F_0(x - \theta).$$

If $F_\theta(x)$ has a density $f_\theta(x)$, then

$$f_\theta(x) = \frac{d}{dx} F_0(x - \theta) = f_0(x - \theta).$$

Therefore, a set of densities of the form $f_0(x - \theta)$ for every θ, defines a location family. This means that all densities in the family have exactly the same shape. They differ only by a shift in position.

If θ is a location parameter and X has distribution $P_\theta(dx)$, then $\mathsf{X} - \theta$ has the standard distribution. If the standard distribution has mean m and a density with maximum at x_0, then $P_\theta(dx)$ has mean $m + \theta$ and maximum density at $x_0 + \theta$. But notice that the variance of $P_\theta(dx)$ is the same as that of the standard distribution. Variance measures concentration around the mean, so certainly it should not be affected by a shift of the entire distribution. In fact, we know that

$$V(\mathsf{X}) = V(\mathsf{X} - \theta) = V(\mathsf{Y}).$$

The concept of a location parameter carries over easily into the multidimensional parameter case.

2.14 ***Definition*** *Given a two-parameter family* $\{P_{(\theta_1,\theta_2)}(dx)\}$, *of distributions,* θ_1 *is called a location parameter if*

a *the cumulative distribution functions are of the form*

$$F(x - \theta_1, \theta_2),$$

or

b *the densities are of the form*

$$f(x - \theta_1, \theta_2),$$

or

c *if* X *has the given distribution with parameters* (θ_1, θ_2), *then* $\mathsf{X} - \theta_1$ *has the distribution with parameters* $(0, \theta_2)$.

It should be clear that μ is a location parameter for the $N(\mu, \sigma^2)$ distributions and that α is a location parameter for both the double exponential and Cauchy. In fact, in all three cases, the parameters μ, α give the location of the midpoints of the distribution.

A one-parameter family of distributions may also be generated by a change of scale: Take Y to have the standard distribution $P_1(dx)$, let $\theta > 0$ be a real number, and define $\mathsf{X} = \theta\mathsf{Y}$. The values of Y are simply scaled down by the factor θ to get the values of X. The distribution P_θ of X is given by

$$P_\theta(I) = P_\theta(\mathsf{X} \in I) = P_1\left(\mathsf{Y} \in \frac{I}{\theta}\right),$$

where, if $I = [a, b]$, then I/θ is the interval $[a/\theta, b/\theta]$. Putting $I = (-\infty, x)$,

$$P_\theta(\mathsf{X} < x) = P_1\left(\mathsf{Y} < \frac{x}{\theta}\right),$$

or

$$F_\theta(x) = F_1\left(\frac{x}{\theta}\right),$$

where $F_1(x)$ is the cumulative distribution function of Y. The density of $F_\theta(x)$ is

$$f_\theta(x) = \frac{d}{dx} F_1\left(\frac{x}{\theta}\right) = \frac{1}{\theta} f_1\left(\frac{x}{\theta}\right),$$

where $f_1(x)$ is the density of $F_1(x)$. The effect of θ is to compress or expand the distribution. If $\theta > 1$, then the distribution $P_\theta(dx)$ has more probability spread further out than the standard distribution, and conversely if $\theta < 1$. This is reflected in the fact that the variance of $P_\theta(dx)$ is θ^2 times the variance of the standard distribution and the standard deviation is increased by a factor of θ. (See Figure 2.15.)

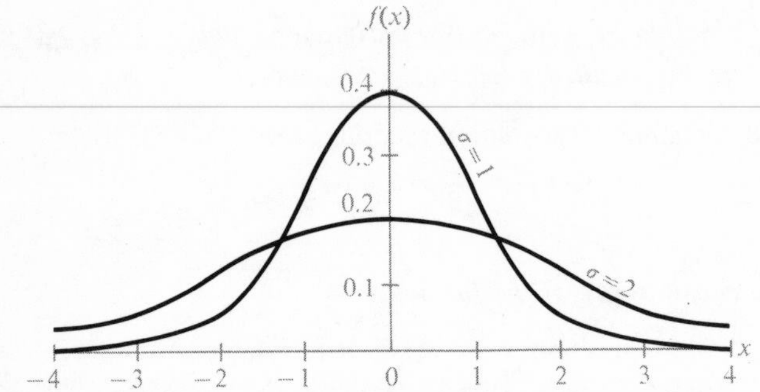

Figure 2.15 *Densities of the normal distribution with* $\mu = 0$*, for* $\sigma = 1$ *and* $\sigma = 2$

Look at the one-parameter family we get from making scale changes of amount $\theta > 0$ on a standard distribution $P_1(dx)$. Call this family $\{P_\theta(dx)\}$ *the scale family generated from the distribution* $P_1(dx)$. Some equivalent definitions are that the cumulative distribution functions for the family have the form $F_1(x/\theta)$, that the set of densities is $(1/\theta)f_1(x/\theta)$, or that if X has distribution $P_\theta(dx)$, then X/θ has the standard distribution $P_1(dx)$.

2.15 ***Definition*** *For a two-parameter family* $\{P_{(\theta_1,\theta_2)}(dx)\}$, θ_2 *is a scale parameter if*

a *the cumulative distribution functions are of the form*

$$F_1(x/\theta_2, \theta_1),$$

or

b *the densities are of the form*

$$\frac{1}{\theta_2} f_1\left(\frac{x}{\theta_2}, \theta_1\right),$$

or

c *if* X *has the distribution with parameters* (θ_1, θ_2), *then* X/θ_2 *has the distribution with parameters* $(\theta_1, 1)$.

It is easy to see that β in the exponential distribution is a scale parameter, as are β in the gamma, Weibull, double exponential, and Cauchy, and σ in the $N(\mu, \sigma^2)$ distribution.

Finally, suppose we generate a two-parameter family of distributions by starting with a standard random variable Y which has a distribution $P_{(0,1)}(dx)$ and making the linear transformations

$$\mathsf{X} = \theta_2 \mathsf{Y} + \theta_1, \qquad \theta_2 > 0.$$

Let $P_{(\theta_1,\theta_2)}(dx)$ denote the distribution of X, and call the family *the location-scale family generated from the standard distribution* $P_{(0,1)}(dx)$. Here θ_1 is a location parameter and θ_2 a scale parameter. Put another way, we get the family $P_{(\theta_1,\theta_2)}(dx)$ by starting with a fixed distribution and considering all distributions we can get from it by changes of location and scale. Location-scale families can be recognized by the following:

a $F_{(\theta_1,\theta_2)}(x)$ *has the form*

$$F_{(0,1)}\left(\frac{x - \theta_1}{\theta_2}\right)$$

or

b $f_{(\theta_1,\theta_2)}(x)$ *has the form*

$$\frac{1}{\theta_2} f_{(0,1)}\left(\frac{x - \theta_1}{\theta_2}\right).$$

Equivalently, if X *has the distribution* $P_{(\theta_1,\theta_2)}(dx)$, *then*

2.16
$$\mathsf{Y} = \frac{\mathsf{X} - \theta_1}{\theta_2}$$

has the standard distribution.

The normal, double exponential, and Cauchy families are location-scale families. Their standard distributions have the densities

a
$$f(x) = \frac{1}{\sqrt{2\pi}} e^{-x^2/2},$$

b
$$f(x) = \tfrac{1}{2} e^{-|x|},$$

c
$$f(x) = \frac{1}{\pi} \cdot \frac{1}{1 + x^2},$$

respectively.

The value of a probability distribution in any location-scale family can be found in tables of the standard distribution. For example, to find the probability $P(\mathsf{X} < x)$ where X is an $N(\mu, \sigma^2)$ variable, use the fact that $\mathsf{X} = \sigma \mathsf{Y} + \mu$, where Y is $N(0, 1)$. Hence

$$P(\mathsf{X} < x) = P\left(\mathsf{Y} < \frac{x - \mu}{\sigma}\right).$$

For a variable T having an exponential distribution with parameter β, use the fact that $T = \beta U$, where U is exponential with parameter 1.

In the last section we saw that a family could be parametrized in many different ways and the choice was somewhat arbitrary. It is possible that the family $\{P_\theta\}$ is actually a location and/or scale family but has been parametrized in such a way that the parameters are not location and/or scale parameters. For example, we could parametrize the exponential family as

$$f(x) = \alpha e^{-\alpha x},$$

where the parameter α ranges over $(0, \infty)$. But α here is not a scale parameter. You may have wondered why we chose to write the exponential density as $(1/\beta)e^{-x/\beta}$. The reason is that the preferred parametrization for a location and/or scale family is such that θ is a location and/or scale parameter, respectively. This is also why the normal is usually parametrized by μ, σ.

This leads us to a reparametrization of the lognormal distribution. Recall that $X = e^{Y}$, where Y is $N(\mu, \sigma^2)$. Write $Y = U + \mu$, where U is $N(0, \sigma^2)$. Then

$$X = e^{U+\mu} = e^{\mu}e^{U}.$$

Put $\theta_2 = e^{\mu}$, $\theta_1 = \sigma^2$. Then

$$X = \theta_2 e^{U}$$

where e^{U} is lognormal with parameters $(0, \theta_1)$, and it is clear that θ_2 is a scale parameter. For simplicity, we use e^{σ^2} as one parameter and, using β, γ to denote parameters, instead of θ_1 and θ_2, get the parametrization

$$\beta = e^{\mu}, \qquad \gamma = e^{\sigma^2}.$$

Do Problem 5 to see that the densities have the right form for β to be a scale parameter.

Problem 7 You are willing to assume that 250 positive numbers were drawn independently from the same underlying distribution. For a first look at the data, a histogram is plotted. It has a maximum at about $x = 4.0$. Your question is whether the data could have come from a gamma or Weibull distribution. How could you proceed to get a fast rough answer?

Problem 8 Each pebble in a shovelful of river bed pebbles is measured along its maximum dimension. You suspect that the log-

$X = e^{\sigma Z + \mu}$

$Z \sim N(0,1)$

normal might be a good choice for underlying distribution. How would you do a rough first check on this possibility?

Problem 9 Reduce the following probabilities by reducing them to the values of a standard distribution (see the tables at the end of the book):

a $P(\mathsf{X} < 3)$ where X is $N(1, 4)$;

b $P(\mathsf{X} < 6)$ where X is lognormal with $\mu = 1$, $\sigma^2 = 1$;

c $P(\mathsf{X} < 2)$ where X is Weibull with $\beta = 2$, $\alpha = 2$;

d $P(\mathsf{X} > 3)$ where X is gamma with $\alpha = 3$, $\beta = 3$.

Problem 10 For any distribution $P_{\mathsf{X}}(dx)$, if there is a unique value of x such that

$$P(\mathsf{X} \le x) = P(\mathsf{X} \ge x) = \tfrac{1}{2},$$

call this value of x the median and denote it by ν. If the density is unimodal and its maximum is at x_0, call x_0 the mode. Let $\{P_\theta(dx)\}$ be a location family. Why is it true that

a if the standard distribution has mean 0, then P_θ has mean θ?

b if the median of the standard distribution is zero, then the median of P_θ is θ?

c if the mode of the standard distribution is 0, then the mode of P_θ is θ?

Notice that one implication of the above is that estimating the mean, the median, or the mode gives an estimate for the location parameter.

d Show that a., b., and c. remain true for θ the location parameter in a location-scale family?

e Do a., b., and c. remain generally true for a two-parameter location family with θ the location parameter?

Problem 11 a Show that if $\{P_\theta\}$ is a scale family, if Y has the standard distribution, and if the constant c is defined by

$$c^k = E|\mathsf{Y}|^k, \qquad k > 0,$$

then for X having the distribution P_θ,

$$(E|X|^k)^{1/k} = c\theta.$$

Conclude from a. that a possible method of estimating θ for a scale family is: For any $k > 0$, estimate the kth absolute moment $E|X|^k$, take its kth root, and divide by c.

b For a location-scale family with parameters θ_1, θ_2, show that when X has distribution $P_{(\theta_1,\theta_2)}$, $E|X|^k$ in general depends on both θ_1 and θ_2; but that

$$(E|X - EX|^k)^{1/k} = d\theta_2,$$

where d is defined by

$$d^k = E|Y - EY|^k$$

and Y has the standard distribution.

Therefore for location-scale families, the scale can be estimated by estimating any kth absolute moment around the mean, taking the kth root, and dividing by d. In particular, the standard deviation $\sqrt{E(X - EX)^2}$ and the absolute deviation $E|X - EX|$ are proportional to θ_2, and are frequently used to get estimates of scale.

Problem 12 By choosing parameters differently, show that a family of uniform distributions is a location-scale family. What can you take as the standard distribution?

The Normal Distribution

For a variety of reasons, the normal family of distributions occurs much more often in statistics than any other parametric family. For us an important reason is that with the words "small" and "approximately" appropriately defined, *the sum of a large number of small independent random variables has an approximately normal distribution.* (See PSP, Chapter 4.) Consequently if $S_n = X_1 + \cdots + X_n$ is a sum of n independent identically distributed variables, small or not, then for n large,

$$\frac{S_n - ES_n}{\sigma(S_n)}$$

has approximately an $N(0, 1)$ distribution.

2.17 ***Notation*** *For the remainder of this book,* Z *without a subscript will denote a random variable having an* $N(0, 1)$ *distribution.*

The normal approximation makes it possible to compute probabilities of the type $P(a \leq S_n < b)$. Subtract ES_n from each member of the inequality $a \leq S_n < b$ and divide by $\sigma(S_n)$ to get

2.18 $$P(a \leq S_n \leq b) \simeq P\left(\frac{a - ES_n}{\sigma(S_n)} \leq Z < \frac{b - ES_n}{\sigma(S_n)}\right).$$

The right-hand side can be looked up in the $N(0, 1)$ table (Table 1) by expressing it as

$$P\left(Z < \frac{b - ES_n}{\sigma(S_n)}\right) - P\left(Z < \frac{a - ES_n}{\sigma(S_n)}\right).$$

You can skip a step in this computation of $P(a \leq S_n < b)$ by realizing that this transformation to the normal variable Z is actually a rescaling of the two endpoints a and b in terms of how many standard deviations they are away from the mean. Thus

2.19 $$a \to \frac{a - ES_n}{\sigma(S_n)}$$
$$b \to \frac{b - ES_n}{\sigma(S_n)},$$

and we can write quickly

$$P(a \leq S_n < b) \simeq P\left(\frac{a - ES_n}{\sigma(S_n)} \leq Z < \frac{b - ES_n}{\sigma(S_n)}\right).$$

There are three important families of distributions closely related to the $N(0, 1)$. We list them briefly here for later reference.

2.20 ***Definition*** *Let* $Z_1, \ldots, Z_n$ *be independent* $N(0, 1)$ *variables. Then the distribution of*

$$Z_1^2 + \cdots + Z_n^2$$

is called the chi-square distribution with n degrees of freedom and is denoted by χ_n^2.

2.21 ***Definition*** *Let* $Z, Z_1, \ldots, Z_n$ *be independent* $N(0, 1)$ *variables. Then the distribution of*

$$\frac{\sqrt{n}\,Z}{\sqrt{Z_1^2 + \cdots + Z_n^2}}$$

is called Student's t distribution with n degrees of freedom and is denoted by t_n.

2.22 Definition *Let $W_1, \ldots, W_m$, $Z_1, \ldots, Z_n$ be independent $N(0, 1)$ variables. Then the distribution of*

$$\frac{\dfrac{W_1^2 + \cdots + W_m^2}{m}}{\dfrac{Z_1^2 + \cdots + Z_n^2}{n}}$$

is called an F distribution with m and n degrees of freedom and is denoted by $F_{m,n}$.

The values of these three distributions are given in Tables 2 to 4. Because of their importance we will usually reserve the symbols C, S, F for random variables having a chi-square, Student's *t*, or *F* distribution, respectively.

Problem 13 Prove that for *n* large, $X_1, \ldots, X_n$ independent and identically distributed, and $S_n = X_1 + \cdots + X_n$, the variable

$$Y = \frac{S_n - ES_n}{\sigma(S_n)}$$

a has mean 0, variance 1;

b is the sum of small independent random variables.

(In b. take *small* to mean *having a small variance.*)

Problem 14 Use 2.18 to compute the probability that in 10000 tosses of a coin with $P(\text{heads}) = .55$, fewer than 5380 heads turn up.

The Multidimensional Normal Distribution

The distributions listed so far are relevant when there is a succession of independent measurements or observations, each resulting in a single number. Now suppose we take a succession of objects and make *J* measurements on each. For instance, suppose *n* individuals are given a battery of *J* tests, or *J* different performance characteristics are measured for each component in a given sample. The data consist of a sequence of *J*-dimensional vectors,

$$\mathbf{x}_1 = (x_1^{(1)}, \ldots, x_J^{(1)}), \mathbf{x}_2 = (x_1^{(2)}, \ldots, x_J^{(2)}), \ldots, \mathbf{x}_n = (x_1^{(n)}, \ldots, x_J^{(n)}),$$

where $\mathbf{x}_k$ is the result of the battery of tests or measurements on the kth object. Assume that $\mathbf{x}_1, \mathbf{x}_2, \ldots, \mathbf{x}_n$ are outcomes of independent random vectors $\mathbf{X}_1, \ldots, \mathbf{X}_n$ having common distribution $P(d\mathbf{x})$.

Successive individuals or objects tested or measured can often be assumed independent. The scores on a battery of tests for one individual should not (if the testing is carefully done) have any effect on the test results for the next individual. But for a given individual, the test scores may be highly dependent. For instance, we might suspect that measurements of weight and blood pressure on individuals may be dependent. Or we might be analyzing samples of the same type of rock collected from different areas for percent content of five different elements. Until proven differently, the a priori suspicion is that these five measurements are correlated.

The multivariate normal distribution is the most useful parametric family of distributions for dependent variables (PSP, Chapter 8). For $\mathbf{X} = (X_1, \ldots, X_J)$ any random vector, the means vector $\boldsymbol{\mu}$ is defined by

$$\boldsymbol{\mu} = (\mu_1, \ldots, \mu_J), \qquad \mu_j = EX_j.$$

The covariance matrix $[\Gamma]$ has entries

$$\Gamma_{ij} = E(X_i - \mu_i)(X_j - \mu_j).$$

If $\mathbf{X}$ has a multivariate normal distribution, it is completely determined by $\boldsymbol{\mu}$ and $[\Gamma]$. In fact, for a nonsingular distribution ($\det [\Gamma] \neq 0$), the joint density of $(X_1, \ldots, X_J)$ is

$$f(\mathbf{x}) = \frac{1}{(2\pi)^{J/2}\sqrt{\det [\Gamma]}}\, e^{-\frac{1}{2}Q(\mathbf{x})},$$

where

$$Q(\mathbf{x}) = \sum_{i,j} \gamma_{ij}(x_i - \mu_i)(x_j - \mu_j),$$

and γ_{ij} are the elements of $[\Gamma]^{-1}$. Thus, the family of J-variate normal distributions is specified by the J parameters $(\mu_1, \ldots, \mu_J)$ and the parameters in $[\Gamma]$. Since $[\Gamma]$ is symmetric, $(\Gamma_{ij} = \Gamma_{ji})$ there are really only $J(J + 1)/2$ parameters in $[\Gamma]$.

In the bivariate case, the parameters are $\mu_1, \mu_2, \sigma_1^2 = V(X_1), \sigma_2^2 = V(X_2)$, and Γ_{12}. Instead of Γ_{12}, it is more convenient to use the parameter

$$\rho = \frac{\Gamma_{12}}{\sigma_1\sigma_2}.$$

Recall that ρ is called the correlation coefficient of X_1, X_2. Then

$$[\Gamma] = \begin{pmatrix} \sigma_1^2 & \sigma_1\sigma_2\rho \\ \sigma_1\sigma_2\rho & \sigma_2^2 \end{pmatrix}$$

and

$$Q(\mathbf{x}) = \frac{1}{1-\rho^2}\left[\left(\frac{x_1-\mu_1}{\sigma_1}\right)^2 - 2\rho\left(\frac{x_1-\mu_1}{\sigma_1}\right)\left(\frac{x_2-\mu_2}{\sigma_2}\right) + \left(\frac{x_2-\mu_2}{\sigma_2}\right)^2\right].$$

We introduce the normalized variables

$$\mathsf{Y}_1 = \frac{\mathsf{X}_1-\mu_1}{\sigma_1}, \qquad \mathsf{Y}_2 = \frac{\mathsf{X}_2-\mu_2}{\sigma_2}.$$

These have 0 means. Their covariance matrix is

$$[\Gamma] = \begin{pmatrix} 1 & \rho \\ \rho & 1 \end{pmatrix}.$$

Hence, $\mathsf{Y}_1, \mathsf{Y}_2$ have a bivariate normal distribution with

$$Q(\mathbf{y}) = \frac{1}{1-\rho^2}\,[y_1^2 - 2\rho y_1 y_2 + y_2^2].$$

This is the standardized form of the bivariate normal. Notice that it depends only on ρ. Now we can write

$$(\mathsf{X}_1, \mathsf{X}_2) = (\mu_1 + \sigma_1\mathsf{Y}_1, \mu_2 + \sigma_2\mathsf{Y}_2).$$

In a sense, μ_1 and μ_2 are location parameters and σ_1, σ_2 scale parameters. The correlation ρ gives a measure of the dependence between X_1 and X_2. They are independent if $\rho = 0$ and linearly dependent if $\rho = \pm 1$.

Problem 15 You are considering modeling a sequence of pairs of measurements $(x_1, y_1), (x_2, y_2), \ldots, (x_n, y_n)$ as the outcomes of independent random vectors $(\mathsf{X}_1, \mathsf{Y}_1), \ldots, (\mathsf{X}_n, \mathsf{Y}_n)$ with a common bivariate normal distribution. How can you separately use the values $x_1, \ldots, x_n$ and the values $y_1, \ldots, y_n$ to get a crude idea of whether joint normality is an unreasonable assumption?

Problem 16 Two different methods are available for estimating the thickness of a metallic film deposited on a plastic strip. A set of film-on-strip samples is prepared and the thickness estimated by both methods for each sample. Assume that the actual thicknesses of the different film samples were drawn from an underlying $N(\mu, \sigma^2)$ distribution, μ and σ^2 unknown. Assume further that if the actual thickness is t, then the outcome of the first method is drawn from an $N(t, \sigma_1^2)$ distribution, and the outcome of the second method from an $N(t, \sigma_2^2)$ distribution, independently of the first. What model does this lead to for the pairs of measurements (x_k, y_k), $k = 1, \ldots, n$? What is the correlation?

Problem 17 For Z having an $N(0, 1)$ distribution, show that Z^2 has a gamma distribution with $\alpha = 1/2$, $\beta = 2$. (Note: $\Gamma(1/2) = \sqrt{\pi}$.) Use this together with the result of Problem 4 to show that χ^2 with J degrees of freedom is a gamma distribution with $\alpha = J/2$, $\beta = 2$.

Summary

What is most important in this chapter is the formulation of a statistical model in terms of a parameter space. The class of possible underlying probabilities $\mathscr{P}$ can be put in one-to-one correspondence with a simpler finite-dimensional set Θ of possible values of a parameter θ. The true state of nature θ^* is assumed to be in Θ. A statistical conclusion is a statement based on the data vector $\mathbf{x}$ regarding the location of θ^*.

In one-sample models, the data are considered to be the outcomes of identically distributed independent random variables $\mathsf{X}_1, \ldots, \mathsf{X}_n$ or random vectors $\mathbf{X}_1, \ldots, \mathbf{X}_n$. In parametric one-sample models, the common distribution is assumed to be in a family of distributions corresponding to a finite dimensional parameter space Θ. Some of the more useful families of distributions are listed, and their properties briefly described. You should remember the concepts of location and scale parameters, how to recognize them, and how to convert a standard distribution into a location and/or scale family.

3 Estimation in Parametric Models

Introduction

This chapter deals with the problem of estimation in parametric models. Before we begin there is a fundamental question: *What do we want to estimate and why*? A standard procedure when faced with a sample from an unknown distribution is to try to estimate its mean and possibly its variance. But this can lead to nonsense. Suppose that the important question is the probability of failure of a satellite component during the planned lifetime of the satellite. Does it help to find the average lifetime of a satellite component?

To a great extent the above question has already been answered in the selection of a parametric model. In a parametric model, the estimation problem is, almost always, to *estimate* θ^*. We want to select a point $\hat{\theta}$ in Θ as our best guess of the value of θ^*. Of course, this selection procedure is based on the data vector $\mathbf{x}$.

Notice that in selecting a point in Θ, we are actually selecting one of the distributions in $\{P_\theta\}$, $\theta \in \Theta$. Thus in the usual estimation problem in parametric models, our answer specifies completely the underlying distribution. For instance, our answers are of the form:

the distribution of this variable is Poisson with parameter $\lambda = 3.7$

or

the distribution of this variable is $N(.3, 4.2)$.

Some thought here may lead to a protest. Suppose we know that it is safe to assume a Weibull distribution for the failure time T of a certain component. But suppose further that all we are interested in is the probability $P(\mathsf{T} \leq 2000)$ of failure before 2000 hours. Why should we go through the complicated procedure of estimating the parameters α and β in a Weibull distribution? If we have tested n components, and m of them have failed in less than 2000 hours, why not estimate $P(\mathsf{T} \leq 2000)$ by m/n? We will come back to this question several times. Our first answer, given later in this chapter, is that in a parametric model, no matter how simple the characteristic of the underlying distribution you want to estimate, the best procedure is to first estimate the value of θ^*, then read off the desired characteristic from the distribution P_θ. This result gives a reason for the statement above that in a parametric model, the important estimation problem usually is: Estimate θ^*.

An Example

Recall the rod example of Chapter 2. We want to estimate the underlying proportion p of defectives in the entire population of rods by testing a batch of n rods. To make the problem numerical, define

$$x_k = \begin{cases} 1 & \text{if the } k\text{th rod is defective} \\ 0 & \text{if the } k\text{th rod is not defective.} \end{cases}$$

An intuitively reasonable estimate for the proportion of defectives in the entire population is the proportion of defectives in the sample. That is, we propose to estimate p by computing

$$\frac{x_1 + \cdots + x_n}{n}.$$

How good is this estimate? This question is important on two counts. First, we simply need to know how close the estimate is likely to be to the actual underlying proportion of defectives. Secondly, we want to compare different methods of estimation.

Suppose that in actuality, in all the millions of rods, $\frac{1}{10}$ are defective. We select a batch of n rods to test. Of course, our estimate does not come out exactly .1; perhaps the estimate we get is .11. If we pull another batch of size n, we may get an estimate of .097. As we repeat, the estimates will be clustered around the value .1. When we ask how good the estimate is, we are really asking how closely or tightly the estimate values are clustered around .1.

Consider $x_1, \ldots, x_n$ to be outcomes of random variables $\mathsf{X}_1, \ldots, \mathsf{X}_n$. Then the estimate is the outcome of the random variable

$$\bar{\mathsf{X}} = \frac{\mathsf{X}_1 + \cdots + \mathsf{X}_n}{n}.$$

We want to get our hands on the distribution of $\bar{\mathsf{X}}$. But the only way to do this is to assume a statistical model that specifies the distribution of the data vector **X**. Use the model of the first chapter in which the class of alternative possible probabilities are given by $\{P_p\}$, $0 \le p \le 1$, and under P_p, the variables $\mathsf{X}_1, \ldots, \mathsf{X}_n$ are independent and identically distributed with

$$P_p(\mathsf{X}_k = 1) = p, \qquad P_p(\mathsf{X}_k = 0) = q, \qquad q = 1 - p.$$

With this model, if we know that $p = .1$, then we know exactly what the distribution of **X** is. The $\mathsf{X}_1, \ldots, \mathsf{X}_n$ are independent, each with probability .1 of equaling 1 and .9 of equaling 0. Knowing this, we can compute the distribution of $\bar{\mathsf{X}}$, that is, we can calculate the probabilities

$P(a \leq \bar{X} \leq b)$. For example, take $n = 20$. Then $X_1 + \cdots + X_{20}$ has a binomial distribution with parameters 20, .1. The proportion of defectives $\bar{X}$ ranges over the values $k/20$, $k = 0, \ldots, 20$, and the probabilities

$$\begin{aligned} p(k) &= P(\bar{X} = k/20) \\ &= P(X_1 + \cdots + X_{20} = k) \end{aligned}$$

are the binomial probabilities defined in (2.3). These probabilities are graphed in Figure 3.1 below. By adding, we get

$$\begin{aligned} P(.05 \leq \bar{X} \leq .15) &= .75 \\ P(.00 \leq \bar{X} \leq .20) &= .96. \end{aligned}$$

Since our estimates are values of $\bar{X}$ which come from this distribution, these equations imply that if the true state of nature is $p = .1$, then 75% of the time our estimate would fall between .05 and .15, 96% of the time between 0 and .20. Obviously, a sample of size 20 is not large enough to give much accuracy. But at any rate, given the true state of nature p and the sample size n, we can compute the distribution of our estimate $\bar{X}$.

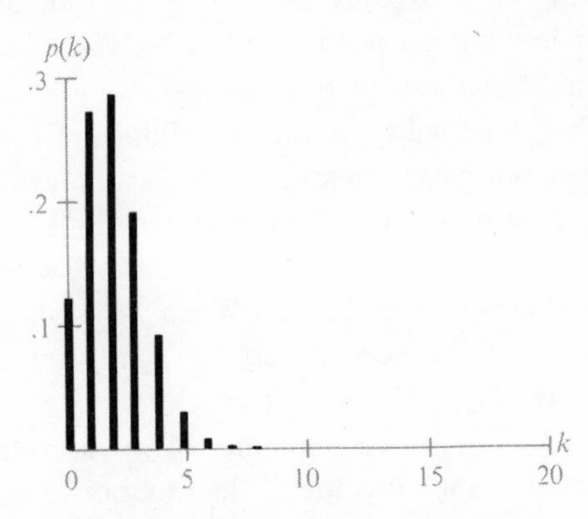

Figure 3.1 *Binomial probabilities with $n = 20$, $p = 0.1$*

Now, how can we measure how closely the distribution of the estimate $\bar{X}$ clusters around the true state of nature p? One way is to consider the difference $\bar{X} - p$ and get some measure of how large this difference is, "on the average." Of course, when the true state of nature is p, then

$$E\bar{X} = \frac{EX_1 + \cdots + EX_n}{n} = p.$$

Then $E(\bar{X} - p) = 0$, so the positive and negative values of $\bar{X} - p$ cancel each other. To find how large $\bar{X} - p$ is "on the average," we could look at the average value of $|\bar{X} - p|$ or perhaps at the squared difference $(\bar{X} - p)^2$. The latter is easier to work with. As our first try at a definition of estimator accuracy we use

3.1 ***Definition*** *The expected squared error, or risk, of the estimator* $\bar{X}$ *at the state of nature p is*

$$R(p) = E(\bar{X} - p)^2.$$

Notice that the risk $R(p)$ depends on the state of nature p. First, p is subtracted from $\bar{X}$ to form the difference $\bar{X} - p$. More important, the expectation above is computed using the distribution P_p. The random variables $X_1, \ldots, X_n$ are 1 or 0 with probability p or $1 - p$, respectively. The expectations of combinations of these random variables will depend on the value of p. To make the dependence on p explicit, we write the expectation as E_p instead of E. Thus, $R(p)$ is denoted by $E_p(\bar{X} - p)^2$.

For the sake of illustration we compute $R(p)$. Since $E_p X_k = p$, $k = 1, \ldots, n$, $E_p\bar{X} = p$, and $E_p(\bar{X} - p)^2$ is simply the variance $V_p(\bar{X})$, where the subscript has the same meaning as in E_p. Since $X_1, \ldots, X_n$ are independent and identically distributed,

$$V_p(\bar{X}) = V_p\left(\frac{X_1 + \cdots + X_n}{n}\right)$$

$$= \frac{1}{n} V_p(X_1) = \frac{p(1 - p)}{n}.$$

Thus

$$R(p) = \frac{p(1 - p)}{n}.$$

Figure 3.2 gives a graph of $R(p)$. Notice that the maximum expected squared error occurs at $p = 1/2$ and is $(1/2)^2/n = 1/4n$. Thus if we test a batch of 500, say, the maximum expected squared error is 1/2000.

The graph of $R(p)$ gives us the picture of the "average accuracy" of the estimator $\bar{X}$ at the various values of p, in this way summarizing the behavior of the estimate. We call this graph *the operating characteristic curve of the estimate using squared error loss.* To compare $\bar{X}$ with any other method of estimation, we compare their operating characteristic curves.

But why use expected squared error as a measure of the accuracy of an estimate? Suppose a firm guarantees that its estimates are accurate to within .05; if not, it pays a fixed penalty. If $\hat{p}$ is the estimate, and p the true state of nature, then the penalty is assessed if

$$|\hat{p} - p| > .05,$$

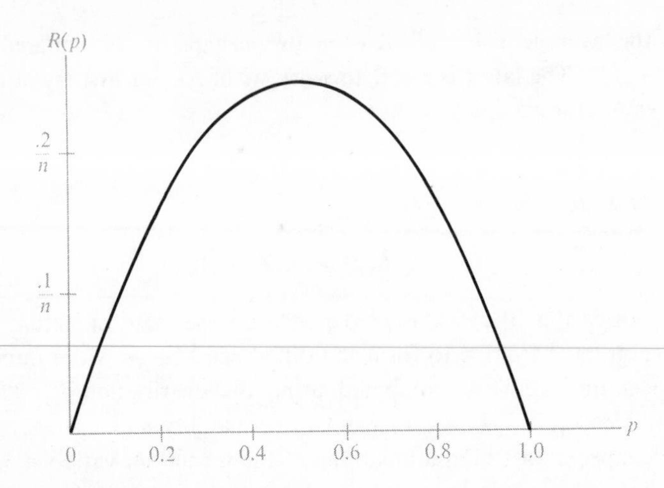

***Figure 3.2** R(p) as a function of p*

and the estimate is satisfactory if

$$|\hat{p} - p| \le .05.$$

How do you gauge the accuracy of this estimator? As long as the estimate $\hat{p}$ is within .05 of p, we don't care how close or how far from p it is. The criterion for judging an estimator here is how frequently it falls in the range $p \pm .05$. Therefore, define the inaccuracy in the estimate $\bar{X}$ at the state of nature p as

$$R(p) = P_p(|\bar{X} - p| > .05).$$

The graph of $R(p)$ is the operating characteristic curve we want. Of course, it differs from the operating characteristic curve we got using squared error as a gauge of inaccuracy. It is more difficult to compute exactly, but for large n we can use the normal approximation. Since $\bar{X}$ is a sum of independent identically distributed random variables and $E_p\bar{X} = p$, $V_p(\bar{X}) = pq/n$,

$$R(p) = P_p\left(\left|\frac{\bar{X} - p}{\sigma_p(\bar{X})}\right| > \frac{.05\sqrt{n}}{\sqrt{pq}}\right)$$

$$\simeq P\left(|Z| > \frac{.05\sqrt{n}}{\sqrt{pq}}\right).$$

For example, if $n = 400$, then we get

$$R(p) \simeq P\left(|Z| > \frac{1}{\sqrt{pq}}\right).$$

From the table of $N(0, 1)$ values, we get

p	$R(p)$
0	0
.1	.0009
.2	.0124
.3	.0293
.4	.0415
.5	.0455

The values are symmetric around $p = .5$.

In these two situations, the common starting point for measuring accuracy is the specification of a *loss function* $L(\hat{p}, p)$, which is a measure of how inaccurate or costly we consider an estimated value of $\hat{p}$ when the true state of nature is p. For example, using squared error the inaccuracy of the estimated value $\hat{p}$ when the true state of nature is p is given by

$$L(\hat{p}, p) = (\hat{p} - p)^2.$$

In the second situation, the inaccuracy is given by

$$L(\hat{p}, p) = \begin{cases} 1 & \text{if } |\hat{p} - p| > .05 \\ 0 & \text{if } |\hat{p} - p| \leq .05. \end{cases}$$

3.2 ***Definition*** *A loss function (for this example) associates a numerical value of inaccuracy or loss to every two numbers $\hat{p}$ and p in $[0, 1]$, p being the true state of nature and $\hat{p}$ the estimate. We denote this function by $L(\hat{p}, p)$.*

The loss function is specified by the statistician. But the specification is not arbitrary. It is tailored to the statistical situation, and in particular, to what the estimate will be used for. Once that information is known, the accuracy requirements on the estimate can be defined and a loss function specified. The loss function is actually a numerical statement of how bad the discrepancy between $\hat{p}$ and p is. For example, for the rod manufacturer, an error of .1 in estimating p may be less important when p is small than when p is large. Then $L(\hat{p}, p)$ should reflect this in the sense that $L(.2, .1)$ should be smaller than $L(.8, .7)$.

Contrast the use of

$$L(\hat{p}, p) = |\hat{p} - p|$$

and

$$L(\hat{p}, p) = |\hat{p} - p|^4$$

as loss functions. Look at the table.

$\hat{p}$	p	$\lvert\hat{p} - p\rvert$	$\lvert\hat{p} - p\rvert^4$
.2	.1	.1	.0001
.4	.1	.3	.0081

Using $|\hat{p} - p|$, the loss due to an error of .3 is 3 times as much as for a .1 error. But using $|\hat{p} - p|^4$, it is 81 times larger. Therefore, we would use the latter loss function only when the loss associated with an error increased very rapidly with the size of the error.

Suppose that the loss function $L(\hat{p}, p)$ is specified. Then to measure how good any estimator is, we compute its expected loss. If $\bar{X}$ is our estimator, then we compute the expected value of $L(\bar{X}, p)$. Use the notation

$$R(p) = E_p L(\bar{X}, p).$$

As in our examples, $R(p)$ is the expected loss or inaccuracy in the estimator $\bar{X}$ when the state of nature is p.

3.3 ***Definition*** *$R(p)$ is called the risk for the estimator $\bar{X}$ at the state of nature p. The graph of $R(p)$ is called the operating characteristic of the estimator $\bar{X}$ with loss function $L(\hat{p}, p)$.*

The smaller the height operating characteristic graph, the more accurate the estimate.

Problem 1 In the above example, how large must n be to insure that the inequality

$$P_p(|\bar{X} - p| \leq .05) \geq .95$$

hold for all possible states of nature, i.e., for all p in $[0, 1)$? (Use the normal approximation.)

Problem 2 For A-type transistors, assume that the underlying distribution of the failure time T is given by

$$P(T > t) = e^{-t/\beta},$$

where β is unknown. The mean of the distribution is

$$ET = \int_0^\infty t\beta^{-1}e^{-t/\beta}\,dt = \beta.$$

A good estimate of β would seem to be the average failure time of the sample tested. Therefore use the estimator

$$\bar{\mathsf{T}} = \frac{\mathsf{T}_1 + \cdots + \mathsf{T}_n}{n}$$

for β. Compute the mean squared error $R(\beta) = E(\bar{\mathsf{T}} - \beta)^2$, and graph it for $n = 100$. How large must n be so that the maximum mean squared error is less than .01 for $0 \leq \beta \leq 5$? Use the normal approximation to graph $R(\beta)$ for the estimator $\bar{\mathsf{T}}$ and the loss function

$$L(\hat{\beta},\beta) = \begin{cases} 0, & |\hat{\beta} - \beta| \leq .1 \\ 1, & |\hat{\beta} - \beta| > .1. \end{cases}$$

The Difficulties of the Statistician's Life

One great difficulty of the statistician's life is that even after he has selected an appropriate loss function there is almost never an estimate that is best for all states of nature. Some may be better if θ is in one range and worse if θ is in another. To illustrate, suppose that in the example of the last section we try using some multiple $\gamma\bar{\mathsf{X}}$ of $\bar{\mathsf{X}}$ as our estimator for p and see if we can select γ advantageously. Using squared error loss, we want to compute

$$R_1(p) = E_p(\gamma\bar{\mathsf{X}} - p)^2.$$

Now

$$E_p(\gamma\bar{\mathsf{X}}) = \gamma E_p\bar{\mathsf{X}} = \gamma p,$$

so write $R_1(p)$ as

$$E_p(\gamma\bar{\mathsf{X}} - \gamma p + \gamma p - p)^2 = E_p(\gamma\bar{\mathsf{X}} - \gamma p)^2 + (\gamma p - p)^2.$$

Here we used the identity: $E(\mathsf{Y} + c)^2 = E\mathsf{Y}^2 + c^2$, if $E\mathsf{Y} = 0$. But

$$E_p(\gamma\bar{\mathsf{X}} - \gamma p)^2 = \gamma^2 E_p(\bar{\mathsf{X}} - p)^2$$

and from a previous computation

$$E_p(\bar{\mathsf{X}} - p)^2 = \frac{p(1 - p)}{n}.$$

Hence

$$R_1(p) = \frac{\gamma^2 p(1 - p)}{n} + p^2(1 - \gamma)^2$$

$$= \frac{\gamma^2 p(1 - p) + np^2(1 - \gamma)^2}{n}.$$

Let us compare this with the risk $R(p) = p(1 - p)/n$ for the estimator $\bar{X}$ by dividing $R_1(p)$ by $R(p)$:

$$\frac{R_1(p)}{R(p)} = \frac{\gamma^2(1 - p) + np(1 - \gamma)^2}{1 - p}$$

$$= \gamma^2 + \frac{np(1 - \gamma)^2}{1 - p}.$$

For $p = 0$, this ratio is equal to γ^2, so $R_1(p)$ is smaller than $R(p)$ for small p, if γ is less than 1. The two are equal when the ratio is 1, or when

3.4 $$\gamma^2 + \frac{np(1 - \gamma)^2}{1 - p} = 1.$$

Solving for the value p_0 where the ratio is 1,

$$p_0 = \frac{1 + \gamma}{n(1 - \gamma) + (1 + \gamma)}.$$

For $p \geq p_0$, $R(p) \leq R_1(p)$, so $\bar{X}$ is a better estimator. But for $p \leq p_0$, $R_1(p) \leq R(p)$, so $\gamma\bar{X}$ is a better estimator in this range. Suppose we know that almost certainly there is a value p_0 such that the true proportion of defectives is less than p_0. Then we would want to select an estimator that is better in the range $p \leq p_0$. To get γ such that $R_1(p) \leq R(p)$ for $p \leq p_0$, we solve 3.4 for γ in terms of p_0, getting

$$\gamma = 1 - \frac{2(1 - p_0)}{(n - 1)p_0 + 1}.$$

Hence γ differs from 1 by a term that decreases like $1/n$.

For large sample size, there is not much advantage in using $\gamma\bar{X}$ instead of $\bar{X}$. But if testing is expensive and you want as small a sample size as possible, then such alternatives should be considered. For example, if $n = 20$ and $p_0 = .1$, then $R_1(p)$ is about $\frac{1}{7}$ the size of $R(p)$ near $p = 0$.

For large sample sizes and p_0 fixed, you can see that the squared error loss of $\gamma\bar{X}$ and $\bar{X}$ are nearly the same. Thus the intuitive guess at $\bar{X}$ as a "good" estimator is not contradicted. You may wonder at this point if the fuss is worth it, because what you think ought to be sensible turns out to really be sensible. But in the next section we give an example whose message is—watch out!

Problem 3 Compare $R(p)$, $R_1(p)$ for $n = 50$ by graphing them on the same axes. Use the value of γ given by $p_0 = .1$.

Intuition Is Not Enough

In this section we will compare two very different estimators for the same situation, both of which are intuitively quite attractive. Suppose n samples are drawn from a distribution which is uniform on $[0, l]$, so that what is unknown is the length of the interval we are drawing from. We could reason this way: Since the sample points $X_1, \ldots, X_n$ are uniformly distributed over the range 0 to l, their average

$$\overline{X} = \frac{X_1 + \cdots + X_n}{n}$$

should be nearly $l/2$. So we estimate l as $2\overline{X}$.

Another way of looking at the situation is this: No matter what l is, the maximum of the readings $X_1, \ldots, X_n$ is less than l and gets closer to l as n increases. Denote $M = \max(X_1, \ldots, X_n)$. We use as our second estimate γM, where γ is a constant we will select to compensate for the fact that M is always less than l by some random factor which tends to get closer to 1 as n increases. Now we compute losses for these two estimators, using squared error loss.

Notice that

$$E_l(2\overline{X}) = 2E_l X_1.$$

Since the density of X_1 is just $1/l$,

$$E_l X_1 = \frac{1}{l}\int_0^l x\,dx = \frac{l}{2},$$

and

$$E_l(2\overline{X}) = l.$$

Therefore

$$E_l(2\overline{X} - l)^2 = V_l(2\overline{X}) = \left(\frac{2}{n}\right)^2 V_l(X_1 + \cdots + X_n)$$

$$= \frac{4}{n} V_l(X_1) = \frac{4}{n} \cdot \frac{1}{l} \int_0^l \left(x - \frac{l}{2}\right)^2 dx$$

$$= \frac{1}{3} \cdot \frac{l^2}{n}.$$

We can compute the distribution of M as follows: $M < x$ only if $X_1 < x, \ldots, X_n < x$ all hold simultaneously. Thus

$$P(M < x) = P(X_1 < x) \cdots P(X_n < x)$$

$$= [P(X_1 < x)]^n = \left(\frac{x}{l}\right)^n$$

for $0 \leq x \leq l$. The density of M is given by

$$\frac{d}{dx} P(\mathsf{M} < x) = \frac{n}{l^n} x^{n-1}$$

for $0 \leq x \leq l$ and is 0 outside this range. Therefore,

$$E_l\mathsf{M} = \int_0^l x f_{\mathsf{M}}(x)\, dx = \int_0^l x \cdot \left(\frac{n}{l^n} x^{n-1}\right) dx$$

$$= \frac{n}{l^n} \int_0^l x^n\, dx = l \cdot \frac{n}{n+1}.$$

The expected value of M, then, is not l, but a little less. However, if we take as our estimator

$$\mathsf{V} = \frac{n+1}{n} \mathsf{M},$$

then $E_l\mathsf{V} = l$ so $(n+1)/n$ looks like a good choice for γ. Now

$$E_l(\mathsf{V} - l)^2 = V_l(\mathsf{V})$$
$$= E_l\mathsf{V}^2 - (E_l\mathsf{V})^2 = E_l\mathsf{V}^2 - l^2,$$

and

$$E_l\mathsf{V}^2 = \left(\frac{n+1}{n}\right)^2 E_l\mathsf{M}^2$$

$$= \left(\frac{n+1}{n}\right)^2 \cdot \int_0^l x^2 \left(\frac{n}{l^n} x^{n-1}\right) dx$$

$$= \left(\frac{n+1}{n}\right)^2 \cdot \frac{n}{l^n} \cdot \frac{l^{n+2}}{n+2} = \frac{(n+1)^2}{n(n+2)} \cdot l^2.$$

Hence

$$E_l(\mathsf{V} - l)^2 = l^2 \left[\frac{(n+1)^2}{n(n+2)} - 1\right] = \frac{1}{n(n+2)} l^2.$$

The point here is that for any $n > 1$, the loss of V is less than the loss of $2\overline{\mathsf{X}}$, and for n large, V is a much better estimate (by a factor of $(n+2)/3$ in terms of the ratio of the losses). On these grounds, $2\overline{\mathsf{X}}$ is highly ineffective as compared to V.

Problem 4 Propose another way of estimating l and compute its expected squared error loss.

Problem 5 Discuss, both in the context of estimating the mean β for exponential distributions

$$P(\mathsf{T} > t) = e^{-t/\beta}, \qquad t \geq 0,$$

and in estimating the range l in a uniform distribution

$$P(\mathrm{X} < x) = \frac{x}{l}, \qquad 0 \le x \le l,$$

why the proportionate error squared loss defined by

$$L(\hat{\theta},\theta) = \left(\frac{\hat{\theta}}{\theta} - 1\right)^2$$

might be preferable to squared error loss.

Some General Principles of Estimation

We can formulate the estimation problem more generally as follows: *Let* $\{P_\theta\}$, $\theta \in \Theta$, *be the class of possible underlying distributions for the random data vector* $\mathbf{X}$. We want to use the observed value of $\mathbf{X}$ to produce a value of θ. An estimation procedure, or an estimator, is some systematic way of using the value of $\mathbf{X}$ to get a value of the parameter. Formally, *an estimator is any function* $\phi(\mathbf{X})$ *of* $\mathbf{X}$ *taking values in the parameter space* Θ. If Θ is 1-dimensional, then $\phi(\mathbf{X})$ is an ordinary random variable. If Θ is k-dimensional, then $\phi(\mathbf{X})$ is a k-dimensional vector random variable.

It is convenient to use the notation $\hat{\boldsymbol{\theta}} = \phi(\mathbf{X})$ for an estimator $\phi(\mathbf{X})$, so $\hat{\boldsymbol{\theta}}$ is always a random variable or vector—some specific estimator of θ. Now reserve the symbol $\hat{\theta}$ for some specific value of $\hat{\boldsymbol{\theta}}$. (This is similar to using X to denote a random variable and x to denote one of its possible values.)

We can just as well consider the estimator as a function of the observed data vector $\mathbf{x}$. If $\hat{\boldsymbol{\theta}} = \phi(\mathbf{X})$, then if the outcome of $\mathbf{X}$ is $\mathbf{x}$, the estimate value is $\hat{\theta} = \phi(\mathbf{x})$. For instance, if

$$\hat{\boldsymbol{\theta}} = \frac{\mathrm{X}_1 + \cdots + \mathrm{X}_n}{n},$$

and the observed data vector is $(x_1, \ldots, x_n)$, then the estimate value is

$$\hat{\theta} = \frac{x_1 + \cdots + x_n}{n}.$$

Because of this equivalence, estimators are sometimes defined in terms of the sample values, that is, in terms of the data vector $\mathbf{x}$. For example, an estimate may be defined as the sample mean

$$\frac{x_1 + \cdots + x_n}{n}.$$

Obviously, in random variable terms this is the estimator

$$\frac{X_1 + \cdots + X_n}{n}.$$

To compare the performance of various estimators, a loss function $L(\hat{\theta}, \theta)$ must be specified. As in our example, this function is a gauge of inaccuracy or loss when the true state of nature is θ and the value of the estimate is $\hat{\theta}$. Define

$$R(\theta) = E_\theta L(\hat{\theta}, \theta).$$

That is, once the loss function is specified we can compute the operating characteristic $R(\theta)$ for any estimator. And once we know their operating characteristics, we can compare the performance of different estimators.

How should the loss function be selected? In the second section of this chapter we pointed out that the selection of the loss function depends on what is required of the estimate, on what use will be made of it. This means that you have to stop and think about what you are really trying to estimate. Suppose that you are really interested in the entire underlying distribution. You want to estimate θ so that the distribution you come up with is not "far" from the true underlying distribution $P_\theta(dx)$. In this case you want your loss function to measure how far apart the two distributions $P_{\hat{\theta}}(dx)$ and $P_\theta(dx)$ are. For example, if $F_\theta(x)$ is the cumulative distribution function of $P_\theta(dx)$, one possible measure of the distance is given by

$$L(\hat{\theta},\theta) = \max_x |F_{\hat{\theta}}(x) - F_\theta(x)|.$$

In other words, the loss is the maximum difference between the two distribution functions. For the exponential family,

$$L(\hat{\beta},\beta) = \max_x |e^{-x/\hat{\beta}} - e^{-x/\beta}|.$$

Finding the maximum is a simple calculus problem. If $\beta < \hat{\beta}$, set their ratio $\beta/\hat{\beta} = r$. Then

$$L(\hat{\beta},\beta) = (1 - r)r^{r/(1-r)}.$$

If $\beta > \hat{\beta}$, we get the same expression but with $\hat{\beta}/\beta = r$. For $\hat{\beta}$ close to β, this loss function takes on a simpler form. As $r \to 1$, $r^{r/(1-r)} \to e^{-1}$; hence for $r \simeq 1$, the loss is approximately $|1 - r|e^{-1}$, or dropping the constant factor e^{-1},

$$L(\hat{\beta},\beta) \simeq \frac{|\hat{\beta} - \beta|}{\beta}.$$

This is essentially percent error loss: To get the percent error in estimating β by $\hat{\beta}$ we would divide the absolute value of the difference by β and multiply by 100.

To go to the other extreme, suppose that in an exponential model we are interested only in estimating

$$P_\beta(\mathrm{T} \le 2000) = 1 - e^{-2000/\beta}.$$

Our estimation procedure is to obtain an estimate $\hat{\beta}$ of β and estimate the above probability by $1 - e^{-2000/\hat{\beta}}$. Suppose we judge our estimate by whether it gets within δ of the true probability or not. That is,

$$L(\hat{\beta},\beta) = \begin{cases} 0, & |e^{-2000/\hat{\beta}} - e^{-2000/\beta}| \le \delta \\ 1, & \text{otherwise.} \end{cases}$$

For $\hat{\beta} \simeq \beta$, this reduces to 0 for $e^{-2000/\beta}|\hat{\beta} - \beta|/\beta^2 \le \delta$ and 1 otherwise. If, furthermore, we are dealing with small failure probabilities; that is, if we are in the range of β such that $P_\beta(\mathrm{T} \le 2000) \ll 1$, then we can approximate the loss as

$$L(\hat{\beta},\beta) = \begin{cases} 0, & |\hat{\beta} - \beta|/\beta^2 \le \delta \\ 1, & \text{otherwise.} \end{cases}$$

Having gone through these two examples, we reverse our tracks and state that henceforth, in 1-parameter problems (Θ 1-dimensional) *we will always use as our loss function the squared error loss*

$$L(\hat{\theta}, \theta) = (\hat{\theta} - \theta)^2.$$

The reasons for this choice in ascending order of importance are:

a In regard to mathematical convenience, expected squared error loss is usually the easiest to compute.

b If, using squared error loss, one estimator $\hat{\boldsymbol{\theta}}_1$ is judged much better than a second estimator $\hat{\boldsymbol{\theta}}_2$, then usually $\hat{\boldsymbol{\theta}}_1$ will be found better than $\hat{\boldsymbol{\theta}}_2$, using any other similar loss function.

Here a similar loss function roughly means a fairly smooth loss function that is small when $\hat{\theta}$ is close to θ. Of course, every problem must be examined to see that b. is applicable. If the appropriate loss function for the problem looks quite unlike squared error loss, you have to think about whether using an estimator on the basis of its performance under squared error loss is justified. Of course b. is vague, and the qualifying word "usually" leaves an uncomfortable escape hatch.

Actually, we can make a more precise but more limited statement. Since $L(\hat{\theta}, \theta)$ is a measure of loss or inaccuracy, we can assume that it is 0 when $\hat{\theta} = \theta$ and that for θ fixed, the value of $\hat{\theta}$ that minimizes the loss is $\hat{\theta} = \theta$. Also we assume that for θ fixed, $L(\hat{\theta}, \theta)$ is a twice-differentiable function of $\hat{\theta}$ with nonzero 2nd derivative at $\hat{\theta} = \theta$. Then (see Problem 13) a partial Taylor expansion reveals that for $\hat{\theta}$ close to θ,

$$L(\hat{\theta}, \theta) \simeq c(\theta)(\hat{\theta} - \theta)^2.$$

Suppose that the sample size is large and that $\hat{\boldsymbol{\theta}}$ is a good estimator of θ in the sense that its outcomes are usually close to θ. Then the expected loss at the state of nature θ is $c(\theta)$ times the expected squared error loss. This implies:

c If the loss function satisfies the above assumptions, then for large sample sizes the ratio of risks $R_1(\theta)/R_2(\theta)$ for any two good estimators is approximately the same as if we used squared error loss to compute the risks.

We can get a useful decomposition of the squared error risk by introducing the concept of bias. Some estimators have the property that their values are centered around the true state of nature. More precisely, notice that in our first example,

$$E_p\bar{X} = p.$$

In our second example

$$E_l(2\bar{X}) = l$$
$$E_l(V) = l.$$

3.5 ***Definition*** *If an estimator $\phi(\hat{\boldsymbol{\theta}})$ of θ has the property that*

$$E_\theta\hat{\boldsymbol{\theta}} = \theta,$$

it is called unbiased. Otherwise, its bias $b(\theta)$ is given by

$$b(\theta) = E_\theta\hat{\boldsymbol{\theta}} - \theta.$$

For any estimator $\hat{\boldsymbol{\theta}}$ write its squared error risk as

$$\begin{aligned} R(\theta) &= E_\theta(\hat{\boldsymbol{\theta}} - \theta)^2 \\ &= E_\theta[\hat{\boldsymbol{\theta}} - E_\theta\hat{\boldsymbol{\theta}} + b(\theta)]^2 \qquad (3.6) \\ &= \sigma_\theta^2(\hat{\boldsymbol{\theta}}) + b^2(\theta). \end{aligned}$$

Thus, the risk breaks down into the sum of the variance of the estimator and the square of the bias. Frequently there is a trade-off between these two—if the absolute value of the bias is decreased, the variance increases, and conversely. For instance, consider the estimator $\gamma\bar{X}$ of p, $0 < \gamma < 1$. Since

$$E_p(\gamma\bar{X}) = \gamma p,$$

it has bias $p(\gamma - 1)$. Further,

$$\sigma_p^2(\gamma\bar{X}) = \gamma^2\sigma_p^2(\bar{X}).$$

Thus, $\bar{X}$ has 0 bias, but $\gamma\bar{X}$, $0 < \gamma < 1$, has a smaller variance than $\bar{X}$.

It is sometimes helpful to know that an estimator is unbiased. But it cannot be considered an essentially desirable property, because some biased estimators may have lower squared error loss than unbiased ones. For instance, we know that if $p \leq p_0$, the biased estimator has a lower risk than the unbiased estimator $\overline{X}$.

In theory, when we are sampling independently from the same underlying distribution, the number of observations is more or less at our disposal. Our data point could consist of the outcome of five observations $(\overline{X}_1, \ldots, \overline{X}_5)$ or of the ten observations $(\overline{X}_1, \ldots, \overline{X}_{10})$ and so on. In any case we are still trying to estimate the same parameter θ_1 of the underlying distribution. But of course, the estimators for different sample sizes will be different functions. If we have 5 observations, then ϕ will be a function of 5 variables. Given 10 observations, any estimator will be a function of 10 variables. *An estimation method, as distinct from an estimator, specifies the estimator for every possible sample size.* In the examples of the first few sections, notice we actually are dealing with estimation methods. When we write: *estimate p by*

$$\frac{X_1 + \cdots + X_n}{n},$$

we specify the estimator for every sample size n, and similarly for the estimators

$$\hat{\theta} = \frac{n+1}{n} \max(X_1, \ldots, X_n).$$

Now we are ready for

3.7 ***Definition*** *If $\hat{\theta}_n$ is a sequence of estimators of θ defined for every sample size, then call the estimation method given by $\hat{\theta}_n$ consistent if the risk*

$$E_\theta(\hat{\theta}_n - \theta)^2 \to 0$$

as $n \to \infty$, for all $\theta \in \Theta$.

Obviously, if we can't make our squared error loss as small as we want by taking the sample size large enough, then our estimation method is seriously awry. Look at the method where we estimate p for all n by $\gamma\overline{X}$. We know from the section that the risk is

$$\frac{\gamma^2 p(1-p)}{n} + p^2(1-\gamma)^2.$$

This method is not consistent. As $n \to \infty$, the risk converges to

$$p^2(1-\gamma)^2.$$

But if we use $\gamma_n\overline{X}$ where $\gamma_n \to 1$ as $n \to \infty$, we get consistency.

Actually there is another definition of consistency that is used more often than 3.7: *An estimation method is called consistent if the probability that $\hat{\theta}_n$ differs from the true state of nature θ by more than any small fixed amount goes to 0 as the sample size becomes large.* In other words, the method is consistent in this sense if

$$P_\theta(|\hat{\theta}_n - \theta| > \delta) \to 0$$

as $n \to \infty$ for every value of $\delta > 0$ and every $\theta \in \Theta$. The intuitive content of the second definition is clear: If our estimates do not get more tightly packed around the true value as the sample size increases, then something is wrong with our approach or assumptions.

How are these two definitions related? If the expected squared distance of $\hat{\theta}_n$ from θ goes to 0, you can see that the values of $\hat{\theta}_n$ must cluster more tightly around θ. That is, if an estimation method is consistent by the first definition, then it is always consistent in terms of the second definition. To prove this, reason as follows: The squared error $|\hat{\theta}_n - \theta|^2$ is greater than δ^2 for all outcomes such that $|\hat{\theta}_n - \theta| > \delta$. Therefore $(\hat{\theta}_n - \theta)^2$ will be greater than δ^2 with probability $P(|\hat{\theta} - \theta| > \delta)$. This leads to the inequality

$$E_\theta(\hat{\theta}_n - \theta)^2 \geq \delta^2 P(|\hat{\theta}_n - \theta| > \delta).$$

Hold δ fixed and let $n \to \infty$. If the left side goes to 0, then so does the right side.

If $P(|\hat{\theta}_n - \theta| > \delta)$ goes to 0 for all $\delta > 0$, then usually the squared error risk goes to 0. If the values of $\hat{\theta}_n$ are tightly clustered around θ and are an appreciable distance away with only small probability, then the expected squared distance of $\hat{\theta}_n$ from θ should be small. The problem comes in those unusual cases where the low probability outcomes for which $\hat{\theta}_n$ is far from θ contribute a disproportionate amount to the expected squared error. Problem 14 is an example. At any rate, for convenience, we use the first definition of consistency, not the second.

Consistency can be regarded as the minimal requirement any reasonable estimation method must satisfy. Virtually any sensible estimation method will be consistent. The requirement that a method be consistent does not help much in choosing the "best" method from among the "good" methods, but it does eliminate the really bad procedures.

In our examples so far the parameter space has always been 1-dimensional. Now suppose the underlying distributions are uniform over some unknown interval. In other words, the set of all possible distributions consists of all distributions which are uniform over some interval. A distribution in this set is specified by the two endpoints θ_1, θ_2, $\theta_1 \leq \theta_2$, of the interval. The parameter space Θ is the 2-dimensional half-plane consisting of all points (θ_1, θ_2), $\theta_1 \leq \theta_2$. To estimate, we need two functions $\hat{\boldsymbol{\theta}}_1$, $\hat{\boldsymbol{\theta}}_2$ of the data vector **X**, one to estimate θ_1 and the other to estimate θ_2. So an estimator here is a vector random variable

$$(\hat{\boldsymbol{\theta}}_1, \hat{\boldsymbol{\theta}}_2).$$

The next now goes exactly like the 1-dimensional case. To see how good an estimator is, we need a loss function

$$L[(\hat{\theta}_1, \hat{\theta}_2), (\theta_1, \theta_2)]$$

that measures the loss due to the discrepancy between the state of nature (θ_1, θ_2) and its estimate $(\hat{\theta}_1, \hat{\theta}_2)$. The risk function is defined by

$$R(\theta_1, \theta_2) = E_{(\theta_1,\theta_2)}L((\hat{\boldsymbol{\theta}}_1, \hat{\boldsymbol{\theta}}_2), (\theta_1, \theta_2)).$$

The only change in the risk is that now it is a function of two variables θ_1, θ_2 and has no easy graphical representation. Take the squared error loss here to be

$$L[(\hat{\theta}_1, \hat{\theta}_2), (\theta_1, \theta_2)] = (\hat{\theta}_1 - \theta_1)^2 + (\hat{\theta}_2 - \theta_2)^2;$$

that is, the loss is just the square of the ordinary Euclidean distance between the two points $(\hat{\theta}_1, \hat{\theta}_2)$ and (θ_1, θ_2). You can see that this generalizes to any number of dimensions.

To define consistency in two or more dimensions, we again demand that the squared error risk go to 0 as the sample size increases. Notice that because of the form of squared error loss in 2 (or more) dimensions, consistency of $(\hat{\boldsymbol{\theta}}_1, \hat{\boldsymbol{\theta}}_2)$ is equivalent to consistency of each of $\hat{\boldsymbol{\theta}}_1$, $\hat{\boldsymbol{\theta}}_2$.

Problem 6 To decide what type of switching equipment to use, a count is made of incoming phone calls over n time intervals. Assume the calls arrive in a Poisson process with parameter λ. To estimate λ use the average number of calls observed per unit time interval. The appropriate loss function $L(\hat{\lambda}, \lambda)$ is 0 if $|\hat{\lambda} - \lambda| \leq .5$. If $\hat{\lambda}$ overestimates λ by .5 or more, the loss is e_1. If $\hat{\lambda}$ underestimates λ by .5 or more, the loss is e_2. Evaluate the loss using the normal approximation if $\lambda = 10$ and $n = 300$. (Ans. $.0031(e_1 + e_2)$.)

Problem 7 Graph $R(\lambda)$ for Problem 6 using the loss function

$$L(\hat{\lambda}, \lambda) = (\hat{\lambda} - \lambda)^2.$$

Problem 8 You are acceptance-testing batches of components with each component classified as defective or nondefective. In each lot you test n items and estimate the parameter p as the proportion of defectives. If there are $\gamma\%$ of defectives in the lot, $\gamma > 5$, then the manufacturer is penalized $e_1(\gamma - 5)$ dollars if he passes the lot on to the user. If the lot is rejected, then the manufacturer loses an estimated amount e_2. It is decided to set a level p_0 and reject the lot if the estimate $\hat{p} \geq p_0$, otherwise accept it.

a Write out the loss function $L(\hat{p}, p)$.

b Write out the risk function in terms of $P_p(\hat{p} \geq p_0)$.

Problem 9 The underlying distribution for a population is normal with unknown mean μ and variance 1. Suppose the loss function depends only on the difference between $\hat{\mu}$ and μ, and we use the estimate

$$\hat{\mu} = \frac{X_1 + \cdots + X_n}{n}.$$

Show that the risk $R(\mu)$ is a constant function.

Problem 10 A body with no forces acting on it is moving through space such that its x-coordinate is

$$x(t) = x_0 + vt$$

with both x_0 and v unknown. At times $0, 1, 2, \ldots, n$ you receive position signals from the body which are assumed to be of the form

$$x(k) + Y_k,$$

where $Y_0, \ldots, Y_n$ are independent $N(0, \sigma^2)$ variables with σ^2 known. You want to estimate the position of the body at time $N > n$. If your estimate is correct to within a distance d, a docking maneuver can be carried out; otherwise it can not. From the data you form an estimate $(\hat{x}_0, \hat{v})$ of (x_0, v). Write out the appropriate form of the loss function $L[(\hat{x}_0, \hat{v}), (x_0, v)]$.

Problem 11 $X_1, \ldots, X_n$ are independent samples from an $N(\mu, \sigma^2)$ distribution, both μ and σ^2 unknown. Denote

$$\hat{\mu} = \frac{X_1 + \cdots + X_n}{n}.$$

Compute

$$R(\mu, \sigma^2) = E_{(\mu,\sigma^2)}(\hat{\mu} - \mu)^2$$

Problem 12 $X_1, \ldots, X_n$ are independent samples from a distribution uniform on the unknown interval $[a, b]$, $a \leq b$. Give a method of estimating (a, b). Can you find unbiased estimators of a and b? Using squared Euclidean distance as loss, compute $R(a, b)$ for your estimators $(\hat{a}, \hat{b})$. Are both $\hat{a}$ and $\hat{b}$ consistent?

Problem 13 For a given loss function $L(\hat{\theta}, \theta)$, hold θ fixed and consider it as a function of $\hat{\theta}$ only. Assume that

a L is a twice-differentiable function of $\hat{\theta}$.

b

$$L(\hat{\theta}, \theta) = 0, \qquad \left.\frac{\partial L(\hat{\theta}, \theta)}{\partial \hat{\theta}}\right|_{\hat{\theta}=\theta} = 0, \qquad \left.\frac{\partial^2 L(\hat{\theta}, \theta)}{\partial \hat{\theta}^2}\right|_{\hat{\theta}=\theta} \neq 0.$$

The second condition is quite sensible. Since $L(\hat{\theta}, \theta)$ is a loss or inaccuracy measure, it should have its minimum (as a function of $\hat{\theta}$) when $\hat{\theta} = \theta$, and if $\hat{\theta} = \theta$, the loss or inaccuracy is 0.

Show that for $\hat{\theta}$ close to θ,

$$L(\theta, \theta) \simeq c(\theta)(\hat{\theta} - \theta)^2,$$

where

$$c(\theta) = \frac{1}{2}\left.\frac{\partial^2 L(\hat{\theta}, \theta)}{\partial \hat{\theta}^2}\right|_{\hat{\theta}=\theta}.$$

Problem 14 For the model of Section 2, we decide that we want to estimate $\log p$ instead of p. We propose to estimate $\log p$ by $\log \bar{X}$

a Show that the squared error risk of $\log \bar{X}$ is infinite for all p.

b Show that $\log \bar{X}$ is a consistent estimator for $\log p$ in the sense of the second definition of consistency.

c What is the difficulty here?

The Maximum Likelihood Method

The point has been made that there is usually no single method of estimation which gives a lower risk than any other method for all values of the parameter. However, there is a method which usually works well. To illustrate this, suppose we toss a coin 10 times independently to estimate $p = P(\text{heads})$ and get 4 heads. Then it seems clear that the best estimate of p is .4. Why? For one thing, .4 is the observed proportion of heads. But here is another way of reasoning: Suppose the observed sequence is

$$\omega_0 = \text{H T T T H T T H H T}.$$

The probability of getting this sequence if the probability of heads is p is simply

$$P_p(\omega_0) = p^4(1 - p)^6.$$

Under some values of p, say $p = .9$, the probability of getting ω_0 is very small. Thus if $p = .9$, we would not be likely to get the outcome ω_0. Actually, then, knowing the outcome, it is very plausible to guess that the true state of nature is that one which give the observed outcome the highest probability. That is, pick as our estimate the value of p that maximizes

$$p^4(1 - p)^6.$$

This function is 0 at $p = 0$ and $p = 1$, and positive inside $[0, 1]$. We can find its maximum by differentiating:

$$\frac{d}{dp}(p^4(1 - p)^6) = 0$$

or

$$4p^3(1 - p)^6 - 6p^4(1 - p)^5 = 0.$$

Transposing and cancelling gives

$$4(1 - p) = 6p;$$

hence the maximum is at $p = .4$. So $p = .4$ is in this case our maximum likelihood estimator.

You can see what the principle is: Given that the outcome of the experiment is $\mathbf{x}$; that is, $\mathbf{X} = \mathbf{x}$, then *estimate θ as that state of nature which maximizes $P_\theta(\mathbf{X} = \mathbf{x})$.*

To illustrate further, suppose that on N successive days batches of 1000 fruit flies were irradiated with X-rays and the number of mutations counted. The results were listed as $n_1, n_2, \ldots, n_N$. The X-ray treatment was the same every day, and the total number of mutations on each day was small compared to the total number (1000) of flies. The problem is to estimate the distribution of the number of mutations. This is first of all a problem in model building. Assume that in each batch of 1000, each fly has some unknown probability p of mutating and $1 - p$ of not mutating, independently of the other flies. Thus the total number of mutations per day is the result of 1000 independent trials with probability p, $1 - p$ of mutation or no mutation respectively. Since n_1, $n_2, \ldots, n_N$ are small compared to the total number of trials, it follows that p must be quite small. Hence a good model for the total number of mutations per day is a Poisson distribution with unknown parameter λ. Thus $n_1, n_2, \ldots, n_N$ are the outcomes of N independent (an additional assumption) random variables $X_1, \ldots, X_N$, each having a Poisson distribution with λ unknown. The probability

$$P_\lambda(X_1 = n_1, \ldots, X_N = n_N)$$

is therefore the product

$$\left(\frac{\lambda^{n_1}}{n_1!} e^{-\lambda}\right)\left(\frac{\lambda^{n_2}}{n_2!} e^{-\lambda}\right) \cdots \left(\frac{\lambda^{n_N}}{n_N!} e^{-\lambda}\right)$$

$$= \frac{1}{n_1! \cdots n_N!} \lambda^{(n_1 + \cdots + n_N)} e^{-N\lambda}.$$

The idea is to find the value of λ that maximizes this probability. It is easier to take the log of this probability and maximize it, reasoning that the log of a function is at a maximum where the function is at a maximum. We have

$$L_\lambda(n_1, \ldots, n_N) = \log P_\lambda(X_1 = n_1, \ldots, X_N = n_N)$$

$$= \log\left(\frac{1}{n_1! \cdots n_N!}\right) + (n_1 + \cdots + n_N) \log \lambda - N\lambda.$$

To find the maximizing λ, differentiate and set equal to 0:

$$\frac{d}{d\lambda} L_\lambda(n_1, \ldots, n_N) = 0$$

or

$$(n_1 + \cdots + n_N) \frac{1}{\lambda} - N = 0,$$

so

$$\lambda = \frac{n_1 + \cdots + n_N}{N}.$$

In this case, then, maximum likelihood leads to the estimator

$$\hat{\lambda} = \frac{X_1 + \cdots + X_N}{N}.$$

This is encouraging, because it is intuitively appealing that the Poisson parameter λ be estimated by the average count per trial.

There is a small difficulty in our path. Suppose the random variables $X_1, \ldots, X_n$ we observe have a continuous distribution, so that for any particular outcomes $x_1, \ldots, x_n$,

$$P_\theta(X_1 = x_1, \ldots, X_n = x_n) = 0$$

for all $\theta \in \Theta$. If these variables have a joint density for every value of θ, then reason this way: If the observed outcomes are $x_1, \ldots, x_n$, the probability that any outcome falls in a little rectangle around the point $(x_1, \ldots, x_n)$ with sides of length $dx_1, \ldots, dx_n$ is

$$P_\theta(X_1 \in dx_1, \ldots, X_n \in dx_n) = f_\theta(x_1, \ldots, x_n)\, dx_1 \cdots dx_n,$$

where $f_\theta(x_1, \ldots, x_n)$ is the joint density. This principle of maximum likelihood leads to: *Estimate θ as that value that makes $f_\theta(x_1, \ldots, x_n)$ a maximum*, thus making the probability of landing in the little rectangle around the point $(x_1, \ldots, x_n)$ a maximum. We illustrate for the transistor problem of Chapter 1. If we assume an exponential failure time distribution with parameter β, then the joint density for the variables $T_1, \ldots, T_n$ is the product of the individual densities:

$$f_\beta(t_1, \ldots, t_n) = \left(\frac{1}{\beta} e^{-t_1/\beta}\right) \cdots \left(\frac{1}{\beta} e^{-t_n/\beta}\right)$$

$$= \frac{1}{\beta^n} e^{-(t_1 + \cdots + t_n)/\beta}.$$

Suppose in a sample of size 100 the observed lifetimes are 1100, 2307, 1986, ..., 1852. To get the maximum likelihood estimate we substitute these numbers for $t_1, \ldots, t_{100}$ in the joint density function. Note that the density only depends on the sum of the t's. Suppose their sum is 178762. Then the evaluation of the density is

$$\frac{1}{\beta^{100}} e^{-178762/\beta}.$$

Taking logarithms gives

$$L_\beta = -100 \log \beta - \frac{178762}{\beta}.$$

The equation

$$\frac{dL_\beta}{d\beta} = 0$$

is

$$-100 \cdot \frac{1}{\beta} + 178762 \cdot \frac{1}{\beta^2} = 0,$$

or $\hat{\beta} = 178762/100$.

To get the general form of the maximum likelihood estimator in the above example for any data vector $\mathbf{t} = (t_1, \ldots, t_n)$, take the logarithm of the density:

$$L_\beta = -n \log \beta - \frac{t_1 + \cdots + t_n}{\beta}.$$

The equation

$$\frac{dL_\beta}{d\beta} = 0$$

leads to

$$\hat{\beta} = \frac{t_1 + \cdots + t_n}{n}.$$

So the maximum likelihood estimator here is

$$\hat{\beta} = \frac{\mathsf{T}_1 + \cdots + \mathsf{T}_n}{n}.$$

In this example notice that we first used specific numbers to get the numerical value of the maximum likelihood estimate, and then proceeded to get the general form of the estimator. This was done to emphasize that it is not necessary to derive the general expression for the estimator to use maximum likelihood—just substitute the outcome values into the density and maximize with respect to β.

Here is an interesting question: For the example in the fourth section of this chapter the outcomes were drawn from a uniform distribution with unknown range $[0, l]$. What is the maximum likelihood estimator for l? The cumulative distribution function of this underlying distribution is given by

$$F_l(x) = \begin{cases} 0, & x \le 0 \\ x/l, & 0 \le x \le l \\ 1, & x \ge l, \end{cases}$$

so the density is

$$f_l(x) = \begin{cases} 0, & x \le 0 \\ 1/l, & 0 \le x \le l \\ 0, & x > l, \end{cases}$$

and the joint density of the sample is

$$f_l(x_1, \ldots, x_n) = \begin{cases} 1/l^n, & 0 \le x_j \le l, j = 1, 2, \ldots, n \\ 0, & \text{otherwise.} \end{cases}$$

Here we have to be careful. If we took—

$$L_l = \log f_l(x_1, \ldots, x_n) = -n \log l$$

and differentiated, we would get

$$-\frac{n}{l} = 0$$

or $\hat{l} = \infty$ as our answer. But $1/l^n$ is *not* the joint density. If we graph the joint density as a function of l, letting $y = \max(x_1, \ldots, x_n)$, we get Figure 3.3,

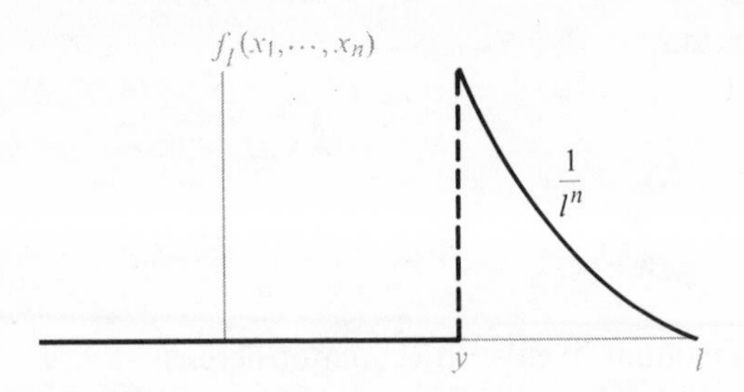

Figure 3.3

because $f_l(x_1, \ldots, x_n) = 0$ for $l \leq y$. Here the maximum of f_l is at $l = y$. Therefore the maximum likelihood estimator for l is given by

$$\hat{l} = \max(X_1, \ldots, X_n).$$

This example illustrates that while there is always (for any reasonable problem) a θ which maximizes L_θ, you cannot always find it by differentiation. Sometimes the maximizing values, as in the above example, will be endpoint values.

Here are two more examples illustrating the use of maximum likelihood. Consider independent identically distributed random variables $X_1, \ldots, X_n$ such that each one can only take on one of a finite number of values, say $1, \ldots, J$. For example, we may be classifying items after a certain amount of usage into J categories, such as, *in good condition*, *needs repair*, and so on.

We want to estimate the probabilities $p_1, \ldots, p_J$ of getting outcomes $1, \ldots, J$. If in n trials, we observe n_1 outcomes 1, n_2 of $2, \ldots, n_j$ of J, then the obvious guess is to estimate p_j, $j = 1, \ldots, J$, by n_j/n. This is also the maximum likelihood estimate, because the probability of getting the sequence $x_1, \ldots, x_n$ of outcomes is given by

$$P_{(p_1,\ldots,p_J)}(X_1 = x_1, \ldots, X_n = x_n) = p_1^{n_1} p_2^{n_2} \cdots \; p_J^{n_J}.$$

So

$$L_{(p_1,\ldots,p_J)} = n_1 \log p_1 + n_2 \log p_2 + \cdots + n_J \log p_J.$$

We want to maximize this expression, but we have to satisfy the restraint $p_1 + \cdots + p_J = 1$. Using a Lagrange multiplier λ gives us the equations

$$\frac{\partial L}{\partial p_j} - \lambda \frac{\partial}{\partial p_j}\left(\sum_1^J p_j\right) = 0, \qquad j = 1, \ldots, J,$$

or

$$\frac{n_j}{p_j} - \lambda = 0, \qquad j = 1, \ldots, J.$$

Hence the maximizing p_j are n_j/λ, and the condition that the $\sum p_j = 1$ gives $\lambda = n$.

Although it is nice to know that the maximum likelihood principle validates our intuitive guesses about the right way to estimate, it hasn't shown us anything really new yet. But consider a more restricted and difficult version of the above problem. Suppose you have a model set up where $p_1, \ldots, p_J$ are to be functions of one or more variables. For example, suppose that the model hypothesizes that the underlying probabilities $p_1, \ldots, p_J$ form a geometric progression with unknown common ratio $r \leq 1$:

$$p_1 = A,\ p_2 = Ar,\ p_3 = Ar^2, \ldots, p_J = Ar^{J-1},$$

where A is chosen to make the sum $p_1 + p_2 + \cdots + p_J = 1$. So

$$A = \frac{1 - r}{1 - r^J}\,.$$

Now find a good estimator for r, given that you observed $n_1, \ldots, n_J$ occurrences of outcomes $1, \ldots, J$, respectively.

One interesting method of solution would be to use a least squares approach. Reason this way: n_1 should be approximately np_1, n_2 should be nearly np_2, and so on. To estimate r, then, minimize

$$\sum_{j=1}^{J} (n_j - nAr^{j-1})^2.$$

But you might just as well argue that if

$$\frac{n_j}{n} \simeq Ar^{j-1},$$

then the numbers $\log (n_j/n)$ as a function of j should fall on the straight line $(j - 1) \log r + \log A$. So it seems as plausible to try to fit a straight line to the log of the data using least squares. Again, since n_{j+1}/n_j should be nearly equal to r for all j from 1 to $j - 1$, you might try to estimate r as the average of these ratios. That is, use the estimator

$$\frac{1}{J - 1} \sum_{j=1}^{J-1} \left(\frac{n_{j+1}}{n_j}\right).$$

Thus, there are a plethora of reasonable methods for estimating r. There is certainly no method which stands out immediately and intuitively as being "the right thing to do."

The maximum likelihood method starts with

$$L_{(p_1, \ldots, p_J)} = \sum_{1}^{J} n_j \log p_j.$$

For $p_j = Ar^{J-1}$, the only unknown parameter is r, and

$$L_r = \sum_1^J n_j[\log A + (j-1)\log r]$$

$$= n \log A + \log r \cdot \sum_1^J (j-1)n_j.$$

Setting the derivative of L_r equal to 0 gives the equation

$$-r \frac{d}{dr} \log A = \sum_1^J (j-1)\left(\frac{n_j}{n}\right).$$

There is no simple way of solving this equation; we have to use numerical methods or approximation techniques. For instance, as a first approximation, we can get a graphical solution. Plot the left-hand side as a function of r. Draw a horizontal line across this graph at a height equal to the value of the right-hand side. The value of r at which the curve and the line intersect is the solution.

The maximum likelihood equation may not have simple solutions. In the previous examples, the probabilities involved were such that solutions were easy. This does not always happen.

Problem 15 To estimate the underlying proportion p of defectives in a large batch, testing is carried out sequentially until 10 defectives are found. If 168 items are checked before stopping, what is the maximum likelihood estimate for p?

Problem 16 Compute the mean square error loss

a for the estimator of λ in a Poisson distribution given by

$$\hat{\lambda} = \frac{X_1 + \cdots + X_n}{n}.$$

b for the maximum likelihood estimator $M = \max(X_1, \ldots, X_n)$ of the range l of a uniform distribution on $[0, l]$.

Problem 17 Particles are emitted from a particle gun and strike a flat target at a distance 1 unit away from the emitting point. In 2 dimensions this looks like Figure 3.4.

Assume that the angle U (see Figure 3.5) that an emitted particle makes with the axis of the gun does not depend on the angle ϕ and has density function $f(u)$. (U is assumed to always be in a

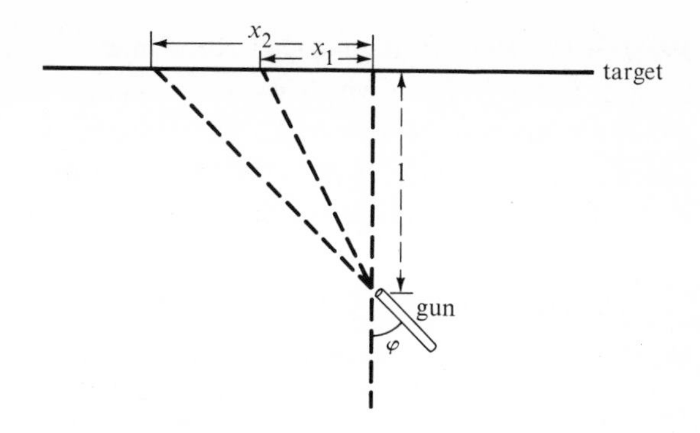

Figure 3.4

range small enough so that all emitted particles hit the target.) What is unknown is the angle ϕ. The striking points $x_1, \ldots, x_n$ of n particles are recorded, as measured from the foot of the perpendicular (see Figure 3.4).

a Show that the maximum likelihood estimate of ϕ is that value of ϕ which maximizes the expression

$$\sum_{k=1}^{n} \log f(y_k - \phi),$$

where $y_k = \text{arc tan } x_k$.

b Assume that U has a truncated normal distribution; that is,

$$f(u) = \begin{cases} ce^{-u^2/2\sigma^2}, & |u| \leq \delta \\ 0, & |u| > \delta, \end{cases}$$

where c is selected so that

$$\int f(u)\, du = 1$$

and $[-\delta, +\delta]$ is the fixed maximum range of U. Show that the maximum likelihood estimate of ϕ is given by

$$\hat{\phi} = \frac{y_1 + \cdots + y_n}{n}.$$

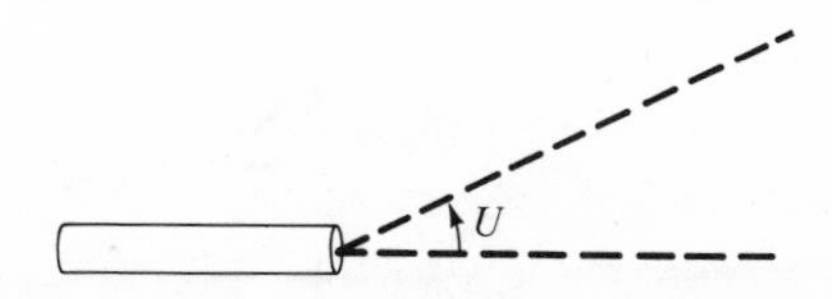

Figure 3.5

Problem 18 Here is a somewhat surprising result. Let the underlying distribution from which we are to sample be uniform on an unknown interval of length 1. That is, we observe $x_1, \ldots, x_n$ selected independently from a uniform distribution on $[m - \frac{1}{2}, m + \frac{1}{2}]$ where m is unknown. Show that the maximum likelihood estimate of m is *any* value of m in the range

$$\max(x_1, \ldots, x_n) - \tfrac{1}{2} \leq m \leq \min(x_1, \ldots, x_n) + \tfrac{1}{2}.$$

What can be said about the size of this range if n is large?

Problem 19 Incoming discrete time noise $X_1, \ldots, X_n$ is assumed to consist of independent $N(\mu, 1)$ variables with μ unknown. However, what you observe is the power in the noise; that is, you observe the values of $X_1^2, \ldots, X_n^2$. From these values, denoted by $y_1, \ldots, y_n$, you want to estimate μ.

a Show that the maximum likelihood estimate of μ.satisfies the equation

$$\mu = \frac{1}{h} \sum_k \sqrt{y_k} \tanh(\sqrt{x_k}\mu).$$

[Recall that $\tanh x = (e^x - e^{-x})/(e^x + e^{-x})$.]

Verify that if $\hat{\mu}$ is a solution to this equation, so is $-\hat{\mu}$. Discuss why this makes sense considering the data.

b Use the expansion (to third order terms)

$$\tanh x = x - \tfrac{1}{3}x^3$$

to find an approximate solution to the equation in a. Discuss in terms of the data $y_1, \ldots, y_n$ when this will be a good approximation to the actual solution.

Problem 20 Components have failure times with the Weibull distribution

$$P(T > t) = e^{-t^{3/2}/\beta}.$$

Find the maximum likelihood estimator of β based on n independent observations. Is this estimator biased or unbiased? Find the squared error risk for this estimator.

(Ans.

$$\hat{\beta} = \frac{1}{n} \sum_1^n t_k^{3/2}$$

$$R(\beta) = \frac{1}{n} \cdot \beta^2.)$$

Maximum Likelihood Applied to Normal Distributions

As our model, assume that the data $x_1, \ldots, x_n$ are drawn independently from an $N(\mu, \sigma^2)$ distribution, with both μ and σ^2 unknown. Thus the parameter space Θ is 2-dimensional, consisting of all points $\theta = (\mu, \sigma^2)$ where $-\infty < \mu < \infty$ and $0 \leq \sigma^2 < \infty$. The joint density is given by the product

$$\left(\frac{1}{\sqrt{2\pi\sigma^2}} \exp\left[-\frac{(x_1 - \mu)^2}{2\sigma^2}\right]\right) \cdots \left(\frac{1}{\sqrt{2\pi\sigma}} \exp\left[-\frac{(x_n - \mu)^2}{2\sigma^2}\right]\right)$$

$$= \frac{1}{(\sqrt{2\pi})^n\sigma^n} \exp\left[-\frac{1}{2\sigma^2}\sum_1^n (x_k - \mu)^2\right]$$

We can write

$$L_\theta(x_1, \ldots, x_n) = -n \log \sigma - n \log (\sqrt{2\pi}) - \frac{1}{2\sigma^2}\sum_1^n (x_k - \mu)^2.$$

To find the maximizing value of θ here, we need the two equations

a $\dfrac{\partial L_\theta}{\partial \mu} = 0$

b $\dfrac{\partial L_\theta}{\partial \sigma} = 0$

or

a $\dfrac{1}{2\sigma^2}\displaystyle\sum_1^n 2(x_k - \mu) = 0$

b $-\dfrac{n}{\sigma} + \dfrac{1}{\sigma^3}\displaystyle\sum_1^n (x_k - \mu)^2 = 0.$

Equation a. can be solved for μ, giving

$$\hat{\mu} = \frac{x_1 + \cdots + x_n}{n}.$$

Using this value for μ in b. we get the maximizing value of σ^2:

$$\hat{\sigma}^2 = \frac{1}{n}\sum_1^n (x_k - \hat{\mu})^2.$$

These are our maximum likelihood estimates for μ and σ^2. We use the notation

$$\bar{X} = \frac{X_1 + \cdots + X_n}{n}$$

$$\hat{\sigma}^2 = \frac{1}{n}\sum_1^n (X_k - \bar{X})^2$$

($\hat{\mu}$ is sometimes used instead of $\bar{X}$). These estimators are called the *sample mean* and *sample variance*, and are used frequently with distributions other than the normal.

The estimate $\hat{\sigma}^2$ is not unbiased. If the state of nature is (μ, σ^2), write

$$Y_k = X_k - \mu$$

$$\bar{Y} = \frac{1}{n} \sum_1^n Y_k.$$

Then $\bar{Y} = \bar{X} - \mu$ and

$$X_k - \bar{X} = (X_k - \mu) - (\bar{X} - \mu) = Y_k - \bar{Y}.$$

Now

$$E(Y_k - \bar{Y})^2 = EY_k^2 + E\bar{Y}^2 - 2EY_k\bar{Y},$$

where the expectation here is really $E_{(\mu,\sigma^2)}$ but we have left out the subscript for simplicity. Under the state of nature (μ, σ^2), Y_k has an $N(0, \sigma^2)$ distribution, so $EY_k^2 = \sigma^2$,

$$E\left(\frac{1}{n}\sum_1^n Y_k\right)^2 = V\left(\frac{1}{n}\sum_1^n Y_k\right) = \frac{1}{n^2}[V(Y_1) + \cdots + V(Y_n)]$$

$$= \frac{1}{n} V(Y_1) = \frac{\sigma^2}{n},$$

and

$$E(Y_k\bar{Y}) = \frac{1}{n} E(Y_k^2) = \frac{\sigma^2}{n}.$$

Therefore

$$E(X_k - \bar{X})^2 = \sigma^2 - \frac{\sigma^2}{n} = \frac{n-1}{n}\sigma^2$$

and

$$E_{(\mu,\sigma^2)}\hat{\sigma}^2 = \frac{n-1}{n}\sigma^2.$$

If you are convinced that unbiased estimates should be used, then instead of $\hat{\sigma}^2$, use

$$\hat{s}^2 = \frac{n}{n-1}\hat{\sigma}^2,$$

which is unbiased and equals

$$\frac{1}{n-1}\sum_1^n (X_k - \bar{X})^2.$$

The maximum likelihood estimator $\hat{\sigma}^2$ has lower risk than $\hat{s}^2$ for all σ^2 (see Problem 22). But for any size n at all, the two risks are very close over a wide range of values of σ^2. Which one you choose depends on how much aesthetic attraction unbiased estimates have for you.

There are some interesting and odd facts about the distributions of the variables $\bar{X}$ and $\hat{\sigma}^2$. First, they are independent (see Problem 23). It's fairly clear that $\bar{X}$ has an $N(\mu, \sigma^2/n)$ distribution. What is harder to see is that

$$\frac{n\hat{\sigma}^2}{\sigma^2} = \sum_1^n \frac{(X_k - \bar{X})^2}{\sigma^2}$$

has a χ^2 distribution with $n - 1$ degrees of freedom. We could reason as follows: The $Y_k = X_k - \bar{X}$ are n normally distributed variables. If $Y_1, \ldots, Y_n$ were independent $N(0, 1)$ variables, then $\sum_1^n Y_k^2$ would have a χ^2 distribution with n degrees of freedom. But the Y_k are not independent —they satisfy one linear relation, namely, $Y_1 + \cdots + Y_n = 0$. Hence we drop one degree of freedom. While this argument produces the right answer and gives an intuitive description of what is going on, it is not generally valid and needs more foundation. Problem 24 gives a more mathematically satisfying method of deriving this result.

Now suppose that the data consist of pairs of readings $(x_1, y_1), \ldots, (x_n, y_n)$ assumed to be drawn independently from a bivariate normal distribution characterized by the 5 unknown parameters $\mu_1, \mu_2, \sigma_1, \sigma_2, \rho$. The product of the densities is

$$\frac{1}{(\sqrt{2\pi\sigma_1^2\sigma_2^2(1 - \rho^2)})^n} e^{-\frac{1}{2}Q(\mathbf{x},\mathbf{y})},$$

where

$$Q(\mathbf{x},\mathbf{y}) = \frac{1}{1 - \rho^2} \sum_k \left[\left(\frac{x_k - \mu_1}{\sigma_1}\right)^2 - 2\rho\left(\frac{x_k - \mu_1}{\sigma_1}\right) \times \left(\frac{y_k - \mu_2}{\sigma_2}\right) + \left(\frac{y_k - \mu_2}{\sigma_2}\right)^2\right].$$

Then

$$L_\theta(\mathbf{x}, \mathbf{y}) = \frac{n}{2} \log\left[2\pi\sigma_1^2\sigma_2^2(1 - \rho^2)\right] - \tfrac{1}{2}Q(\mathbf{x}, \mathbf{y}).$$

To find the maximizing value of $\theta = (\mu_1, \mu_2, \sigma_1, \sigma_2, \rho)$, solve the 5 equations

$$\frac{\partial L_\theta}{\partial \mu_1} = 0, \qquad \frac{\partial L_\theta}{\partial \mu_2} = 0, \qquad \frac{\partial L_\theta}{\partial \sigma_1} = 0, \qquad \frac{\partial L_\theta}{\partial \sigma_2} = 0, \qquad \frac{\partial L_\theta}{\partial \rho} = 0.$$

The computations are tedious (try it if you have the time). But the results are simple and intuitively gratifying:

$$\hat{\mu}_1 = \frac{1}{n} \sum_k x_k, \qquad \hat{\mu}_2 = \frac{1}{n} \sum_k y_k$$

$$\hat{\sigma}_1^2 = \frac{1}{n} \sum_k (x_k - \hat{\mu}_1)^2, \qquad \hat{\sigma}_2^2 = \frac{1}{n} \sum_k (y_k - \hat{\mu}_2)^2$$

$$\hat{\rho} = \frac{1/n \sum_k (x_k - \hat{\mu}_1)(y_k - \hat{\mu}_2)}{\hat{\sigma}_1 \hat{\sigma}_2}.$$

The estimates $\hat{\mu}_1$, $\hat{\mu}_2$, $\hat{\sigma}_1^2$, $\hat{\sigma}_2^2$ are exactly what we might expect from looking at the univariate normal case. The novel feature of the bivariate model is the appearance of $\hat{\rho}$, called *the sample correlation coefficient*, as the maximum likelihood estimate for ρ. In view of the fact that if X and Y have a bivariate normal distribution with parameters $(\mu_1, \mu_2, \sigma_1, \sigma_2, \rho)$, then

$$\rho = \frac{E(X - \mu_1)(Y - \mu_2)}{\sigma_1 \sigma_2},$$

the form of the estimate $\hat{\rho}$ above is not surprising. The computation of the squared error risk of $\hat{\rho}$ is formidable. We shall come back to it in a later chapter.

These examples show how maximum likelihood works if the parameter θ is multidimensional. If $\theta = (\theta_1, \ldots, \theta_k)$, then the maximum likelihood equations are simply

$$\frac{\partial L_\theta}{\partial \theta_1} = 0, \frac{\partial L_\theta}{\partial \theta_2} = 0, \ldots, \frac{\partial L_\theta}{\partial \theta_k} = 0.$$

These give k equations for the k unknowns $\theta_1, \ldots, \theta_k$. They may be complicated and difficult to solve, perhaps requiring a machine search for the maximizing value of θ.

Problem 21 Given samples $X_1, \ldots, X_n$ from a normal $N(\mu, \sigma^2)$ distribution with both μ and σ^2 unknown, find the mean squared error loss $R(\mu, \sigma^2)$ in estimating μ by the sample mean.

Problem 22 (More difficult!) In the situation of Problem 21 consider the two estimators for σ^2:

a $$\hat{\sigma}^2 = \frac{1}{n} \sum_1^n (X_k - \bar{X})^2,$$

where $\bar{X}$ is the sample mean, and

b
$$\hat{s}^2 = \frac{n}{n-1}\hat{\sigma}^2,$$

($\hat{s}^2$ is an unbiased estimator for σ^2).

Show that the expected squared error loss for $\hat{s}^2$ is

$$\frac{2\sigma^4}{n-1},$$

and for $\hat{\sigma}^2$,

$$\frac{2n-1}{n^2}\sigma^4.$$

Show that among all estimators of the form $r_n\hat{\sigma}^2$, the one that gives the smallest risk for all n is

$$\frac{n}{n+1}\hat{\sigma}^2.$$

(Use the fact that if Y has an $N(\mu, \sigma^2)$ distribution, then $E(Y - \mu)^4 = 3\sigma^4$.)

Problem 23 If $X_1, \ldots, X_n$ are independent with an $N(\mu, \sigma^2)$ distribution, why do the $n + 1$ variables

$$X_1 - \bar{X}, \ldots, X_n - \bar{X}, \bar{X}$$

have a joint normal distribution? Let $Y_k = X_k - \bar{X}$. Show that Y_k and $\bar{X}$ are uncorrelated. Now justify the conclusion that

$$\bar{X}, \sum_1^n Y_k^2$$

are independent variables.

Problem 24 (Difficult!) Let $\mathbf{e}_1, \mathbf{e}_2, \ldots, \mathbf{e}_n$ be perpendicular unit vectors in n-dimensional space. That is, their dot (or inner) products satisfy

$$\mathbf{e}_i \cdot \mathbf{e}_j = \begin{cases} 0, & i \neq j \\ 1, & i = j, \end{cases}$$

where if $\mathbf{e}_i = (e_1^{(i)}, \ldots, e_n^{(i)})$, then

$$\mathbf{e}_i \cdot \mathbf{e}_j = e_1^{(i)}e_1^{(j)} + e_2^{(i)}e_2^{(j)} + \cdots + e_n^{(i)}e_n^{(j)}.$$

Any vector $\mathbf{y} = (y_1, \ldots, y_n)$ can be written as

$$\sum_i z_i \mathbf{e}_i$$

where $z_i = \mathbf{y} \cdot \mathbf{e}_i$. The equality

$$\sum_1^n y_i^2 = \sum_1^n z_n^2,$$

always holds (why?). Select any such n perpendicular unit vectors with the single proviso that

$$\mathbf{e}_n = \left(\frac{1}{\sqrt{n}}, \ldots, \frac{1}{\sqrt{n}}\right).$$

For $X_1, \ldots, X_n$ independent and $N(\mu, \sigma^2)$ distributed, the variables $Y_k = (X_k - \bar{X})/\sigma$, $k = 1, \ldots, n$, have a joint normal distribution with 0 means. Define new joint normal variables $Z_1, \ldots, Z_n$ by

$$Z_i = \mathbf{Y} \cdot \mathbf{e}_i.$$

From the above we have that

$$\sum_1^n Y_i^2 = \sum_1^n Z_i^2.$$

Notice that

$$Z_n = \frac{1}{n} \sum_1^n Y_i = 0.$$

Now if we could prove that the remaining variables $Z_1, \ldots, Z_{n-1}$ are independent $N(0, 1)$, this would give a real proof that $\sum_1^n Y_k^2$ has a χ^2 distribution with $n - 1$ degrees of freedom. Actually, the proof now is not difficult; complete it by showing that

$$EZ_iZ_j = \begin{cases} 0, & i \neq j \\ 1, & i = j \end{cases}$$

Problem 25 The following list of numbers was taken from a table of outputs of a random number generator which was designed to model independent sampling from an $N(0, 1)$ distribution:

1.799	−0.148
−0.244	0.495
−0.224	0.890
−0.183	0.696
0.770	1.075
0.972	−1.955
−0.368	0.985
0.325	−0.486
2.309	−0.679
−1.516	1.710
−1.451	0.821
−0.544	−1.440

Pair off the first and second columns, and treat the pairs as a sample of size 12 from a bivariate normal distribution. Compute $\hat{\mu}_1, \hat{\mu}_2, \hat{\sigma}_1, \hat{\sigma}_2, \hat{\rho}$.

Other Examples

One might say that up to this point maximum likelihood has produced very little that is surprising. If X is $N(\mu, \sigma^2)$, then $\mu = E\mathrm{X}$ so "naturally" we would estimate μ by the sample mean:

$$\hat{\mu} = \frac{x_1 + \cdots + x_n}{n}.$$

Similarly, $\sigma^2 = E(\mathrm{X} - E\mathrm{X})^2$ so "naturally" σ^2 would be estimated by the sample variance

$$\hat{\sigma}^2 = \frac{1}{n} \sum_1^n (x_k - \hat{\mu})^2.$$

We now give an example that shows that these "natural" estimates are natural only if we are using the normal family as the underlying distribution.

Example a Let $x_1, \ldots, x_n$ be the outcomes of independent observations on a double exponential distribution

$$f(x) = \frac{1}{2\beta} e^{-|x-\alpha|/\beta}, \qquad \beta > 0.$$

Just as with the normal, the location parameter α is given by

$$\alpha = E\mathrm{X}.$$

The square of the scale parameter satisfies

$$\beta^2 = V(\mathrm{X})/2.$$

This might encourage the belief that the "natural" estimates for α and β^2 are the sample mean and one-half of the sample variance. We compute the maximum likelihood estimates:

$$L_\theta = -n \log (2\beta) - \frac{1}{\beta} \sum_1^n |x_k - \alpha|.$$

For any value of β, the α that maximizes L_θ is the value of α that minimizes

$$l(\alpha) = \sum_1^n |x_k - \alpha|. \tag{3.8}$$

Order the numbers $x_1, \ldots, x_n$: let $x_{(1)}$ be the smallest one, that is, $x_{(1)} = \min(x_1, \ldots, x_n)$, $x_{(2)}$ the next smallest and so on, with $x_{(n)}$ the largest reading.

3.9 ***Definition*** *The numbers $x_{(1)}, \ldots, x_{(n)}$ are called the ordered sample or the sample order statistics.*

Assume that no two of the points $x_1, \ldots, x_n$ are the same, so that we get a strict ordering $x_{(1)} < x_{(2)} < \cdots < x_{(n)}$. Actually, if they are readings from a continuous underlying distribution, there is probability 0 of getting two identical outcomes. Suppose $x_{(m)} < \alpha < x_{(m+1)}$. Then m of the x_k's are less than α and $n - m$ are greater than α. If we increase α by a small $\Delta\alpha$, then for every $x_k < \alpha$, $|x_k - \alpha|$ increases by $\Delta\alpha$, and for $x_k > \alpha$, $|x_k - \alpha|$ decreases by $\Delta\alpha$. Hence

$$\begin{aligned} &\text{if } m < n - m, \ l(\alpha) \text{ is decreasing;} \\ &\text{if } m > n - m, \ l(\alpha) \text{ is increasing;} \\ &\text{if } m = n - m, \ l(\alpha) \text{ is constant.} \end{aligned}$$

Starting with α less than any of the sample points and increasing it, $l(\alpha)$ decreases as long as there are more of the numbers $x_1, \ldots, x_n$ above α than below, remains constant when there are as many above as below, and then increases when there are fewer above than below. So a minimizing value of α is a number that "splits" the sample. This leads to

3.10 ***Definition*** *A number $\hat{\nu}$ is called a sample median of $x_1, \ldots, x_n$ if there are as many of the $x_1, \ldots, x_n$ less than $\hat{\nu}$ as there are greater than $\hat{\nu}$.*

If n is odd ($n = 2m + 1$), then $x_{(m+1)}$ is the unique median value. For instance, if $n = 5$, then the ordered sample is

$$x_{(1)} < x_{(2)} < x_{(3)} < x_{(4)} < x_{(5)}$$

and $x_{(3)}$ is clearly the number that "splits" the sample. But if n is even ($n = 2m$), then any number between $x_{(m)}$ and $x_{(m+1)}$ "splits" the sample; there is no unique median value. What happens is that $l(\alpha)$ is constant at its minimum value for α in the range $x_{(m)} \leq \alpha \leq x_{(m+1)}$.

We settle on one value for the sample median by the arbitrary decree that if $n = 2m$ is even, $\hat{\nu}$ is halfway between $x_{(m)}$ and $x_{(m+1)}$.

We now have the result that for a double exponential distribution, the maximum likelihood estimate of the location parameter α is the sample median $\hat{\nu}$. To estimate β, write

$$\frac{\partial L_\theta}{\partial \beta} = -\frac{n}{\beta} + \frac{1}{\beta^2} \sum_1^n |x_k - \alpha| = 0.$$

The solution is

$$\beta = \frac{1}{n} \sum_{1}^{n} |x_k - \alpha|.$$

But we know that the value of α that minimizes L_θ is $\hat{\nu}$; hence the maximum likelihood estimate for the scale parameter β is

$$\hat{\beta} = \frac{1}{n} \sum_{1}^{n} |x_k - \hat{\nu}|.$$

This expression is called the absolute deviation around the median.

The fact that the sample median and not the sample mean is the maximum likelihood estimator for the location parameter in the double exponential family is not useful unless we know that maximum likelihood estimates are particularly good. This does turn out to be generally true, and, in fact, for large n, the sample mean has twice the squared error loss of the sample median if the underlying distributions are in the double exponential family.

The point is that estimates that are best or "almost" best for a small parametric family of distributions are specifically tailored to that family and may do very poorly when applied to other families. Using the sample mean on the double exponential distribution is bad enough, but as we will see, using it to estimate location in the Cauchy family is much worse.

Example b Take $x_1, \ldots, x_n$ independent draws from a Cauchy distribution. Then

$$L_\theta = -n \log \pi\beta - \sum_k \log\left(1 + \left(\frac{x_k - \alpha}{\beta}\right)^2\right).$$

The maximizing equations for α and β are

$$\frac{\partial L_\theta}{\partial \alpha} = \frac{2}{\beta^2} \sum_k \frac{x_k - \alpha}{1 + [(x_k - \alpha)/\beta]^2} = 0$$

$$\frac{\partial L_\theta}{\partial \beta} = -\frac{n}{\beta} + \frac{2}{\beta^2} \sum_k \frac{(x_k - \alpha)^2}{1 + [(x_k - \alpha)/\beta]^2} = 0.$$

These can be simplified in various ways, but they yield no general closed form solution for α and β in terms of $x_1, \ldots, x_n$. The alternatives are either to use a machine maximization program to compute α and β, or to use estimators for α and β that are easier to compute. Then the question becomes: How much do you lose (or possibly gain) in terms of squared error loss by using some simpler estimates? Having something you can compute by hand or by brief use of a desk calculator is often worth an increase in squared error loss.

However, use caution! We will see that using maximum likelihood estimates is usually very sensible. If we decide in the Cauchy example above that computational difficulties make it undesirable, then we might consider estimating the location parameter α by the sample mean. But the squared error loss for the sample mean in the Cauchy situation turns out to be infinite for all values of the parameters. Why? What makes the sample mean such a poor estimator in the Cauchy family?

Problem 26 Order each of the two columns in Problem 25, and compute their medians. Compute their absolute deviation from the median.

The Method of Moments

Maximum likelihood is not the only general estimation method. There are many others. As an example, we discuss the *moment method* which had widespread use in the early days of statistics. Here parameters are estimated by matching sample moments with the theoretical moments. For instance, suppose θ is 1-dimensional. The first moment of the underlying distribution is a function of θ,

$$\mu(\theta) = \int x P_\theta(dx).$$

Matching this to the first sample moment gives the equation

$$\mu(\theta) = \frac{x_1 + \cdots + x_n}{n}.$$

Solve this for θ to get the estimate. For p in coin-tossing distributions, λ in Poisson, β in exponential, and μ in normal distributions, these moment estimates are the sample means. Hence the moment method in these cases gives the same answer as maximum likelihood. But notice that the first moment of a uniform distribution on $[0, l]$ is $l/2$. Hence the moment method estimates l as twice the sample mean. This we know is a considerably less accurate estimate than the maximum likelihood estimate.

If θ is multidimensional, the moment method generalizes this way: Denote

$$\mu_k(\theta) = \int x^k P_\theta(dx),$$

$$\hat{\mu}_k = \frac{x_1^k + \cdots + x_n^k}{n}.$$

Then if θ is 2-dimensional, say, use the two equations

$$\mu_1(\theta) = \hat{\mu}_1$$
$$\mu_2(\theta) = \hat{\mu}_2$$

to solve for θ.

Sometimes the method of moments will give equations that are considerably easier to solve than the maximum likelihood equations. But you have to treat these estimates with caution. In some models, they may be good estimates with low risk. But in other situations they may have extremely high risk as compared to other estimates.

In general, it is hard to say when these estimates are going to be good. Using the method of moments is of dubious value when maximum likelihood estimators are known to have generally desirable properties and to always be as good as or better than the moment estimates for large sample sizes.

Problem 27 Estimate the parameters of the double exponential family by the moment method.
(Ans.

$$\hat{\alpha} = \hat{\mu}_1, \qquad \hat{\beta} = \sqrt{\frac{\hat{\mu}_2 - \hat{\mu}_1^2}{2}}\,.)$$

Desirable Qualities of Maximum Likelihood Estimation

Obviously, since we have spent a fair amount of time illustrating the maximum likelihood method, it must have some desirable properties. We discuss the more important of these in ascending order of difficulty.

3.11 *If $\hat{\boldsymbol{\theta}}$ is the maximum likelihood estimator for a parameter θ, then the maximum likelihood estimator for any parameter which is a function $h(\theta)$ of θ is the function $h(\hat{\boldsymbol{\theta}})$ of $\hat{\boldsymbol{\theta}}$.*

This property tells us a great deal. For instance, if components have an exponential failure time with scale parameter β, how do we estimate the probability that a component will fail in less than 2000 hours? The data vector is the observed failure times $(t_1, \ldots, t_n)$. Recall that one simple way to estimate $P([0, 2000])$ is to use the proportion of the sample that failed in less than 2000 hours. However, for an exponential distribution

$$P([0, 2000]) = 1 - e^{-2000/\beta}.$$

This is a function of the unknown parameter β, and hence the maximum likelihood estimate of $P([0, 2000])$ is

$$1 - e^{-2000/\hat{\beta}},$$

where $\hat{\beta}$ is the maximum likelihood estimate of β.

In a parametric problem, the underlying distribution is completely specified by the value of θ. Therefore any characteristic of the underlying distribution, such as $P_\theta([0, 2000])$ or the variance of $P_\theta(dx)$, or its mean, is a function of θ. The maximum likelihood estimate of any such characteristic is not usually obtained by a direct estimation of the characteristic. Instead, as (3.11) indicates, a simpler procedure is to get the maximum likelihood estimate $\hat{\theta}$ of θ. This gives $P_{\hat{\theta}}(dx)$ as the maximum likelihood estimate for the underlying distribution. Then the maximum likelihood estimates for any characteristics of the underlying distribution are the corresponding characteristics of $P_{\hat{\theta}}(dx)$. For example, in any family $\{P_\theta\}$, the maximum likelihood estimate for the probability of a given interval I is $P_{\hat{\theta}}(I)$, the maximum likelihood estimate for the variance of the underlying distribution is the variance of $P_{\hat{\theta}}(dx)$, and so on. As another illustration, in the case of exponential failure times, suppose we wanted to estimate the variance of the underlying distribution. Why not form the sample variance

$$\hat{\sigma}^2 = \frac{1}{n} \sum_1^n (t_k - \bar{t})^2,$$

where

$$\bar{t} = \frac{t_1 + \cdots + t_n}{n}$$

is the sample mean? The reason is that the variance of an exponentially distributed variable with scale parameter β is β^2. Since the maximum likelihood estimate for β is the sample mean $\bar{t}$, the maximum likelihood estimate for the variance is $\bar{t}^2$. Actually (see Problem 31), $\hat{\sigma}^2$ is a poor estimate for the variance of an exponential distribution.

Property 3.11 also implies that no matter how you parametrize your problem, you come up with the same answer; the estimated underlying distribution is the same. For example, in the set of distributions uniform over an unknown interval, we took the parameters $a \le b$ to be the endpoints of the interval. We could just as well have used parameters (m, l), where m is the midpoint of the interval and l its length,

$$m = \frac{a + b}{2}, \qquad l = b - a.$$

If we use (m, l) as parameters and find their maximum likelihood estimates, it turns out that

$$\hat{m} = \frac{\hat{a} + \hat{b}}{2}, \qquad \hat{l} = \hat{b} - \hat{a},$$

where $\hat{a}$, $\hat{b}$ are the maximum estimates of a, b.

More formally, this property can be stated as: If we transform from the parameter $\theta = (\theta_1, \ldots, \theta_k)$ to the new parameters $(\gamma_1, \gamma_2, \ldots, \gamma_k)$

$$\begin{aligned} \gamma_1 &= h_1(\theta_1, \ldots, \theta_k) \\ &\vdots \\ \gamma_k &= h_k(\theta_1, \ldots, \theta_k) \end{aligned}$$

then the maximum likelihood estimates of $\gamma_1, \ldots, \gamma_k$ and those of $\theta_1, \ldots, \theta_k$ are also connected by the above functional relations.

The next two properties are much more difficult to prove and the statements of them are purposely loose. In all but a few pathological examples:

3.12 *The maximum likelihood method is consistent.*

For large sample sizes, maximum likelihood estimates usually compare well with any other estimation method. In fact, for the class of problems where the probabilities are "smooth" functions of θ and the maximum likelihood estimator is uniquely defined, the following holds true:

3.13 *The maximum likelihood method is asymptotically (as $n \to \infty$), as good as or better than any other estimation method.*

In what sense is maximum likelihood better? We need to formulate this important result in terms of the risk $R_n(\theta)$. The desired formulation is connected with the notion of efficiency of an estimator and with confidence intervals. We will present this in the next section.

Problem 28 Take the underlying distribution to be $N(\mu, \sigma^2)$, and sample independently n times.

a You decide to parametrize the problem by setting $\gamma_1 = \mu$, $\gamma_2 = \mu^2/\sigma^2$. Compute the maximum likelihood estimates of γ_1 and γ_2 *directly*, and then show that $\hat{\gamma}_1 = \hat{\mu}$, $\hat{\gamma}_2 = (\hat{\mu})^2/\hat{\sigma}^2$.

b Find the maximum likelihood estimator of the point x such that the probability that an observation from the underlying population will be greater than x is .13.

Problem 29 If the underlying distributions are $N(\mu, \sigma^2)$, use 3.11 to deduce that the maximum likelihood estimate for σ is given by

$$\hat{\sigma} = \sqrt{\frac{1}{n} \sum_{1}^{n} (x_i - \hat{\mu})^2},$$

where $\hat{\mu}$ is the sample mean.

Problem 30 Take $x_1, \ldots, x_n$ to be independent draws from the distribution

$$P_p(X = 1) = p, \qquad P_p(X = 0) = 1 - p.$$

What is the maximum likelihood estimate for p^2? What is the bias of this estimate?

Problem 31 **a** Compute the loss for $\hat{\sigma}^2$ as an estimator fort he variance of an exponential family.

b Compute the loss for the maximum likelihood estimate of the variance.

c What is their ratio for large sample size?

(In **a.** use the fact that for any underlying distribution $P(dx)$, the loss in using $\hat{\sigma}^2$ to estimate the variance σ^2 of $P(dx)$ is

$$\left(\frac{n-1}{n}\right)^2 \frac{1}{n}\left(\mu_4 - \frac{(n-4)}{(n-1)}\sigma^4 + \frac{1}{(n-1)^2}\sigma^4\right),$$

where

$$\mu_4 = \int (x - \mu)^4 P(dx)$$

and μ is the mean of the distribution.)

Efficiency of Estimation Methods

In virtually all of the common estimation models, if $R_n(\theta)$ is the risk using maximum likelihood, and $R_n'(\theta)$ the risk using any other consistent estimation method, then

$$\frac{R_n(\theta)}{R_n'(\theta)} \leq 1 + o(1), \tag{3.14}$$

where $o(1)$ denotes a quantity that goes to 0 as $n \to \infty$, for θ fixed. This is a more precise statement of what we were getting at in the last section. For large sample size, maximum likelihood usually is as good as or better than any other estimation method, as measured by expected squared error loss. In problems where this is so, we define the asymptotic efficiency $e(\theta)$ at θ of the method with loss $R_n'(\theta)$ as

3.15 ***Definition***

$$e(\theta) = \lim_{n \to \infty} \frac{R_n(\theta)}{R_n'(\theta)}$$

if this limit exists.

There are some interesting and useful questions prompted by the idea of efficiency. How fast does the squared error loss decrease? For what class of statistical models is maximum likelihood asymptotically a best estimation method?

We will call a statistical model or problem *smooth* if the densities $f_\theta(x)$ (or the probabilities $P_\theta(x_j)$ in the discrete case) are well-behaved functions of the parameter θ. For instance, we have to be able to differentiate them twice with respect to θ and the derivatives have to be reasonably bounded. We will state the precise conditions for smoothness later. First, we want to summarize what happens in a smooth problem.

We know that we can make very bad estimates. In other words, by choosing a poor estimator, we can get large expected loss. What we do not know is how small we can make the expected loss. Suppose we want to use some estimator $\hat{\boldsymbol{\theta}}$ and compute its risk. If we have no idea of how small we can make the risk by using other estimators, then we have no standard to judge $\hat{\boldsymbol{\theta}}$. Furthermore, note that in every example except for the uniform distributions on $[0, l]$, the risk went down like $1/n$. Is this behavior typical? We need information about the behavior of the risks for large n. In a smooth problem we have some remarkable results. In the case where θ is a 1-dimensional parameter and $\{P_\theta\}$ are the corresponding distributions:

3.16 ***Definition*** *The information content $I(\theta)$ in $P_\theta(dx)$ about θ is defined as*

$$I(\theta) = \sum_j \left[\frac{\partial}{\partial\theta} \log P_\theta(x_j)\right]^2 P_\theta(x_j) \qquad (\textit{discrete case})$$

or

$$I(\theta) = \int \left[\frac{\partial}{\partial\theta} \log f_\theta(x)\right]^2 f_\theta(x)\, dx \qquad (\textit{density case}).$$

For example, in the exponential case

$$\log f_\beta(x) = -\log \beta - \frac{x}{\beta}$$

$$\frac{\partial}{\partial \beta} \log f_\beta(x) = -\frac{1}{\beta} + \frac{x}{\beta^2} = \frac{1}{\beta^2}(x - \beta),$$

so

$$I(\beta) = \frac{1}{\beta^4} \int (x - \beta)^2 f_\beta(x)\, dx.$$

The integral is just the variance of an exponential with parameter β, so

$$I(\beta) = \frac{1}{\beta^2}.$$

If an estimation method is consistent, then its bias $b_n(\theta)$ must go to 0 as $n \to \infty$. (Why?) With any kind of smoothness this implies that the derivative

$$\frac{d[b_n(\theta)]}{d\theta}$$

of the bias also goes to 0 as n becomes large. Otherwise, the functions $b_n(\theta)$ would have an increasingly large number of small wiggles. For the purpose of the following theorem, assume that consistency implies that the derivative of the bias goes to 0. Then

3.17 ***Theorem*** *For any consistent method of estimation in a smooth problem, the squared error loss satisfies*

3.18 $$R_n(\theta) \geq \frac{1}{nI(\theta)} + o\left(\frac{1}{n}\right).$$

(Here $o(1/n)$ denotes a quantity which goes to 0 faster than $1/n$.)

The exciting thing about this theorem is that it gives a lower bound on the squared error loss: No consistent estimation method can have a squared error loss lower than the right side of 3.18.

Remarkably, in a smooth problem the maximum likelihood estimates achieve this minimum.

3.19 ***Theorem*** *Suppose that for every sample size the maximum likelihood equation*

$$\frac{\partial L_\theta}{\partial \theta} = 0$$

has a unique solution. Then the maximum likelihood estimator has the property that

$$R_n(\theta) = \frac{1}{nI(\theta)} + o\left(\frac{1}{n}\right).$$

Together, 3.17 and 3.19 give the result that in a smooth problem maximum likelihood is a "best" estimation method in the sense of 3.13. Notice that we say *a best* rather than *the best*. There may be other estimation methods whose efficiency is 1 for all θ.

This theorem implies that in a smooth problem the squared error loss of the maximum likelihood estimation method decreases like $1/n$. This gives a simple relationship between efficiency and sample size. For instance, if we use an estimation method whose efficiency at θ is $1/2$, say, then to get the same squared error risk as a method of efficiency 1, we have to *double our sample size*. Often increased sample size is either inherently difficult to come by or expensive. Under these circumstances the message is: *Use high efficiency estimation methods*. However, there is another situation in which one deliberately uses a lower efficiency estimation method as insurance against "contaminated" data points. We come back to this in Chapter 7.

Because of the importance of the information inequality 3.18 we indicate a simple and interesting proof in Problem 39. Theorem 3.19, while also important, has a proof that is too long and difficult to give here.

We are left with defining exactly what we mean by a smooth problem. The variable we are observing does not necessarily take on all real values. For instance, in an exponentially distributed model, X takes on positive values, and in a Poisson model, only nonnegative integer values. In the discrete case, define the *range* R of outcomes in a model to be all those points $\{x_j\}$ which have positive probability for some value of θ. In the density case, take R to be the smallest interval which has probability 1 for all values of θ.

Our first requirement for a smooth problem is

3.20 *For every fixed value of x in the range of outcomes, $f_\theta(x)$ (or $P_\theta(x)$ in the discrete case) has a continuous second derivative in θ.*

The second requirement is that as functions of x the partial θ-derivatives decrease rapidly enough so that various integrals or sums in which they appear are well-behaved:

For any small interval of θ-values J, the two functions of x

3.21
$$\max_{\theta \in J} \left| \frac{\partial f_\theta(x)}{\partial \theta} \right|, \qquad \max_{\theta \in J} \left| \frac{\partial^2 f_\theta(x)}{\partial^2 \theta} \right|$$

have finite integrals. For every θ,

$$\max_{x \in R} |f_\theta(x)| < \infty.$$

In the discrete case, simply read sums instead of integrals.

These two requirements are met by every family of distributions we have worked with except two:

a the uniform distributions on $[0, l]$

b the double exponential family with known scale parameter.

In both cases the first requirement is violated. For the uniform distributions we know that the squared error risk goes down like $1/n^2$. Hence, the information inequality and the various conclusions that it implies cannot be true. The trouble is that the outcome space is $[0, l]$; therefore the range of outcomes varies with the parameter value. Problems in which this is true typically do not satisfy the information inequality results, and the maximum likelihood estimators may not be asymptotically "best" (see Problem 40).

But the information inequality results are true for the double exponential family. The above conditions a. and b. are actually too stringent and eliminate a number of situations in which the conclusions hold. Some generality has been sacrified for the sake of simplicity.

Problem 32 Verify that for estimating the means of the following distributions by the sample mean, $1/[nI(\theta)]$ gives the correct form of the squared error loss.

a $P(X = 1) = p$, $P(X = 0) = 1 - p$.

b X is Poisson with parameter λ.

c X is exponential with parameter $1/\theta$.

d X is $N(\mu, \sigma^2)$.

Problem 33 Suppose you decide to use the unbiased estimator

$$\hat{s}^2 = \frac{1}{n-1} \sum_1^n (X_k - \hat{\mu})^2$$

to estimate the variance of the following distributions:

a Poisson with parameter λ.

b $P(X = 1) = p$, $P(X = 0) = 1 - p$.

For each case, compute the efficiency of the method. (Use the fact that for any underlying distribution,

$$V(\hat{s}^2) = \frac{\mu_4 - \sigma^4}{n},$$

where μ_4 is the fourth moment around the mean.)

Problem 34 Some failure data $t_1, \ldots, t_n$ are to be fitted by a Weibull distribution of the form

$$P(T > t) = e^{-t^\alpha}, \qquad t \geq 0,$$

where α is to be estimated by maximum likelihood. Evaluate the information $I(\alpha)$. (Use the value

$$\int_0^\infty [1 + \log u - u \log u]^2 e^{-u}\, du \simeq 1.82.)$$

Problem 35 Find a 1-parameter model where the moment estimator has efficiency less than 1.

Problem 36 The sample median $\hat{\nu}$ is used to estimate the location parameter α in a double exponential family. The scale parameter β is assumed equal to 1. Find the squared error loss for large n. What is the efficiency of the sample mean as an estimator for α?

Problem 37 Find the asymptotic expression for the risk of the maximum likelihood estimate of the location parameter in the Cauchy family where β is assumed known and equal to 1. (Use the value

$$\frac{\pi}{8} = \int_{-\infty}^{\infty} \frac{x^2}{(1 + x^2)^3}\, dx.)$$

Problem 38 **a** Show that for a location parameter family, the maximum likelihood risk is approximately c/n, where c is a constant.

b Show that in a scale parameter family, the risk is approximately maximum likelihood $d\theta^2/n$, where d is a constant.

Problem 39 We have a model with data vector $\mathbf{x}$ and joint densities $f_\theta(\mathbf{x})$. For any estimator $\phi(\mathbf{x})$ of θ, the bias $b(\theta)$ is

$$b(\theta) = E_\theta\phi(\mathbf{X}) - \theta = E_\theta(\phi(\mathbf{X}) - \theta)$$
$$= \int (\phi(\mathbf{x}) - \theta) f_\theta(\mathbf{x})\, d\mathbf{x}.$$

Assuming we can take the derivative inside the integral sign,

$$\frac{d[b(\theta)]}{d\theta} = \frac{d}{d\theta}\int (\phi(\mathbf{x}) - \theta) f_\theta(\mathbf{x})\, d\mathbf{x}$$
$$= \int \frac{\partial}{\partial\theta}[(\phi(\mathbf{x}) - \theta) f_\theta(\mathbf{x})]\, d\mathbf{x}$$
$$= -\int f_\theta(\mathbf{x})\, d\mathbf{x} + \int (\phi(\mathbf{x}) - \theta)\frac{\partial}{\partial\theta} f_\theta(\mathbf{x})\, d\mathbf{x}.$$

Writing

$$\frac{\partial}{\partial\theta} f_\theta(\mathbf{x}) = \left[\frac{1}{f_\theta(\mathbf{x})}\frac{\partial}{\partial\theta} f_\theta(\mathbf{x})\right] f_\theta(\mathbf{x})$$
$$= \left[\frac{\partial}{\partial\theta}\log f_\theta(\mathbf{x})\right] f_\theta(\mathbf{x}),$$

we have, since the integral of a density is always 1,

$$1 + b'(\theta) = \int \Big(\phi(\mathbf{x}) - \theta\Big)\left(\frac{\partial}{\partial\theta}\log f_\theta(\mathbf{x})\right) f_\theta(\mathbf{x})\, d\mathbf{x}.$$

Applying the Schwarz inequality in the form

$$\left[\int g(x)h(x)P(dx)\right]^2 \le \left(\int g^2(x)P(dx)\right)\left(\int h^2(x)P(dx)\right)$$

to the above equation gives

$$E_\theta(\phi - \theta)^2 \ge \frac{[1 + b'(\theta)]^2}{\int [(\partial/\partial\theta)\log f_\theta(\mathbf{x})]^2 f_\theta(\mathbf{x})\, d\mathbf{x}}.$$

In the case of independence use the fact that

$$f_\theta(\mathbf{x}) = f_\theta(x_1) \cdots f_\theta(x_n)$$

to show that

$$\int \left[\frac{\partial}{\partial\theta} \log f_\theta(\mathbf{x})\right]^2 f_\theta(\mathbf{x})\, d\mathbf{x} = n \int \left[\frac{\partial}{\partial\theta} \log f_\theta(x)\right]^2 f_\theta(x)\, dx.$$

This proves that the risk $R(\theta)$ of the estimate ϕ satisfies

$$R_n(\theta) \geq \frac{[1 + b_n'(\theta)]^2}{nI(\theta)}.$$

Problem 40 **a** Find out why the family of distributions uniform on $[0, l]$ is not a smooth statistical model. How about the double exponential family with known scale parameter?

b For the family of distributions uniform on $[0, l]$; compare the risk of the maximum likelihood estimator M_n to that of the unbiased estimator

$$\mathsf{U} = \frac{n+1}{n} \mathsf{M}_n.$$

Summary

In a parametric statistical model $\{P_\theta\}$, $\theta \in \Theta$, the usual estimation problem is to find the true value of θ, or equivalently, to select one distribution from the family $\{P_\theta\}$ as the actual underlying distribution. To judge the performance of any estimator, the statistician specifies a loss function and then computes the expected loss $R(\theta)$ for the given estimator at the state of nature θ. The risk function $R(\theta)$ gives the operating characteristic of the estimator. For a variety of reasons, we restrict ourselves to squared error loss. Unfortunately there is usually no single way of estimating that is best for all θ. However, the maximum likelihood method of estimation has some very attractive properties. It is a consistent method of estimation in all but very pathological problems. Its risk in smooth problems is generally about as small as possible when the sample size is large. The things to remember about maximum likelihood estimates are

a the concept: estimate θ as that value of the parameter which maximizes the probability of getting the data vector actually observed,

b the applications: how to compute the maximum likelihood estimates,

c the properties: in particular, the maximum likelihood estimate for any function $h(\theta)$ is $h(\hat{\theta})$, where $\hat{\theta}$ is the maximum likelihood estimate of θ.

Also keep in mind the concept of the efficiency of an estimation method. This will serve well as a measure of comparing estimation methods.

4 Confidence Intervals

Confidence Intervals

Along with any method of estimation comes the natural question, How close to the true value is the estimate? This question has to be asked whenever an estimate is used. We can always compute the operating characteristic of the estimate using squared error risk. But this does not give an immediate sense of how large the error is likely to be.

What we want is a more primitive, less sophisticated way of indicating the accuracy of an estimate. To do this, we use *confidence intervals.* Suppose that we are given a statistical model with $\{P_\theta\}$, $\theta \in \Theta$, the probabilities for the outcome of a random vector **X**. Suppose further that θ is a 1-dimensional parameter and $\hat{\boldsymbol{\theta}}$ our estimator for θ. Now we know that $\hat{\theta}$ will rarely equal θ exactly. For instance, if $\hat{\boldsymbol{\theta}} = 7.0$ we suspect that θ will be near 7.0, but not exactly equal to it, perhaps 6.9, perhaps 7.1. What we would really like to find is some interval around the estimated value, say $7.0 \pm .2$, such that with high probability the true value θ falls in the given interval. Then, of course, the size of the interval we need to contain θ is a convincing measure of the accuracy of $\hat{\boldsymbol{\theta}}$. The length of the interval depends on $\hat{\theta}$, because the accuracy of the estimator depends on the value of θ. For example, if we are estimating the parameter λ of a Poisson distribution by using

$$\hat{\boldsymbol{\lambda}} = \frac{\mathsf{X}_1 + \cdots + \mathsf{X}_n}{n},$$

the expected squared error is $R(\lambda) = \lambda/n$. Thus, the larger λ is, the more error we expect in our estimator. If we come up with a large value of $\hat{\lambda}$, we suspect that the "interval of high probability" around $\hat{\lambda}$ needs to be much larger than if our estimated value of λ is small.

In general, then, what we want to find are two functions $g(\hat{\theta})$ and $h(\hat{\theta})$ with $g(\hat{\theta}) \leq \hat{\theta} \leq h(\hat{\theta})$ such that if the estimated value of θ is $\hat{\theta}$, then with high probability, θ is in the interval $[g(\hat{\theta}), h(\hat{\theta})]$. Notice that what we are setting up is a procedure, rather than a single answer. We are interested in finding the *functions* $g(\hat{\theta})$ and $h(\hat{\theta})$ which provide the intervals. In other words, no matter what value of $\hat{\theta}$ the data may produce, we want a procedure that will use that value of $\hat{\theta}$ to produce a corresponding interval around it. This leads to the following:

4.1 **Definition** *A confidence procedure for θ based on the estimator $\hat{\theta}$ is any computational scheme which, starting from $\hat{\theta}$ computes an interval $J(\hat{\theta})$ containing $\hat{\theta}$.*

We will see that it is the whole procedure we must consider, not just the individual intervals it produces.

If $\hat{\theta}$ is the estimated value, can we choose an interval $J(\hat{\theta}) = [g(\hat{\theta}), h(\hat{\theta})]$ so that with high probability the parameter value θ is in $J(\hat{\theta})$? Suppose we are sampling from an $N(\mu, 1)$ population and get the estimate $\hat{\mu} = 3$. What meaning can we attach to a statement such as, The probability that μ is in [2.5, 3.5] is .95, or, The probability that μ is in [2.8, 3.2] is .9? Take a sample from one $N(\mu, 1)$ population and estimate μ, then from another $N(\mu', 1)$ population with, possibly, a different value μ' of the mean, and so on repeating the sampling process a large number of times. Then with our usual frequency interpretation, the first statement is read as: In 95% of those samples for which the estimated mean was around 3, the true population mean μ was in the interval [2.5, 3.5]. But this is nonsense. Suppose nature has set all the populations means equal to 0. We still occasionally get the estimate value $\hat{\mu} \simeq 3$. The probability that $\hat{\mu} \simeq 3$ if $\mu = 0$ is small, but not 0; it will still happen as a chance fluctuation now and then. Is it true that in 95% of the times we get $\hat{\mu} \simeq 3$, μ is in [2.5, 3.5]? Of course not.

The point to be emphasized is that there is no way of gauging the effectiveness of a single interval corresponding to a single given value of $\hat{\theta}$. *But we can judge the effectiveness of our entire procedure.* Suppose now that we look at a confidence procedure that associates an interval $J(\hat{\theta}) = [g(\hat{\theta}), h(\hat{\theta})]$ to every possible value of the estimator $\hat{\theta}$. We play a game with nature as follows: Nature picks a first value of θ, say θ_1. We sample, get an estimate $\hat{\theta}_1$ from our data point $\mathbf{x}_1$, and form the interval $J(\hat{\theta}_1)$. We win if $\theta_1 \in J(\hat{\theta}_1)$; otherwise we lose. Now nature picks another value of θ, say θ_2, and we sample again (independently of the first game), get the estimate $\hat{\theta}_2$, compute the interval $J(\hat{\theta}_2)$, and win if $\theta_2 \in J(\hat{\theta}_2)$. You can see that a long series of such games may involve the entire procedure, because we have to compute intervals corresponding to all the different values $\hat{\theta}_1, \hat{\theta}_2, \ldots$. Now we make a first try at formulating the problem:

4.2 **Definition** *Call the procedure for computing $J(\hat{\theta})$ a $100\gamma\%$ confidence procedure if no matter how nature selects the successive values $\theta_1, \theta_2, \ldots$, you (the statistician) win at least $100\gamma\%$ of the time.*

How do we formulate this condition in terms of the quantities that define a statistical problem? Recall that our data point $\mathbf{x}$ is the outcome of a random vector $\mathbf{X}$ whose distribution is assumed to be in $\{P_\theta\}$, $\theta \in \Theta$.

On the basis of **x**, we form our estimate $\hat{\theta}(\mathbf{x})$ of θ and then we compute $J(\hat{\theta})$. Now I claim that the condition for a $100\gamma\%$ confidence procedure is

4.3 $$P_\theta[\theta \in J(\hat{\boldsymbol{\theta}})] \geq \gamma, \qquad \text{for all } \theta \in \Theta.$$

For instance, a 95% confidence procedure must satisfy

$$P_\theta[\theta \in J(\hat{\boldsymbol{\theta}})] \geq .95, \qquad \text{for all } \theta \in \Theta.$$

It is easy to see why 4.3 is equivalent to Definition 4.2. If 4.3 holds, then no matter what value of θ nature selects, there is probability at least γ that $\theta \in J(\hat{\boldsymbol{\theta}})$. But if 4.3 fails to hold for any value of θ, for instance, if

$$P_{\theta_0}[\theta_0 \in J(\hat{\boldsymbol{\theta}})] = .99\gamma,$$

then if nature selects the value θ_0 every time, you win only $99\gamma\%$ of the time. We now have:

4.4 ***Definition*** *A confidence procedure for computing* $J(\hat{\theta})$ *is a* $100\gamma\%$ *confidence procedure if*

$$P_\theta[\theta \in J(\hat{\boldsymbol{\theta}})] \geq \gamma, \qquad \text{for } all\ \theta \in \Theta.$$

Therefore, a 95% confidence procedure is not "an interval that contains the true state of nature with probability .95 or more," it is a method of computing intervals in a given problem such that if you use this method repeatedly in the same problem, you will be right at least 95% of the time. Usually, such a procedure is called a 95% (or generally, a $100\gamma\%$) *system of confidence intervals*, or more simply, a 95% *confidence interval*. As you have seen in this section, the expression "confidence interval" can be misleading, and I much prefer to talk in terms of "confidence procedures." But since it is the accepted nomenclature, we will live with it.

Simple Confidence Intervals for Large Sample Sizes

Actually the question of confidence intervals hinges on the distribution of the estimate. If we know how $\hat{\boldsymbol{\theta}}$ is distributed around the true state of nature θ, then we can get some idea of how far away $\hat{\boldsymbol{\theta}}$ is liable to be from θ.

In most applications there is a simple way of computing confidence intervals, based on the fact that for large sample sizes, $\hat{\boldsymbol{\theta}}$ has a fairly simple distribution. To begin with, suppose that $\hat{\boldsymbol{\theta}}$ is unbiased. Denote its variance under the state of nature θ by $\sigma^2(\theta)$. That is,

$$\sigma^2(\theta) = E_\theta(\hat{\boldsymbol{\theta}} - \theta)^2.$$

Suppose further that $\hat{\theta}$ *is normally distributed.* Then, 95% of the time the values of $\hat{\theta}$ will fall in the range

$$\theta \pm 2\sigma(\theta).$$

Now look at the interval

$$J = \hat{\theta} \pm 2\sigma(\theta).$$

Every time $\hat{\theta}$ falls within $2\sigma(\theta)$ of θ, the interval J contains the true state of nature θ. Unfortunately, we cannot give our error bounds as $\hat{\theta} \pm 2\sigma(\theta)$, since θ is unknown to us. But for large sample size, if we are using a consistent method, then $\hat{\theta}$ will usually be close to θ. If $\sigma(\theta)$ is a smooth function of θ, then for $\hat{\theta}$ close to θ,

$$\sigma(\hat{\theta}) \simeq \sigma(\theta).$$

So we can say that no matter what the state of nature, about 95% of the time it will fall in the range

$$\hat{\theta} \pm 2\sigma(\hat{\theta}).$$

In other words, for all $\theta \in \Theta$,

$$P_\theta[\hat{\theta} - 2\sigma(\hat{\theta}) \leq \theta \leq \hat{\theta} + 2\sigma(\hat{\theta})] \simeq .95.$$

Therefore

$$J(\hat{\theta}) = \hat{\theta} \pm 2\sigma(\hat{\theta})$$

forms an approximate 95% confidence interval for θ.

More generally, select z such that

4.5 $$P(-z \leq \mathsf{Z} \leq z) = \gamma.$$

Then

$$P_\theta[\theta - z\sigma(\theta) \leq \hat{\theta} \leq \theta + z\sigma(\theta)] = \gamma.$$

Transpose θ and $\hat{\theta}$ to get

$$P_\theta[\hat{\theta} - z\sigma(\theta) \leq \theta \leq \hat{\theta} + z\sigma(\theta)] = \gamma.$$

Replace $\sigma(\theta)$ by $\sigma(\hat{\theta})$ to conclude that $\hat{\theta} \pm z\sigma(\hat{\theta})$ forms a $100\gamma\%$ confidence interval for θ.

For $\hat{\theta}$ normally distributed and unbiased, the above discussion gives a quick way of getting confidence intervals for large sample sizes. Unfortunately, to date we know only one estimator that is unbiased and normally distributed. But what we need for these simple confidence intervals to work is that

$$\frac{\hat{\theta} - \theta}{\sigma(\theta)}$$

be *approximately* $N(0, 1)$ for large sample sizes. This will be true if the estimator is *asymptotically normal and asymptotically relatively unbiased.* We define these terms separately. Let $\mu(\theta)$ be the expected value $E_\theta\hat{\theta}$ of the estimator. Then call the estimator *asymptotically normal* if

4.6
$$\frac{\hat{\theta} - \mu(\theta)}{\sigma(\theta)}$$

is approximately $N(0, 1)$ for n large. Secondly, we would like to replace $\mu(\theta)$ by θ in 4.6. Write $\mu(\theta) = \theta + b(\theta)$, where $b(\theta)$ is the bias. Then

$$\frac{\hat{\theta} - \mu(\theta)}{\sigma(\theta)} = \frac{\hat{\theta} - \theta}{\sigma(\theta)} - \frac{b(\theta)}{\sigma(\theta)}.$$

Define the estimator to be *asymptotically relatively unbiased* if $b(\theta)/\sigma(\theta)$ goes to 0 as $n \to \infty$.

4.7 ***Proposition*** *If $\hat{\theta}$ is asymptotically normal and asymptotically relatively unbiased, then approximate* $100\gamma\%$ *confidence intervals for large sample size are given by*

$$\hat{\theta} \pm z\sigma(\hat{\theta}),$$

where z is computed from an $N(0, 1)$ *table as in* 4.5.

Most of the estimators we have met are asymptotically normal and asymptotically relatively unbiased. For instance, look at the estimate of the mean of an exponential given by

$$\hat{\beta} = \frac{T_1 + \cdots + T_n}{n}.$$

Since $E_\beta\hat{\beta} = \beta$, $\hat{\beta}$ is exactly unbiased. Furthermore, writing $\hat{\beta}$ as

$$\hat{\beta} = \frac{T_1}{n} + \cdots + \frac{T_n}{n}$$

shows that it is the sum of many small independent identically distributed components. We invoke the central limit theorem to conclude that $\hat{\beta}$ is nearly normally distributed.

In fact, we have the quite important result:

4.8 ***Theorem*** *In a smooth statistical problem the maximum likelihood estimators are asymptotically normal and relatively unbiased.*

The proof of this theorem is based on the central limit theorem plus some complicated computations. It is a happy result and justifies more trust, for large sample sizes, in maximum likelihood.

Now squared error risk, which we discarded in favor of confidence intervals, is ready to come in by the back door. From the last chapter, recall the decomposition 3.6

$$R(\theta) = \sigma^2(\theta) + b^2(\theta).$$

Write this as

$$R(\theta) = \sigma^2(\theta)\left(1 + \left(\frac{b(\theta)}{\sigma(\theta)}\right)^2\right).$$

In this form you can see that if the estimator is asymptotically relatively unbiased, then for large sample sizes

$$R(\theta) \simeq \sigma^2(\theta).$$

Thus $100\gamma\%$ confidence intervals can be given in terms of the expected squared error loss as

$$\hat{\theta} \pm z\sqrt{R(\hat{\theta})}.$$

Using the results of the last chapter, we have

4.9 ***Theorem*** *In a smooth model, if the estimator $\hat{\theta}$ is a maximum likelihood estimator, then for large sample sizes, $100\gamma\%$ confidence intervals are given by*

4.10
$$\hat{\theta} \pm \frac{z}{\sqrt{n}} \cdot \frac{1}{\sqrt{I(\hat{\theta})}}.$$

In a smooth problem, the lengths of these $100\gamma\%$ confidence intervals decrease like $1/\sqrt{n}$. This is an unpleasant fact: To cut the lengths of the confidence intervals in half, for instance, we have to quadruple our sample size.

For applications, the important question that remains is how large the sample size has to be so that we can use these simple intervals. Usually, it is safe to use them for samples of size 20. The important thing to realize is that the size of the confidence intervals is not that critical. Suppose your estimate of the confidence interval is in error by 20% or even 30%. You want confidence intervals in order to get an idea of how accurate your estimate is, of how far it might be from the true value. Perhaps you want to know if there is enough data on hand to give the kind of accuracy you want. But whether your $100\gamma\%$ confidence interval is $7.00 \pm .20$ or the 20% larger interval, $7.00 \pm .24$ is usually not critical in deciding whether the estimate is accurate enough for your purposes.

A good rule-of-thumb is to compute the bias and variance of any estimate you use. Usually, in a smooth problem, almost every estimate you use will be asymptotically normal and asymptotically relatively unbiased. If the sample size is at least 20, then the interval $\hat{\theta} \pm z\sigma(\hat{\theta})$ will be close enough to give you a reasonable measure of your accuracy.

Problem 1 Successive one-minute counts of the number of cars on a freeway passing a given point are:

23, 36, 33, 33, 31, 23, 27, 30, 37, 36,
34, 26, 26, 30, 43, 28, 35, 35, 31, 33,
28, 28, 40, 26, 33, 23, 32, 35, 24, 29,
31, 29, 28, 32, 27, 37, 32, 30, 23, 24,
24, 26, 26, 32, 31, 30, 27, 35, 25, 31.

Assume these counts are independent samples from a normal distribution with $\sigma = 5$. Find 95% confidence intervals for the mean based on the maximum likelihood estimator.

Problem 2 Your boss, who knows no statistics, comes to you and asks you to design an experiment to estimate the mean failure time of a certain component to within about 5% accuracy. After some investigation you decide that it is reasonable to assume that the failure times of the components are exponentially distributed. Formulate the problem in terms of confidence intervals, choose a suitable level, and then find the sample size required.

Problem 3 The number of electrons produced by the passage of an alpha particle through a container of a given medium is known to have a Poisson distribution. One hundred of these particles are observed at various times passing through the container, the electron production is measured, and the parameter λ estimated as 9.4. Another group repeats the experiment and estimates λ as 8.6. Is this surprising? Why?

Problem 4 A certain component is manufactured by two different subcontractors. One hundred components from each manufacturer are tested to failure. The first batch has a sample mean of 8.9 days; the second batch, has a sample mean of 11.8 days. Assuming exponential failure times, use confidence intervals to discuss the first manufacturer's assertion that the difference may have been due to random fluctuations.

(If β_1 is the mean of the first distribution and β_2 the mean of the second distribution, $\hat{\beta}_1$ and $\hat{\beta}_2$ the sample means, then $\hat{\beta}_1 - \hat{\beta}_2$ is the maximum likelihood estimate of $\beta_1 - \beta_2$. (Why?) Now find $100\gamma\%$ confidence intervals for $\beta_1 - \beta_2$ by computing

$$E(\hat{\boldsymbol{\beta}}_1 - \hat{\boldsymbol{\beta}}_2 - \beta_1 + \beta_2)^2,$$

remembering that $\hat{\boldsymbol{\beta}}_1$ and $\hat{\boldsymbol{\beta}}_2$ are independent.)

Problem 5 The set of possible underlying distributions consists of all gamma distributions having $\alpha = 3$. Find 95% confidence intervals for β using the maximum likelihood estimate, and assuming n large.

Problem 6 On a Presidential preference poll with forced choice between two candidates the sample size is 1600 and 53% of the people polled prefer candidate N to candidate H. Find 95% confidence limits for the estimate, assuming that the maximum likelihood estimator is being used by the pollsters. (Ans. 50.55, 55.45.)

Problem 7 Suppose you believe that a roulette wheel has a slight bias favoring a certain group of numbers. In particular, suppose you want to estimate the probability of the group of numbers 15, 16, 17, 18 to within 1% at the 90% level. Assuming that the probability of each number is about 1/38 and that a spin of the wheel takes about 30 seconds, how many hours' observation would be required? (Ans. 1930.)

Problem 8 It is planned to estimate $p = P(\mathsf{X} < 1)$ for the underlying distribution by taking $\hat{p}$ equal to the proportion of all points in the sample $x_1, \ldots, x_n$ that are less than 1. Find a $100\gamma\%$ confidence interval for p.

Confidence Intervals for a Multidimensional Parameter

What if the parameter θ is multidimensional? For instance, suppose $\theta = (\theta_1, \theta_2)$ and we want to gauge the accuracy of an estimate $\hat{\boldsymbol{\theta}}_1$ of θ_1. More specifically, suppose we are sampling from an $N(\mu, \sigma^2)$ distribution with both μ and σ^2 unknown. From a sample $x_1, \ldots, x_n$ we get estimators $\hat{\mu}$ and $\hat{\sigma}^2$. Suppose $\hat{\mu} = 0$. If $\hat{\sigma}^2$ is very small, for example, $\hat{\sigma}^2 = 10^{-10}$, then the values $x_1, \ldots, x_n$ are tightly clustered around 0, and we suspect that the estimate $\hat{\mu} = 0$ is accurate. But if $\hat{\sigma}^2$ is very large, say $\hat{\sigma}^2 = 10^{10}$, then the values $x_1, \ldots, x_n$ are widely spread out and substantial error in the estimate $\hat{\mu} = 0$ is likely. Thus, if σ^2 is unknown, then the sample variance $\hat{\sigma}^2$ is indispensable in judging the accuracy of $\hat{\mu}$. Therefore $100\gamma\%$ confidence intervals for μ will depend on $\hat{\sigma}$ as well as $\hat{\mu}$. The larger $\hat{\sigma}^2$ is, the larger the intervals will be. What we are looking for, then, are intervals $J(\hat{\mu}, \hat{\sigma})$ around the estimated value $\hat{\mu}$ such that

4.11 $$P_{(\mu,\sigma)}[\mu \in J(\hat{\mu}, \hat{\sigma})] \geq \gamma, \qquad \text{for all } (\mu, \sigma).$$

Analogous to the 1-dimensional case, no matter what sequence of values of (μ, σ) nature chooses, if we use intervals $J(\hat{\mu}, \hat{\sigma})$ satisfying 4.10, then we are right at least $100\gamma\%$ of the time.

In general, if $\theta = (\theta_1, \theta_2)$ and we have an estimator $\hat{\theta} = (\hat{\theta}_1, \hat{\theta}_2)$ for $\theta = (\theta_1, \theta_2)$, then intervals $J(\hat{\theta})$ around the point $\hat{\theta}_1$ are said to be $100\gamma\%$ confidence intervals for θ_1 if

4.12 $$P_{(\theta_1,\theta_2)}[\theta_1 \in J(\hat{\theta})] \geq \gamma, \qquad \text{for all } (\theta_1, \theta_2).$$

The way to get $100\gamma\%$ confidence intervals for large sample size is almost exactly the same as in the 1-dimensional case. Assume that $\hat{\theta}_1$ is asymptotically normal and unbiased. The variance of $\hat{\theta}_1$, denoted by $\sigma_1^2(\theta)$, may depend on both θ_1 and θ_2. Then computing z from an $N(0, 1)$ table in the usual way, large sample $100\gamma\%$ confidence intervals for θ_1 are given by

$$\hat{\theta}_1 \pm z\sigma_1(\hat{\theta}).$$

Example a Let $X_1, \ldots, X_n$ be samples from an $N(\mu, \sigma^2)$ distribution, μ and σ^2 unknown. Use

$$\hat{\mu} = \frac{X_1 + \cdots + X_n}{n}$$

and

$$E_{(\mu,\sigma)}(\hat{\mu} - \mu)^2 = \frac{\sigma^2}{n}.$$

Replace σ^2 by its maximum likelihood estimate

$$\hat{\sigma}^2 = \frac{1}{n}\sum_1^n (X_i - \hat{\mu})^2$$

to get the $100\gamma\%$ confidence interval

$$\hat{\mu} - z\frac{\hat{\sigma}}{\sqrt{n}}, \hat{\mu} + z\frac{\hat{\sigma}}{\sqrt{n}}.$$

The theorem concerning the distribution of maximum likelihood estimators also holds in the multidimensional case. If smoothness is defined in the appropriate multidimensional sense and the maximum likelihood estimates are the unique solutions of the likelihood equations, then the estimates are asymptotically normal and asymptotically relatively unbiased. Further, if $\theta = (\theta_1, \theta_2)$, then $\hat{\boldsymbol{\theta}}_1$ and $\hat{\boldsymbol{\theta}}_2$ have a *joint normal distribution* asymptotically. Therefore, in the large sample case it is almost as easy to get confidence intervals based on the maximum likelihood estimates when Θ has 2, 3, or more dimensions as when Θ is 1-dimensional. Of course, saying "almost as easy" assumes that the variances of the estimators can be evaluated. We will come back to this problem shortly.

So far we have focused on the problem of getting a confidence interval for only one of the parameters, say θ_1, in a many-parameter problem. But often we are interested in estimating *all* parameters, or at any rate, more than one parameter. For example, suppose that $\theta = (\theta_1, \theta_2)$ and we want 95% intervals for both parameters. Suppose we go through the above procedure twice, once for θ_1 and once for θ_2. The resulting two intervals

$$J_1(\hat{\theta}) = \hat{\theta}_1 \pm 2.0\sigma_1(\hat{\theta})$$

$$J_2(\hat{\theta}) = \hat{\theta}_2 \pm 2.0\sigma_2(\hat{\theta})$$

are such that

$$P_\theta[\theta_1 \in J_1(\hat{\boldsymbol{\theta}})] \simeq .95$$

$$P_\theta[\theta_2 \in J_2(\hat{\boldsymbol{\theta}})] \simeq .95.$$

But is this what we want? Suppose that the 5% of the time that θ_1 is not in $J_1(\hat{\boldsymbol{\theta}})$ does not overlap with the 5% of the time that θ_2 is outside of $J_2(\hat{\boldsymbol{\theta}})$. Altogether, then, one or the other of θ_1 or θ_2 is not in the specified range 10% of the time.

If we are looking for a 2-dimensional confidence region that contains the point (θ_1, θ_2) 95% of the time, then separate 95% intervals are not what we want. It makes sense to require *simultaneous confidence intervals*; that is, intervals $J_1(\hat{\theta})$, $J_2(\hat{\theta})$ such that the combined statement

$$\theta_1 \in J_1(\hat{\theta}),\ \theta_2 \in J_2(\hat{\theta})$$

holds at least 95% of the time. In other words, at least 95% of the time, θ_1 and θ_2 *simultaneously* fall into their respective intervals. We say that $J_1(\hat{\theta})$ *and* $J_2(\hat{\theta})$ *are simultaneous* 95% *confidence intervals if*

$$P_\theta[\theta_1 \in J_1(\hat{\boldsymbol{\theta}}),\ \theta_2 \in J_2(\hat{\boldsymbol{\theta}})] \geq .95.$$

More generally,

4.13 ***Definition*** *In a k-parameter problem, the intervals* $J_1(\hat{\theta}), \ldots, J_k(\hat{\theta})$ *are said to form simultaneous* $100\gamma\%$ *confidence intervals if*

$$P_\theta[\theta_1 \in J_1(\hat{\boldsymbol{\theta}}), \ldots, \theta_k \in J_k(\hat{\boldsymbol{\theta}})] \geq \gamma, \qquad \text{for all } \theta \in \Theta.$$

The difficulty in getting simultaneous intervals is that the estimates $\hat{\boldsymbol{\theta}}_1, \ldots, \hat{\boldsymbol{\theta}}_k$ may have any amount of interdependence. However, there is a crude but effective method of getting $100\gamma\%$ intervals that always works. It is based on the following simple result.

4.14 ***Proposition*** *If* $J_1(\hat{\theta}), \ldots, J_k(\hat{\theta})$ *individually are* $100\gamma\%$ *confidence intervals, then they form simultaneous* $100[1 - k(1 - \gamma)]\%$ *confidence intervals.*

A hint of the proof was given above. If each of $\theta_1, \ldots, \theta_k$ falls outside its given interval at most $100(1 - \gamma)\%$ of the time, then at worst, the percentage of the time that at least one of $\theta_1, \ldots, \theta_k$ will be outside its range is k times $100(1 - \gamma)\%$. All of the parameters will be simultaneously inside their respective intervals at least $100[1 - k(1 - \gamma)]\%$ of the time.

For example, to use the above method to get simultaneous 95% intervals for θ_1 and θ_2, $J_1(\hat{\theta})$ and $J_2(\hat{\theta})$ individually have to be 97.5% intervals; to get a simultaneous 90% interval, both intervals individually have to be 95% intervals.

Now we go back to the problem of computing the variances of estimates of the various parameters. One approach is simple: Write out the expressions for the estimators and try to compute their variances directly. If the estimators are maximum likelihood estimates, then you might consider trying to compute their variances or equivalently (for large n) their risks by using the information expression. But when there are two or more parameters, the 1-dimensional result is no longer generally true. For instance, if $\theta = (\alpha, \beta)$ and $\hat{\boldsymbol{\alpha}}, \hat{\boldsymbol{\beta}}$ are the maximum likelihood estimators, it is not true that you can simply define

$$I(\alpha, \beta) = \int \left[\frac{\partial}{\partial \alpha} \log f_{(\alpha,\beta)}(x)\right]^2 f_{(\alpha,\beta)}(x)\, dx$$

and have

$$E_{(\alpha,\beta)}(\hat{\boldsymbol{\alpha}} - \alpha)^2 \simeq \frac{1}{nI(\alpha, \beta)}.$$

This does hold if you have a model in which β is assumed known and fixed. But where β is unknown, there is more variability in the problem; and $\hat{\boldsymbol{\alpha}}$ generally has a higher risk.

The appropriate generalization of 3.19 is as follows: Let $\theta = (\theta_1, \ldots, \theta_k)$ with $\hat{\boldsymbol{\theta}} = (\hat{\boldsymbol{\theta}}_1, \ldots, \hat{\boldsymbol{\theta}}_k)$ the maximum likelihood estimators.

Form the *risk matrix* $[R(\theta)]$ with elements

$$R_{ij}(\theta) = E_\theta(\hat{\theta}_i - \theta_i)(\hat{\theta}_j - \theta_j),$$

and the *information matrix* $[I(\theta)]$ with elements

$$I_{ij}(\theta) = \int \left[\frac{\partial}{\partial \theta_i} \log f_\theta(x)\right]\left[\frac{\partial}{\partial \theta_j} \log f_\theta(x)\right] f_\theta(x)\, dx.$$

Then

4.15 ***Theorem*** *In a "smooth" problem*

$$[R(\theta)] = \frac{1}{n}[I(\theta)]^{-1} + o\left(\frac{1}{n}\right).$$

The risk of $\hat{\theta}_k$ is the kth diagonal element of $[R(\theta)]$;

$$E_\theta(\hat{\theta}_k - \theta)^2 = R_{k,k}(\theta).$$

Therefore for large sample size, the approximation to the squared error loss of $\hat{\theta}_k$ is obtained by computing the kth diagonal element of the inverse of $I(\theta)$ and dividing by n.

There is a special case where this somewhat complicated procedure simplifies. Let $\{P_{(\alpha,\beta)}(dx)\}$ be a location-scale family where the standard distribution is symmetric around the origin. Then a computation shows that the off-diagonal elements of $[I(\alpha, \beta)]$ vanish. In consequence

$$E_{(\alpha,\beta)}(\hat{\alpha} - \alpha)^2 \simeq \frac{1}{nI_{11}(\alpha, \beta)}$$

$$E_{(\alpha,\beta)}(\hat{\beta} - \beta)^2 \simeq \frac{1}{nI_{22}(\alpha, \beta)}.$$

But these are exactly the expressions we would have gotten had we used the 1-dimensional version 3.19 on each parameter separately, assuming that the other parameter were known. What is happening is that the estimates $\hat{\alpha}$ and $\hat{\beta}$ have asymptotically a joint normal distribution. Since $[I]$ is diagonal, then for n large $[R]$ is approximately diagonal. This implies that $\hat{\alpha}$ and $\hat{\beta}$ are asymptotically independent, and there is no "cross-interference" between the two estimators.

Problem 9 The sample mean and variance in a sample of size 100 drawn from a normal population are $\hat{\mu} = 69.3$, $\hat{\sigma}^2 = 25.8$. Compute simultaneous 95% confidence intervals.

Problem 10 Suppose that

$$\frac{\hat{s}}{\sqrt{2}} = \sqrt{\frac{1}{2(n-1)} \sum_{1}^{n} (x_i - \bar{x})^2}$$

is used as an estimator of the scale parameter of a double exponential distribution. What is its efficiency? (Recall that the variance σ^2 of a double exponential is $2\beta^2$. Write

$$(\hat{s}^2 - \sigma^2)^2 = (\hat{s} - \sigma)^2(\hat{s} + \sigma)^2.$$

Now $\hat{s}$ is a consistent estimator for σ; hence $(\hat{s} + \sigma)^2 \simeq 4\sigma^2$, and

$$4\sigma^2 E(\hat{s} - \sigma)^2 \simeq E(\hat{s}^2 - \sigma^2)^2.$$

Look at Chapter 3, Problem 33 for an evaluation of the right-hand side.)

Problem 11 Evaluate the squared error losses for the maximum likelihood estimates of α, β for large n in a Weibull family, using the values

$$\int (1 + \log v - v \log v)^2 e^{-v}\, dv = 1.82$$

$$\int (1 - v)(1 + \log v - v \log v) e^{-v}\, dv = .42.$$

Problem 12 Assume that $(X_1, Y_1), \ldots, (X_n, Y_n)$ are samples from a bivariate normal distribution with 0 means and equal but unknown variances. Use $c = EX_kY_k$ to denote the covariance.

a Show that the maximum likelihood estimate for the covariance c is

$$\hat{c} = \frac{1}{n} \sum_{1}^{n} X_iY_i.$$

(Estimate σ^2 by

$$\hat{\sigma}^2 = \frac{1}{2}\left(\frac{1}{n} \sum_{1}^{n} X_i^2 + \frac{1}{n} \sum_{1}^{n} Y_i^2\right),$$

and verify that $\hat{\sigma}^2$ and the given $\hat{c}$ satisfy the maximum likelihood equations.)

b Find 90% confidence intervals for c.

$$\left(\text{Ans.} \qquad \hat{c} \pm \frac{1.65}{\sqrt{n}} \sqrt{\hat{c}^2 + (\hat{\sigma}^2)^2}.\right)$$

Problem 13 In Problem 12, make the transformation from the parameter $c = E\mathsf{XY}$ to the correlation coefficient ρ defined as c/σ^2.

a Why is the maximum likelihood estimate for ρ given by $\hat{\rho} = \hat{c}/\hat{\sigma}^2$?

b (Difficult!) Write $\hat{\sigma}^2(\hat{\rho} - \rho) = \hat{c} - \hat{\sigma}^2\rho$ and observe that for n large, $\hat{\boldsymbol{\sigma}}^2 \simeq \sigma^2$, so

$$E[\hat{\boldsymbol{\sigma}}^2(\hat{\boldsymbol{\rho}} - \rho)]^2 \simeq \sigma^4 E(\hat{\boldsymbol{\rho}} - \rho)^2.$$

Note that $\mathsf{U}_i = \mathsf{X}_i - \rho\mathsf{Y}_i$ satisfies $E\mathsf{U}_i\mathsf{Y}_i = 0$, so that U_i is independent of Y_i, and therefore X_i can be written as $\mathsf{U}_i + \rho\mathsf{Y}_i$. Apply this to the expression $\hat{\mathsf{c}} = \hat{\sigma}^2\rho$ to evaluate $E(\hat{\mathsf{c}} - \hat{\boldsymbol{\sigma}}^2\rho)^2$, and then find 95% confidence intervals for ρ.

$$\left(\text{Ans.} \qquad \hat{\rho} \pm \frac{2}{\sqrt{n}}(1 - \hat{\rho}^2).\right)$$

Problem 14 (Difficult!) Evaluate $E(\hat{\boldsymbol{\rho}} - \rho)^2$ approximately in Problem 13 by computing the information matrix.

$$\left(\text{Ans.} \qquad E(\hat{\boldsymbol{\rho}} - \rho)^2 = \frac{(1 - \rho^2)^2}{n}.\right)$$

Rough and Ready Confidence Intervals

In all of our general work on confidence intervals to date, we have made some strong assumptions, namely: $\hat{\boldsymbol{\theta}}$ is asymptotically normal and unbiased, the problem is "smooth," there is a unique likelihood estimate, and the sample size is large enough for all of the asymptotic results to hold. Trying to show that these conditions are satisfied for a particular problem is sometimes difficult.

There is a method of getting confidence intervals that has universal applicability—asymptotic normality or no, large or small sample size. Unfortunately it yields confidence intervals that are usually much too large. But still, these are usually order-of-magnitude correct. This method employs a very useful inequality.

4.16 *Chebyshev's Inequality* *For any random variable* Y, *and any number* $\delta > 0$,

$$P(|Y| \geq \delta) \leq \frac{EY^2}{\delta^2}.$$

The proof of this is simple: Let A be the set on which $|Y| \geq \delta$. On A, $Y^2 \geq \delta^2$, so

$$EY^2 \geq \delta^2 P(A)$$

Dividing by δ^2 completes the proof.

In Chebyshev's inequality let

$$Y = \frac{\hat{\theta} - \theta}{r(\theta)},$$

where $r(\theta)$ is the square root of $R(\theta)$. Note that since

$$r^2(\theta) = R(\theta) = E_\theta(\hat{\theta} - \theta)^2,$$

$E_\theta Y^2 = 1$. Therefore, by 4.16, for any δ,

$$P_\theta\left(\left|\frac{\hat{\theta} - \theta}{r(\theta)}\right| \geq \delta\right) \leq \frac{1}{\delta^2},$$

or

$$P_\theta\left(\left|\frac{\hat{\theta} - \theta}{r(\theta)}\right| \leq \delta\right) \geq 1 - \frac{1}{\delta^2},$$

and

4.17 $$P_\theta[\hat{\theta} - \delta r(\theta) \leq \theta \leq \hat{\theta} + \delta r(\theta)] \geq 1 - \frac{1}{\delta^2}.$$

Now let δ satisfy

$$1 - \frac{1}{\delta^2} = \gamma$$

If the estimation method is consistent and the sample size moderately large, we can replace $r(\theta)$ in 4.17, by $r(\hat{\theta})$ getting

4.18 $$P_\theta[\hat{\theta} - \delta r(\hat{\theta}) \leq \theta \leq \hat{\theta} + \delta r(\hat{\theta})] \geq \gamma.$$

The $100\gamma\%$ intervals are therefore given by

$$J(\hat{\theta}) = \hat{\theta} \pm \delta r(\hat{\theta}).$$

The assumption made in deriving 4.17 holds universally. Usually, $r(\theta)$ can be replaced by $r(\hat{\theta})$ for very moderate sample sizes; $n \geq 10$ is usually sufficient. The drawback of this widely applicable method is that the resulting confidence intervals are usually much too large. For instance, for 95% confidence intervals, we have

$$1 - \frac{1}{\delta^2} = .95,$$

so $\delta = 4.5$ and the intervals are $\hat{\theta} \pm 4.5r(\hat{\theta})$. But if the unbiased normal approximation holds, we know that the smaller intervals $\hat{\theta} \pm 2.0r(\hat{\theta})$ are sufficient. Thus the "rough and ready" method leads to confidence intervals more than two times too large. Still, it gives the intervals to the right order of magnitude.

These intervals can be obtained just as easily in the multidimensional case. For instance, in Chebyshev's inequality take Y to be

$$\mathrm{Y} = \frac{\hat{\theta}_1 - \theta_1}{r^{(1)}(\theta)},$$

where $r^{(1)}(\theta)$ denotes the risk $E_\theta(\hat{\theta}_1 - \theta_1)^2$. Notice that $E\mathrm{Y}^2 = 1$, and replace $r^{(1)}(\theta)$ by $r^{(1)}(\hat{\theta})$ to get confidence intervals for θ_1.

General Confidence Intervals

If the sample size is small, if the estimator is not asymptotically normal, if the rough and ready intervals seem too large, then what? In any problem there are an infinite number of different $100\gamma\%$ confidence intervals. For example, suppose we have a confidence interval that satisfies

$$P_\theta[\theta \in J(\hat{\boldsymbol{\theta}})] \geq .95, \qquad \text{for all } \theta \in \Theta.$$

Now for every value of $\hat{\theta}$ take $J'(\hat{\theta})$ to be any interval larger than $J(\hat{\theta})$, that is, $J(\hat{\theta}) \subset J'(\hat{\theta})$. Then certainly $P_\theta[\theta \in J'(\hat{\boldsymbol{\theta}})] \geq .95$, for all $\theta \in \Theta$. One general principle is that we always want our confidence intervals as small as possible. Since $J(\hat{\theta})$ is our estimate of the interval in which θ is located, the smaller the intervals $J(\hat{\theta})$ are, the more precisely we have located θ. We will say that a $100\gamma\%$ confidence interval $J(\hat{\theta})$ is *tight* if there is no other $100\gamma\%$ confidence procedure with intervals $J'(\hat{\theta})$ such that $J'(\hat{\theta})$ is always smaller than (contained in) $J(\hat{\theta})$. Suppose that the confidence intervals satisfy

$$P_\theta[\theta \in J(\hat{\boldsymbol{\theta}})] = \gamma, \qquad \text{for all } \theta \in \Theta, \tag{4.19}$$

(or come as close to γ as possible if the $\{P_\theta\}$ are discrete distributions). Then if we try to shrink all of the intervals $J(\hat{\theta})$ a little, the probabilities $P_\theta[\theta \in J(\hat{\boldsymbol{\theta}})]$ decrease and become less than γ and we no longer have $100\gamma\%$ confidence procedure. Therefore, to find tight procedures, we will look for procedures satisfying 4.19. (We ignore the comparatively unusual cases for which 4.19 has no solutions, not even approximate solutions.)

We can almost always find tight $100\gamma\%$ intervals in the following way. Our estimator $\hat{\theta}$ is a random variable whose distribution depends on θ. For every value of θ, find an upper bound $b(\theta)$ and a lower bound $a(\theta)$ such that

4.20 $$P_\theta[a(\theta) \le \hat{\theta} \le b(\theta)] = \gamma.$$

If $a(\theta)$ and $b(\theta)$ are monotone functions of θ, the inequality

$$a(\theta) \le \hat{\theta} \le b(\theta)$$

can be written in the form

$$g(\hat{\theta}) \le \theta \le h(\hat{\theta}).$$

Then 4.20 can be written as

$$P_\theta[g(\hat{\theta}) \le \theta \le h(\hat{\theta})] = \gamma, \qquad \text{for all } \theta,$$

which shows that the intervals $[g(\hat{\theta}), h(\hat{\theta})]$ define a $100\gamma\%$ confidence procedure. Here is an illustration.

Example b $X_1, \ldots, X_n$ are samples from a distribution uniform on $[0, l]$. We want to find 95% confidence intervals for l based on the maximum likelihood estimator

$$\hat{l} = \max(X_1, \ldots, X_n) = M_n.$$

The first step is to find two numbers $a(l)$ and $b(l)$ such that

$$P_l[a(l) \le M_n \le b(l)] = .95,$$

or

$$P_l[M_n < b(l)] - P_l[M_n \le a(l)] = .95.$$

From p. 55, of the last chapter, we have

$$P_l(M_n \le x) = \begin{cases} (x/l)^n, & 0 \le x \le l \\ 1, & x > l. \end{cases}$$

So assuming that $b(l) \le l$, we get the equation

$$(b(l)/l)^n - (a(l)/l)^n = .95, \qquad \text{for all } l \ge 0.$$

This single equation for two undetermined functions $a(l)$ and $b(l)$ has an infinity of solutions; one simple set of solutions is $a(l) = \alpha l$, $b(l) = \beta l$. These will satisfy the equation for all l if

$$\beta^n - \alpha^n = .95.$$

This is also one equation in two unknowns—so there are still an infinite number of solutions. For example, let $\beta = 1$, getting

$$\begin{aligned} 1 - \alpha^n &= .95 \\ \alpha^n &= .05 \\ n \log \alpha &= \log .05 \\ &= -3.0. \end{aligned}$$

For n large, $\log \alpha$ is near 0, hence $\alpha \simeq 1$. Put $\alpha = 1 - \delta$, where $\delta \ll 1$; then

$$\log \alpha = \log (1 - \delta) \simeq -\delta.$$

The above equation becomes

$$-n\delta = -3.0$$

or

$$\delta = 3/n,$$

so we have the solution

$$\beta = 1, \qquad \alpha = 1 - 3/n,$$

which leads to

$$a(l) = (1 - 3/n)l, \qquad b(l) = l;$$

hence, going back to the original equation,

$$P_l[(1 - 3/n)l \leq \mathsf{M}_n \leq l] = .95, \qquad \text{for all } l.$$

The final step is to rewrite the two inequalities defining the above region, namely

$$(1 - 3/n)l \leq \mathsf{M}_n, \qquad \mathsf{M}_n \leq l,$$

as an inequality on l. The first one can be rewritten as

$$l \leq \frac{\mathsf{M}_n}{1 - (3/n)}.$$

The second inequality is already in the correct form. Combining these two gives

$$\mathsf{M}_n \leq l \leq \frac{\mathsf{M}_n}{1 - (3/n)}.$$

Since this region is exactly the same as the one defined by the original inequalities, we can write

$$P_l\left(\mathsf{M}_n \leq l \leq \frac{\mathsf{M}_n}{1 - (3/n)}\right) = .95.$$

Hence, 95% confidence intervals for l are given by $[\mathsf{M}_n, \mathsf{M}_n/(1 - 3/n)]$. For example, if $n = 100$, and the sample $x_1, \ldots, x_{100}$ has a maximum value of 2.8, then the corresponding 95% interval is [2.8, 2.8/.97], or [2.8, 2.9].

One interesting aspect of this example is the length of the confidence intervals as a function of sample size. For n large,

$$\frac{1}{1 - (3/n)} \simeq 1 + \frac{3}{n}.$$

The length of the confidence intervals is about $3\mathsf{M}_n/n$. As $n \to \infty$, $\mathsf{M}_n \to l$, and the size of the confidence interval is about $3l/n$. Therefore, the accuracy of the estimate increases linearly with the sample size. Double the sample size and the length of the confidence interval is cut in half. As we know, this problem is not smooth and the information results do not hold. In fact, as Problem 15 shows, the maximum likelihood estimator M_n is not asymptotically normal.

Problem 15 Find the distribution, for large n, of

$$\frac{\mathsf{M}_n - l}{\sigma_n(l)},$$

where $\sigma_n^2(l)$ is the variance of M_n. That is, evaluate

$$P\left(\frac{\mathsf{M}_n - l}{\sigma_n(l)} \le x\right).$$

$$\left(\text{Ans.} \qquad P\left(\frac{\mathsf{M}_n - l}{\sigma_n(l)} \le x\right) \simeq \begin{cases} 1, & x > 0 \\ e^x, & x < 0. \end{cases}\right)$$

Problem 16 **a** Find rough and ready 95% confidence intervals using the estimator M_n for the problem of Example **b**.

$$(\text{Ans.} \qquad \mathsf{M}_n\left(1 \pm \frac{4.5\sqrt{2}}{\sqrt{(n+2)(n+1)}}\right) \simeq \mathsf{M}_n\left(1 \pm \frac{6.4}{n}\right),$$

but since we know that $l \ge \mathsf{M}_n$, we can use the intervals

$$\mathsf{M}_n \le l \le \mathsf{M}_n\left(1 + \frac{6.4}{n}\right).$$

Compare this with the tight intervals computed in Example **b**.)

b Use the method of Example **b**. to compute 95% confidence intervals based on the unbiased estimator $(n + 1/n)\mathsf{M}_m$.

Exact Confidence Intervals for Location and Scale Parameters

It would be useful to have confidence intervals that are tight and, at the same time, valid for all sample sizes. Besides being useful for small sample sizes, these would also give us an idea of the sample size needed to make our normal approximations accurate. Exact confidence intervals can be computed for location and scale parameters. We start by giving some examples.

Example c Take $X_1, \ldots, X_n$ to be exponential with scale parameter β. The maximum likelihood estimate of β is the sample mean

$$\hat{\beta} = \frac{X_1 + \cdots + X_n}{n}.$$

Now $X_k = \beta Y_k$, where Y_k has the standard exponential distribution with parameter 1. Thus, for all β, $\hat{\beta}/\beta$ has the distribution given by

$$\frac{Y_1 + \cdots + Y_n}{n},$$

which is gamma with parameters n, $1/n$. Use a table of the gamma distribution to find numbers y_0, y_1 such that

$$P\left(y_0 \leq \frac{Y_1 + \cdots + Y_n}{n} \leq y_1\right) = \gamma.$$

The numbers y_0, y_1 depend only on n and γ. Then

$$P_\beta\left(y_0 \leq \frac{\hat{\beta}}{\beta} \leq y_1\right) = \gamma,$$

and solving the inequalities for β gives the $100\gamma\%$ intervals

$$\frac{\hat{\beta}}{y_1} \leq \beta \leq \frac{\hat{\beta}}{y_0}.$$

Since $\hat{\beta}$ is asymptotically normal and asymptotically relatively unbiased, for large sample size approximate 95% confidence intervals are given by

$$\hat{\beta}\left(1 - \frac{2}{\sqrt{n}}\right) \leq \beta \leq \hat{\beta}\left(1 + \frac{2}{\sqrt{n}}\right).$$

There is rapid convergence of the exact intervals to the approximate intervals. For 95% intervals and $n = 15$, the total difference between the endpoints of the two intervals is less than 7% of the length of the shortest exact interval.

Example d Take $X_1, \ldots, X_n$ to be independent $N(\mu, \sigma^2)$ variables and estimate σ^2 by

$$\hat{s}^2 = \frac{1}{n-1} \sum_1^n (X_k - \bar{X})^2.$$

Problem 24, page 79 outlines a proof that

$$V^2 = \frac{1}{\sigma^2} \sum_1^n (X_k - \bar{X})^2$$

has a χ^2 distribution with $n - 1$ degrees of freedom. In consequence $\hat{s}^2/\sigma^2$ has the distribution of a χ^2 variable divided by $n - 1$. From a χ^2 table find two numbers y_0, y_1 such that

$$P\left(y_0 \leq \frac{1}{n-1} C \leq y_1\right) = \gamma.$$

Then for all (μ, σ^2),

$$P_{(\mu,\sigma^2)}\left(y_0 \leq \frac{\hat{s}^2}{\sigma^2} \leq y_1\right) = \gamma,$$

and solving the inequalities gives

$$\frac{\hat{s}^2}{y_1} \leq \sigma^2 \leq \frac{\hat{s}^2}{y_0}$$

as exact $100\gamma\%$ confidence intervals.

By a direct computation

$$E(\hat{s}^2 - \sigma^2)^2 = \frac{2}{n-1} \sigma^4,$$

so that the approximate 95% confidence intervals based on asymptotic normality are

$$\hat{s}^2 \left(1 - \frac{4}{\sqrt{n-1}}\right) \leq \sigma^2 \leq \hat{s}^2 \left(1 + \frac{4}{\sqrt{n-1}}\right).$$

The convergence here is considerably slower than in Example a. For $n = 20$ the approximate interval differs by 30% from the shortest exact interval. At $n = 40$, this figure has dropped to about 15%.

Example e The best known example of exact confidence intervals for a location parameter occurs in the estimation of μ when sampling from an $N(\mu, \sigma^2)$ distribution. A famous statistical name is attached to this example.

Let $X_1, \ldots, X_n$ be independent $N(\mu, \sigma^2)$ variables. Take

$$\hat{\mu} = \frac{1}{n} \sum_{1}^{n} X_k,$$

and use the unbiased sample variance

$$\hat{s}^2 = \frac{1}{n-1} \sum_{1}^{n} (X_k - \hat{\mu})^2.$$

We know

$$R^{(1)}(\mu, \sigma) = E_{(\mu,\sigma)}(\hat{\mu} - \mu)^2 = \frac{\sigma^2}{n}.$$

Hence the normalized version of the estimator $\hat{\mu}$ is given by

$$\sqrt{n}\left(\frac{\hat{\mu} - \mu}{\sigma}\right).$$

This has an $N(0, 1)$ distribution. For large n, $\hat{s} \simeq \sigma$; hence the variable

$$T = \sqrt{n}\left(\frac{\hat{\mu} - \mu}{\hat{s}}\right)$$

is approximately $N(0, 1)$. But for small n, we know little about the distribution of T. However, an important fact about the distribution of T that holds true for all sample sizes is

4.21 ***Proposition*** *The distribution of* T *does not depend on* μ *or* σ.

To verify 4.21, if $X_1, \ldots, X_n$ have an $N(\mu, \sigma^2)$ distribution, then they can be written as

$$X_m = \sigma Y_m + \mu,$$

where $Y_1, \ldots, Y_n$ are independent $N(0, 1)$. We denote by $\hat{\mu}(\mathbf{X})$ and $\hat{s}^2(\mathbf{X})$ the sample means and unbiased sample variance obtained from the variables $X_1, \ldots, X_n$, and by $\hat{\mu}(\mathbf{Y})$, $\hat{s}^2(\mathbf{Y})$ the same quantities obtained from the $N(0, 1)$ variables $Y_1, \ldots, Y_n$. First of all,

$$\begin{aligned}\frac{\hat{\mu}(\mathbf{X}) - \mu}{\sigma} &= \frac{1}{\sigma}\frac{1}{n} \sum_{1}^{n} (X_k - \mu) \\ &= \frac{1}{n} \sum_{1}^{n} Y_k \\ &= \hat{\mu}(\mathbf{Y}).\end{aligned}$$

Secondly,

$$\begin{aligned}X_m - \hat{\mu}(\mathbf{X}) &= (X_m - \mu) - (\hat{\mu}(\mathbf{X}) - \mu) \\ &= \sigma Y_m - \sigma\hat{\mu}(\mathbf{Y}).\end{aligned}$$

Therefore

$$\hat{s}^2(\mathbf{X}) = \sigma^2 \hat{s}^2(\mathbf{Y}),$$

and

$$\hat{s}(\mathbf{X}) = \sigma \hat{s}(\mathbf{Y}).$$

Putting these together gives

$$\sqrt{n}\left(\frac{\hat{\mu}(\mathbf{X}) - \mu}{\hat{s}(\mathbf{X})}\right) = \sqrt{n}\,\frac{\hat{\mu}(\mathbf{Y})}{\hat{s}(\mathbf{Y})}.$$

Since $\mathbf{Y}$ consists of independent $N(0, 1)$ components, the distribution of the variable on the right certainly does not depend on μ, σ. Having gone this far, we state (see p. 41):

4.22 ***Proposition*** *For independent* $N(0, 1)$ *variables* $\mathsf{Y}_1, \ldots, \mathsf{Y}_n$, *the distribution of*

$$\mathsf{S} = \sqrt{n}\,\frac{\hat{\mu}(\mathbf{Y})}{\hat{s}(\mathbf{Y})}$$

is given by Student's t distribution with $n - 1$ *degrees of freedom.*

Now $\hat{s}(\mathbf{Y})$ is a consistent estimator for the standard deviation of the underlying distribution of the $\mathsf{Y}_1, \ldots, \mathsf{Y}_k$, so it is usually nearly equal to 1 for n large, say $n \geq 30$. Then

$$\mathsf{S} \simeq \sqrt{n}\bar{\mathsf{Y}},$$

and has approximately an $N(0, 1)$ distribution. For smaller values of n, see Table 3.

To get $100\gamma\%$ confidence intervals, use the table values of the distribution of S to select t so that

4.23 $$P(-t \leq \mathsf{S} \leq t) = \gamma.$$

By 4.22 we know that for any μ, σ,

$$P_{(\mu,\sigma)}\left(-t \leq \sqrt{n}\,\frac{\hat{\mu} - \mu}{\hat{s}} \leq t\right) = \gamma,$$

so the $100\gamma\%$ confidence intervals for μ are given by

4.24 $$\left[\hat{\mu} - t\frac{\hat{s}}{\sqrt{n}}, \hat{\mu} + t\frac{\hat{s}}{\sqrt{n}}\right].$$

The only difference between these intervals and the large sample size intervals is that t appears instead of z. Since the distribution of S depends on n, so will the value of t. For 95% intervals, z computed from the normal approximation is 1.96. Here are the exact values of t for various sample sizes:

n	t
5	2.78
10	2.62
15	2.14
20	2.09
30	2.04
60	2.00
120	1.98

Notice that for small sample sizes, the normal approximation computes erroneously short confidence intervals and thereby gives a false sense of accuracy. But even for n as small as 15, the normal approximation is surprisingly good.

What is going on in these examples? In the first two examples we used the fact that in a scale family or in a location-scale family, we had estimates $\hat{\beta}$ of the scale such that the distribution of $\hat{\beta}/\beta$ did not depend on the values of (α, β). In the second example, we used the fact that the estimates $\hat{\mu}$ and $\hat{s}$ of the location and scale parameters combined in such a way that the distribution of

$$\frac{\hat{\mu} - \mu}{\hat{s}}$$

did not depend on the value of (μ, σ). Actually, for most location and scale estimates $\hat{\alpha}$ and $\hat{\beta}$ used in a location-scale family these two facts hold true; namely, *the distributions of*

4.25
$$\frac{\hat{\beta}}{\beta}, \quad \frac{\hat{\alpha} - \alpha}{\hat{\beta}}$$

are the same for all parameter values of (α, β).

The reasoning is very similar to that used in physics to get dimensionless quantities. Subtracting α from $\hat{\alpha}$, and dividing by β hopefully gets rid of the dependence on the parameters. It's more easily seen for $\hat{\beta}/\beta$. Since $\hat{\beta}$ is an estimate of the scale, its distribution should not depend on the value of the location parameter. Furthermore, in some sense $\hat{\beta}$ should be directly proportional to β; hence $\hat{\beta}/\beta$ should be "dimensionless."

An equivalent way of stating 4.25 is:

4.26
$$\text{For all } (\alpha, \beta), \quad \frac{\hat{\alpha}(\mathbf{X}) - \alpha}{\hat{\beta}(\mathbf{X})} = \frac{\hat{\alpha}(\mathbf{Y})}{\hat{\beta}(\mathbf{Y})}$$

and

4.27
$$\frac{\hat{\beta}(\mathbf{X})}{\beta} = \hat{\beta}(\mathbf{Y}),$$

where $\mathbf{Y} = (\mathrm{Y}_1, \ldots, \mathrm{Y}_n)$, each Y_k having the standard $P_{(0,1)}(dx)$ distribution.

If 4.26 and 4.27 hold, we can get exact $100\gamma\%$ confidence intervals for α and β based on $\hat{\boldsymbol{\alpha}}$ and $\hat{\boldsymbol{\beta}}$. Select z_0, z_1 such that

$$P\left(z_0 \leq \frac{\hat{\alpha}(\mathbf{Y})}{\hat{\beta}(\mathbf{Y})} \leq z_1\right) = \gamma$$

and y_0, y_1 such that

$$P(y_0 \leq \hat{\beta}(\mathbf{Y}) \leq y_1) = \gamma.$$

Then, denoting $\hat{\boldsymbol{\alpha}}(\mathbf{X})$, $\hat{\beta}(\mathbf{X})$ by $\hat{\boldsymbol{\alpha}}$, $\hat{\boldsymbol{\beta}}$, 4.26 and 4.27 imply that

$$P_{(\alpha,\beta)}\left(z_0 \leq \frac{\hat{\boldsymbol{\alpha}} - \alpha}{\hat{\boldsymbol{\beta}}} \leq z_1\right) = \gamma$$

$$P_{(\alpha,\beta)}\left(y_0 \leq \frac{\hat{\boldsymbol{\beta}}}{\beta} \leq y_1\right) = \gamma.$$

Solving the inequalities in these probabilities for α and β gives the $100\gamma\%$ confidence intervals

4.28 $$\hat{\alpha} - z_1\hat{\beta} \leq \alpha \leq \hat{\alpha} - z_0\hat{\beta},$$

4.29 $$\frac{\hat{\beta}}{y_1} \leq \beta \leq \frac{\hat{\beta}}{y_0}$$

for α and β, respectively. The problem is now reduced to evaluating the distributions of the variables $\hat{\alpha}(\mathbf{Y})/\hat{\beta}(\mathbf{Y})$ and $\hat{\beta}(\mathbf{Y})$.

In any location and/or scale parameter problem we can justifiably say that every "sensible" location estimator satisfies 4.26 and that 4.27 holds for every "sensible" scale estimator. We use "sensible" here to mean that our estimates are well-behaved under a transformation of the data. Formally,

4.30 ***Definition*** *If an estimator $\hat{\alpha}$ satisfies*

$$\hat{\alpha}(a + bx_1, \ldots, a + bx_n) = a + b\hat{\alpha}(x_1, \ldots, x_n)$$

for all $\mathbf{x}$, *a, and* $b > 0$, *call $\hat{\alpha}$ an invariant location estimator.*

Similarly, if $\hat{\beta}$ satisfies

$$\hat{\beta}(a + bx_1, \ldots, a + bx_n) = b\hat{\beta}(x_1, \ldots, x_n)$$

call $\hat{\beta}$ an invariant scale estimator.

Notice that invariance of the estimators does not depend on the underlying family of distributions, but only on the form of the estimators themselves. Virtually every common location and scale estimate is invariant. Why are invariant estimates "sensible?" Suppose $\hat{\alpha}$ and $\hat{\beta}$ are estimates of α and β on the basis of the data $x_1, \ldots, x_n$. This is equivalent to the conclusion that these points were drawn from $\mathsf{X} = \hat{\alpha} + \hat{\beta}\mathsf{Y}$, where Y has the standard distribution of the family. Then "sensibly," you should conclude on the basis of the data $a + bx_1, \ldots, a + bx_n$ that they were drawn from the variable

$$a + b\mathsf{X} = (a + b\hat{\alpha}) + b\hat{\beta}\mathsf{Y}.$$

In other words, the estimate for α based on $a + bx_1, \ldots, a + bx_n$ should be $a + b\hat{\alpha}$, where $\hat{\alpha}$ is the estimate based on $x_1, \ldots, x_n$. Similarly, the estimate for scale based on $a + bx_1, \ldots, a + bx_n$ should be $b\hat{\beta}$.

The fact that most of the usual estimators for location and scale parameters are invariant is emphasized by:

4.31 ***Theorem*** *If the maximum likelihood estimators in any location-scale family are unique, then they are invariant.*

The proof is not difficult (see Problem 22).

Problem 17 A chemical analysis for the percent-by-weight of an element in a compound was repeated 10 times with the results:

57.27, 57.23, 56.09, 56.43, 57.18,
58.12, 57.41, 57.45, 56.78, 56.88.

a Assuming these are from an $N(\mu, \sigma^2)$ distribution, find 95% confidence intervals for μ using Student's t distribution.

b Find 95% confidence intervals for σ using the χ^2 distribution.

Problem 18 Which of the following are invariant location estimators?

a $\sum_{1}^{n} a_k x_k$

b $\dfrac{x_{(k+1)} + x_{(n-k)}}{2}$

($x_{(1)}, \ldots, x_{(n)}$ are the order statistics for the sample)

c $\dfrac{1}{n - 2m} \sum_{m+1}^{n-m} x_{(k)}, \qquad 2m < n$

d $x_{(k)}$

e $\gamma\bar{x}$, ($\bar{x}$ is the sample mean and $0 \leq \gamma < 1$).

Problem 19 Which of the following are invariant scale estimators? (Note: c is a fixed constant, $\bar{x}$ is the sample mean.)

a
$$c \sum_{1}^{n} |x_k - \bar{x}|$$

b $c[\max(x_1, \ldots, x_n) - \min(x_1, \ldots, x_n)]$ (the quantity in brackets is called the *range* of the sample)

c
$$c(x_{(n-k)} - x_{(k+1)})$$

d
$$\left(\frac{1}{n} \sum_{1}^{n} (x_k - \bar{x})^2\right)^{1/2}$$

e
$$c\left(\frac{1}{n} \sum_{1}^{n} (x_k - \bar{x})^r\right)^{1/r}, \qquad r > 0.$$

Problem 20 (Difficult). Consider the family of distributions uniform over

$$\left[m - \frac{l}{2},\ m + \frac{l}{2}\right].$$

a Show that the location and scale estimates given by

$$\hat{m} = \frac{\max(x_1, \ldots, x_n) + \min(x_1, \ldots, x_n)}{2}$$

$$\hat{l} = \max(x_1, \ldots, x_n) - \min(x_1, \ldots, x_n)$$

are invariant.

b Compute the cumulative distribution function of

$$\frac{\hat{m}(\mathbf{X}) - m}{\hat{l}(\mathbf{X})},$$

given that for $Y_1, \ldots, Y_n$ uniform on $[-1/2, 1/2]$ and $Y_{(1)} = \min(Y_1, \ldots, Y_n)$, $Y_{(n)} = \max(Y_1, \ldots, Y_n)$, the joint density of $Y_{(n)}, Y_{(1)}$ is

$$f(x, y) = n(n-1)(y - x)^{n-2}.$$

c Find 95% confidence intervals for m using **b**.

Problem 21 **a** Show that if $\hat{\alpha}$, $\hat{\beta}$ are invariant location and scale estimators, then for any given location-scale family, the squared error losses are given by

$$E_{(\alpha,\beta)}(\hat{\alpha} - \alpha)^2 = c_1 \beta^2$$
$$E_{(\alpha,\beta)}(\hat{\beta} - \beta)^2 = c_2 \beta^2,$$

where

$$c_1 = E[\hat{\alpha}(\mathbf{Y})]^2, \qquad c_2 = E[\hat{\beta}(\mathbf{Y}) - 1]^2$$

and $\mathsf{Y}_1, \ldots, \mathsf{Y}_n$ have the standard distribution of the family.

b Use **a.** to show that the ratio of risks for any two invariant location or scale estimates is constant. Deduce that the efficiency $e(\alpha, \beta)$ of such estimates is the same for all values of α and β.

c What can you deduce about the bias of invariant location estimators in a location-scale family?

Problem 22 In a location-scale family with standard density $f(x)$, the maximum likelihood estimates $\hat{\alpha}$ and $\hat{\beta}$ are those values of α and β that maximize

$$f_{(\alpha,\beta)}(\mathbf{x}) = \frac{1}{\beta^n} f\left(\frac{x_1 - \alpha}{\beta}\right) \cdots f\left(\frac{x_n - \alpha}{\beta}\right)$$

or

$$L_{\alpha,\beta}(\mathbf{x}) = -n \log \beta + \sum_1^n l\left(\frac{x_k - \alpha}{\beta}\right),$$

where $l(x) = \log f(x)$. For $x_k' = a + bx_k$, a, b constant, the new maximum likelihood estimates $\hat{\alpha}'$, $\hat{\beta}'$ are the values of α', β' that maximize

$$L_{(\alpha',\beta')}(\mathbf{x}') = -n \log \beta' + \sum_1^n l\left(\frac{x_k' - \alpha'}{\beta'}\right).$$

Make the substitution $\beta' = b\beta$, $\alpha' = a + b\alpha$; then $L_{(\alpha',\beta')}(\mathbf{x}')$ becomes

$$L_{(\alpha',\beta')}(\mathbf{x}') = -n \log b - n \log \beta - \sum_1^n l\left(\frac{x_k - \alpha}{\beta}\right)$$
$$= -n \log b + L_{(\alpha,\beta)}(\mathbf{x}).$$

Use this to show that

$$\hat{\alpha}' = a + b\hat{\alpha}, \qquad \hat{\beta}' = b\hat{\beta}.$$

Confidence Intervals—A Higher Vision

We now know how to compute $100\gamma\%$ confidence intervals in a number of statistical situations. But how can the concept of confidence intervals be formulated to fit into our general definition of a statistical model? To get confidence intervals, we observe the data point $\mathbf{x}$. On the basis of $\mathbf{x}$ we compute the estimate $\hat{\theta}$, then from $\hat{\theta}$ we compute $J(\hat{\theta})$. If we put these steps together, then what we have done is to associate an interval J with the data point $\mathbf{x}$. This leads to the more general idea of *confidence regions.* Suppose we are given a statistical model $\{P_\theta\}$, $\theta \in \Theta$. We observe an outcome $\mathbf{x}$. On the basis of $\mathbf{x}$ we select a *region in* Θ which we believe contains the true state of nature θ^*. Notice how this differs from an estimation procedure. We use the data $\mathbf{x}$ to get a region in Θ, rather than a single point in θ.

4.32 ***Definition*** *A* $100\gamma\%$ *confidence region is a procedure that associates with every outcome* $\mathbf{x}$ *of* $\mathbf{X}$ *a region* $R(\mathbf{x}) \subset \Theta$ *such that for every value of* θ, $\theta \in \Theta$,

4.32 $$P_\theta[\theta \in R(\mathbf{X})] \geq \gamma.$$

Again, note that a confidence region is not a single region but a procedure describing how to use the data $\mathbf{x}$ to get a region, which of course depends on $\mathbf{x}$. Secondly, the statement

$$P_\theta[\theta \in R(\mathbf{X})] \geq \gamma, \qquad \text{for all } \theta \in \Theta,$$

has a simple intuitive meaning: If the experiment is repeated over and over again independently with different values of θ, and if the procedure $R(\mathbf{x})$ is always used, $100\gamma\%$ *of the time the true state of nature is contained in the computed region.*

Summary

Foremost in this chapter is the concept of a $100\gamma\%$ confidence interval as a procedure for computing intervals $J(\hat{\theta})$ such that

$$P_\theta[\theta \in J(\hat{\theta})] \geq \gamma, \qquad \text{for all } \theta \in \Theta.$$

The rest of the chapter consists of methods of finding confidence intervals. These methods, ranging from the very generally applicable to the highly specialized, are as follows:

a *Applicable to large sample sizes and asymptotically normal and relatively unbiased estimates:*

$$J(\hat{\theta}) = \hat{\theta} \pm z\sigma(\hat{\theta}),$$

where z *is computed from the* $N(0, 1)$ *distribution.*

Recall that in "smooth" problems maximum likelihood estimates are asymptotically normal and unbiased. This method also works when the parameter is multidimensional.

b *Generally applicable to moderate sample sizes, but yielding intervals larger than necessary:*

Use Chebyshev's inequality to get

$$J(\hat{\theta}) = [\hat{\theta} - \delta r(\hat{\theta}), \hat{\theta} + \delta r(\hat{\theta})],$$

where δ satisfies

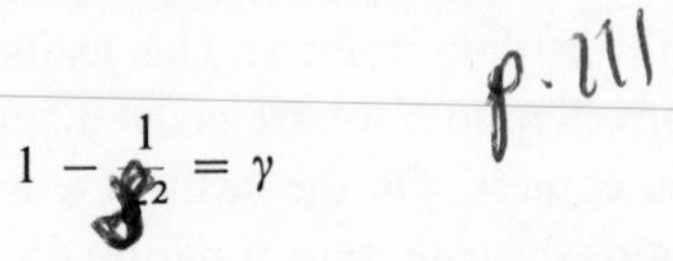

$$1 - \frac{1}{\delta^2} = \gamma$$

c *Generally applicable and usually tight, but hard to compute:*

Find $a(\theta)$, $b(\theta)$ such that

$$P_\theta[a(\theta) \leq \hat{\boldsymbol{\theta}} \leq b(\theta)] = \gamma, \qquad \text{for all } \theta \in \Theta.$$

Put the inequalities $a(\theta) \leq \hat{\theta} \leq b(\theta)$ in the form

$$g(\theta) \leq \hat{\theta} \leq h(\theta).$$

d *Applicable to location-scale families and tight for all sample sizes:*

If $\hat{\alpha}$ and $\hat{\beta}$ are invariant estimators of location and scale parameters, $100\gamma\%$ intervals for α and β are given by

$$[\hat{\alpha} - x_1\hat{\beta}, \hat{\alpha} - x_0\hat{\beta}], \qquad [\hat{\beta}/y_1, \hat{\beta}/y_0],$$

where x_0, x_1 and y_0, y_1 are such that

$$P(x_0 \leq \alpha(\mathbf{Y})/\beta(\mathbf{Y}) \leq x_1) = \gamma$$

$$P(y_0 \leq \beta(\mathbf{Y}) \leq y_1) = \gamma$$

and $\mathbf{Y}$ has the standard distribution of the family.

For the normal family, $(n - 1)\hat{s}^2(\mathbf{Y})$ has a χ^2 distribution and

$$\sqrt{n}\,\frac{\hat{\mu}(\mathbf{Y})}{\hat{s}(\mathbf{Y})}$$

has Student's t distribution.

Procedures similar to the above work in the multidimensional parameter case. But frequently, what may be wanted in the multidimensional problem are simultaneous confidence intervals. That is, if $\theta = (\theta_1, \theta_2)$, we want two intervals $J_1(\hat{\theta})$, $J_2(\hat{\theta})$ such that

$$P_\theta[\theta_1 \in J_1(\hat{\boldsymbol{\theta}}), \theta_2 \in J_2(\hat{\boldsymbol{\theta}})] \geq \gamma, \qquad \theta \in \Theta.$$

This will be completely satisfied if $J_1(\hat{\theta})$ and $J_2(\hat{\theta})$ are individually $100[1 - 2(1 - \gamma)]\%$ confidence intervals.

5 Hypothesis Testing in Parametric Models

Introduction

In an estimation problem, you are asked to find the true value of the parameter θ^*, and your answer may be any point in θ. In a confidence region problem, you are asked to find a region containing θ^* and your answer may be any region in Θ. In a hypothesis testing problem, on the other hand, there are only two possible answers, Yes or No. The questions asked are of the form: Are these variables independent? Is this coin biased? Of course, for any such problem we must specify a statistical model: a random data vector $\mathbf{X}$ and a set $\{P_\theta\}$, $\theta \in \Theta$, of possible underlying distributions for $\mathbf{X}$. Then, a hypothesis testing problem always has the following form: A subset $\Theta_1 \subset \Theta$ is specified and the question is whether θ^* is in Θ_1. Actually, since the data vector $\mathbf{x}$ is subject to chance, the answer Yes, θ^* is in Θ_1, for example, really means that on the basis of $\mathbf{x}$ it is more reasonable to believe that θ^* is in Θ_1 than not in Θ_1.

The selection of a model and a hypothesis set Θ_1, as well as the assumptions involved in this selection, are of primary importance. Here is an example:

Example a A nonverbal language-free intelligence test has been standardized on large numbers of American children so that the scores are normally distributed with a mean of 100 and a standard deviation of 10. This test is given to 50 African tribal children near Dakar, and their average score is 97.1. Is this average far enough below 100 to indicate a significance between the test scores of the African children and the American test scores? Actually what we are really trying to find out is whether the underlying distribution of scores for these 50 children deviates significantly from an $N(100, 100)$ distribution.

This problem can be treated on different levels. Assume that the random data vector $\mathbf{X} = (X_1, \ldots, X_{50})$ consists of independent and identically distributed components. For the simplest model, assume that the underlying distribution for the African children is normal with the same standard deviation of 10, but possibly a different mean. That is, the class of possible underlying distributions are $N(\mu, 100)$. For intelligence tests, μ has to be positive—that is, the parameter space $\Theta = (0, \infty)$. What is Θ_1? The question given above is formulated in this model by:

Does $\mu^* = 100$ or not? We could take Θ_1 to be the point $\{100\}$; then the hypothesis is that the underlying distribution is $N(100, 100)$. Or we could take Θ_1 to be the set of all positive points except 100, with the corresponding hypothesis that the mean of the African distribution is not 100.

On a next step up, we might model this situation as follows: Suppose there is no reason to suspect that the 50 scores do not come from some normal distribution. There is no detectable skew and their histogram looks normal. But also there is no reason to believe that their standard deviation is 10. In fact, their sample variance is 59.5. The appropriate model here takes the family $N(\mu, \sigma^2)$, $0 < \mu < \infty$, $0 < \sigma < \infty$, as the possible underlying distributions. The hypothesis that their distribution does not detectably differ from $N(100, 100)$ takes the form $(\mu, \sigma) = (100, 10)$. Thus Θ_1 is the single point $(100, 10)$.

We might reason that the standard deviation of the distribution might not be 10, but that this does not concern us. We only want to test whether the mean is 100. This gets translated into the hypothesis set consisting of all points $(100, \sigma)$, $\sigma \geq 0$.

On a different level of complexity, we might feel that the assumption of normality for the test scores in these new circumstances is seriously suspect. This leads us into new possibilities that we explore in Chapter 8.

The formulation of a hypothesis testing problem is clear:

5.1 ***Definition*** *A hypothesis testing problem consists of a statistical model* $\{P_\theta\}$, $\theta \in \Theta$, *and a specified subset* Θ_1 *of* Θ. *The hypothesis is* $\theta^* \in \Theta_1$ *and* Θ_1 *is called the hypothesis set.*

Either Θ_1 or its complement can be taken as the hypothesis set. But sometimes there is a glaring discrepancy in simplicity between the two sets. For instance, if we take as hypothesis the unbiasedness of a coin, the hypothesis set is the single point $p = 1/2$. The complement of this is the much larger and more complicated set $[0, 1/2) \cup (1/2, 1]$. Similarly, in the first formulation of Example **a.** we have our choice of taking as hypothesis set $\{100\}$ or the complement of this point.

This type of lack of symmetry between the two complementary hypothesis sets is common in hypothesis testing. We will usually, in these situations, take the smaller, simpler set of parameter values to be the hypothesis set. The set Θ_1 thus chosen is called the *null hypothesis.*

Problem 1 Thirty trios of infant mice, 90 in all, were selected such that within each trio the mice came from the same litter and were matched in size and weight. In each trio, one mouse was given a control diet, the second a moderate amount of a drug being tested for its effect in enlarging the thyroid gland, and the third, a large amount of the drug. At the end of 30 days, the mice were killed and their thyroid glands weighed. Denote the weights for the kth trio by $w_k^{(1)}, w_k^{(2)}, w_k^{(3)}$ for the control, moderate dose, and heavy dose, respectively. This is later reduced to pairs of readings

$$x_k = w_k^{(2)} - w_k^{(1)}, \qquad y_k = w_k^{(3)} - w_k^{(1)}.$$

a Formulate a statistical model for the hypothesis that the drug has no effect on thyroid gland enlargement.

b Formulate a model for the hypothesis that the heavy dose and the moderate dose have the same effect on thyroid gland enlargement.

Problem 2 Consider the statement: For any given birth, the probability that the infant will be female is slightly higher than the probability of a male. Describe the data, model, and hypothesis set you would use to test this statement.

Problem 3 The specifications that a manufacturer must meet on a certain component include a requirement that the failure probability during 100 hours of use be less than .001. To check this requirement, n components are selected at random from the first run and tested until failure. Describe the model and hypothesis set assuming

a an exponential failure time distribution

b a Weibull failure time distribution.

Problem 4 There is some question as to whether the arrival times of user calls in a shared time computer system form a Poisson process. It is decided to test the Poisson hypothesis as follows: The times between successive arrivals are to be assumed independent and identically distributed, with a gamma distribution. Set up a model and state the hypothesis.

Problem 5 The diameter of strands of plastic produced by a process follows a lognormal distribution. The process is considered in need of adjustment when more than 1% of the strands have a diameter exceeding a given value d_0. Set up a model and hypothesis set, assuming that n successive strands are measured.

Tests and Their Operating Characteristics

In a hypothesis testing model we want to decide on the basis of the data vector $\mathbf{x}$ whether or not the hypothesis is true. Furthermore, we are interested in rules that specify the decision for all possible outcomes of the experiment. Any such decision procedure has this form: Let Ω be the set of all possible values of the data vector. Select a set $S \subset \Omega$. Then if $\mathbf{x} \in S$, accept the hypothesis, and if $\mathbf{x} \notin S$, reject the hypothesis.

The specification of the set S determines the decision for any possible outcome. Therefore we call the set S a decision rule for the problem. Alternatively, such a set S is also called a test of the hypothesis Θ_1. To summarize:

5.2 ***Definition*** *In a parametric statistical model $\{P_\theta\}$, $\theta \in \Theta$, with hypothesis set Θ_1, a decision rule, or a test of the hypothesis, consists of specifying a set S of data vectors such that*

if $\mathbf{x}$ falls in S, accept the hypothesis $\theta^* \in \Theta_1$;

if $\mathbf{x}$ does not fall in S, reject the hypothesis $\theta^* \in \Theta_1$.

Sometimes we will call S the *acceptance region.* Usually the complement of S is called the *critical region.*

There are usually an infinity of different decision rules one could use in any given hypothesis testing problem. In fact, there are as many different decision rules as there are different subsets of Ω. To compare these various rules, to select a "good" rule, we need a measure of how well a decision rule performs. A natural measure is the probability with which a rule S leads to the wrong decision. If the true state of nature θ is in the hypothesis set Θ_1, then we make a wrong decision if we reject the hypothesis. The probability of rejecting is

$$P_\theta(\mathbf{X} \notin S) = 1 - P_\theta(\mathbf{X} \in S).$$

If the true state of nature θ is not in Θ_1, then we make an error if we accept the hypothesis. The probability of this is

$$P_\theta(\mathbf{X} \in S).$$

What we want from a good decision rule is that $P_\theta(\mathbf{X} \notin S)$ be small for $\theta \in \Theta_1$ and that $P_\theta(\mathbf{X} \in S)$ be small when $\theta \notin \Theta_1$.

Define a function $\phi(\theta)$ to be the probability of accepting the hypothesis

$$\phi(\theta) = P_\theta(\mathbf{X} \in S).$$

Then $\phi(\theta)$ gives the error probability for θ outside of Θ_1. For $\theta \in \Theta_1$, the error probability is $1 - \phi(\theta)$. Therefore all of the error probabilities can be read from the graph of $\phi(\theta)$.

5.3 ***Definition*** *The function $\phi(\theta)$ defined by*

$$\phi(\theta) = P_\theta(\mathbf{X} \in S)$$

is called the operating characteristic of the test S.

A perfect test S has the property that

$$\phi(\theta) = \begin{cases} 1, & \theta \in \Theta_1 \\ 0, & \theta \notin \Theta_1. \end{cases}$$

This $\phi(\theta)$ is achieved if $\mathbf{X}$ always falls in S when $\theta \in \Theta_1$ and $\mathbf{X}$ never falls in S when $\theta \notin \Theta_1$. In either case we invariably draw the correct conclusion on the basis of our data vector. Looking at the graph of $\phi(\theta)$ and seeing how much it departs from ideal behavior gives us the measure of how good the test is.

Level of Significance, Detecting, and Discrimination

We have pointed out that there are two possible types of error in hypothesis testing:

> *first*, rejecting the hypothesis when in fact $\theta^* \in \Theta_1$,
> *second*, accepting the hypothesis when $\theta^* \notin \Theta_1$.

In each situation the statistician tries to choose S so that these two different types of errors are at acceptable levels. The acceptable levels are fixed by the demands of the situation.

In some problems one might want the probability of rejecting the hypothesis when it is actually true to be extremely small, while not being as concerned about the probability of accepting it as true when it does not hold. For example, consider a situation in which the hypothesis is that a certain medical treatment has harmful side effects. One would certainly not want to reject this hypothesis if it were true. But the consequences of accepting it as harmful even when it is not may not be so disastrous.

There is a terminology in common use in hypothesis testing which singles out one of these errors. Take S to be a test of the hypothesis $\theta \in \Theta_1$.

5.4 ***Definition*** *If one selects* α, $0 < \alpha < 1$ *and enforces the condition*

$$\min_{\theta \in \Theta_1} P_\theta(\mathbf{X} \in S) \geq 1 - \alpha,$$

then S *is called a test at significance level* α.

The common procedure in hypothesis testing is to first select the level α of the test. This is the upper bound for the probability of rejecting the hypothesis if it is true. Once α has been selected and S chosen accordingly, $\phi(\theta)$ is computed to see how the test behaves for values of θ outside of Θ_1.

This brings us to the concept of *detection capability*. Let S be a test region and θ be a state of nature outside of the null hypothesis Θ_1.

5.5 ***Definition*** *We say that* θ *can be detected at level* α *if* S *is a level* α *test and if* $\phi(\theta) \leq \alpha$.

How far we have to go from Θ_1 to get to states that can be detected at level α gives a measure of the discrimination ability of the test.

5.6 ***Definition*** *Suppose we have a distance measure such that* θ *is detectable at level* α *if and only if its distance from* Θ_1 *is greater than or equal to some value* Δ. *Then we say* Δ *is the minimum detectable distance at level* α *and all distances greater than* Δ *can be detected at level* α.

As an example, suppose Θ is 1-dimensional and the hypothesis is $\theta = \theta_0$; that is Θ consists of the single point θ_0. A special case of the above definition is

5.7 ***Definition*** *A given level* α *test has the minimum detectable distance* $\Delta\theta$ *from* θ_0 *at level* α *if every* θ *that can be detected at level* α *satisfies*

$$|\theta - \theta_0| \geq \Delta\theta$$

that is, if every detectable θ *is at least the distance* $\Delta\theta$ *from* θ_0.

If θ is in Θ_1 or at a distance greater than or equal to Δ from Θ_1 the test functions reliably. Its error probability is less than or equal to α. If θ is outside of Θ_1 but less than a distance Δ away, we may erroneously accept the hypothesis with probability greater than α. The test cannot reliably discriminate between those values of θ for which the hypothesis is true and those values for which it is false but whose distance from Θ_1 is less than Δ.

Whenever you accept a hypothesis on the basis that $\mathbf{x} \in S$, you should have some idea of the detection capability of your test. There may be a substantial probability of accepting states of nature far removed from your hypothesis set. Another reason for always trying to get some idea of the detection capability of your test is to decide what sample size will be required. We will clarify this in the next section.

An Example

Suppose you are handed a coin and asked to see whether it is biased. You should not accept this job until the question is more fully specified. How much bias are you being asked to detect? Are you being asked to set up a test that will detect a 1% bias or a 5% bias, or is it satisfactory for the purposes of this problem to be able to detect a 20% bias? Without this information, you cannot decide how many times to toss the coin. A moderate sample size may be enough to detect a 20% bias, but a much larger sample will be required to detect a 5% bias.

More precisely, take the null hypothesis to be $p = .5$, and decide how much of a bias $\Delta p = |p - .5|$ you want to be able to detect, and at what level. Then, in order to find the required sample size, you need to know the detection capability of your test as a function of sample size.

This is typical of what is done in designing an experiment to test some hypothesis. You decide how much of a departure from the null hypothesis you want to detect, compute the detection capability of your test, and then find the sample size necessary to give you the discrimination you want.

Suppose that we decide to use the following test: We select an integer $N > 0$, toss the coin n times, and let m be the number of heads. If

$$\left| m - \frac{n}{2} \right| \leq N,$$

we accept the hypothesis; otherwise we reject it. Let x_k be 0 if the kth toss is tails, 1 if it is heads. Then S consists of all $\mathbf{x} = (x_1, \ldots, x_n)$ such that

$$\left| x_1 + \cdots + x_n - \frac{n}{2} \right| \leq N.$$

The operating characteristic of this test is

$$\phi(p) = P_p\left(\left| \mathsf{X}_1 + \cdots + \mathsf{X}_n - \frac{n}{2} \right| \leq N\right).$$

We can compute $\phi(p)$ for small n by using tables of the binomial distribution. For large n, $\phi(p)$ can be computed by using the fact that

$$\frac{\mathsf{X}_1 + \cdots + \mathsf{X}_n - np}{\sqrt{npq}}$$

has approximately a $N(0, 1)$ distribution under P_p.

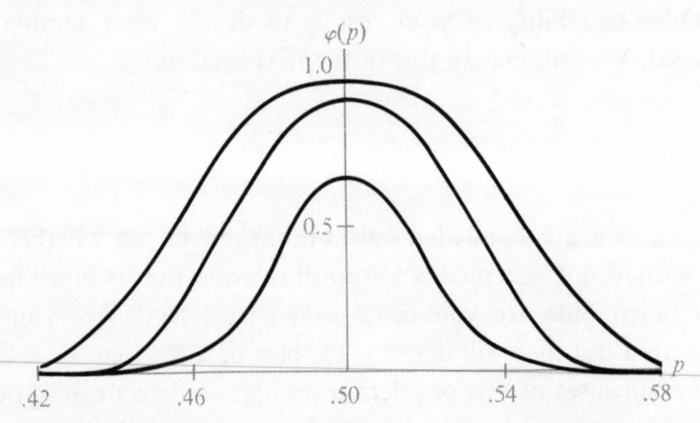

***Figure 5.1** The operating characteristics of the tests with N = 16, N = 32, N = 48*

The following tables give some values of $\phi(p)$ for $n = 1000$ and $N = 16, 32, 48$. (See also Figure 5.1.)

$N = 16$

p	.46	.48	.50	.52	.54
$\phi(p)$	.07	.39	.68	.39	.07

$N = 32$

p	.44	.46	.48	.50	.52	.54	.56
$\phi(p)$	.04	.31	.77	.95	.77	.31	.04

$N = 48$

p	.42	.44	.46	.48	.50	.52	.54	.56	.58
$\phi(p)$	.02	.23	.60	.96	.997	.96	.60	.23	.02

A large value of N gives a large acceptance region. This is desirable on the following grounds: If we take N small, say $N = 16$, then $P_{.5}(S) = .68$. This implies that even if the coin is unbiased, 32% of the time we will get an outcome not in S, leading us to the false conclusion that the coin is biased. On the other hand, if we take $N = 48$, then $P_{.5}(S) = .997$, so we will make the above mistake only $\frac{3}{10}$ of a percent of the time. But now even if the coin has probability .46 of heads, we will decide it is unbiased 60% of the time.

For fixed sample size, if we make N (equivalently, S) too large, then we lose detection capability. But if we make N (or S) too small, the significance level of the test increases. We try to compromise, and select the N that best fits the requirements of the problem.

If we are designing the experiment, then both the sample size n and N are determined by the amount of bias we wish to detect and by the level of significance. For instance, suppose you want to be able to detect a .06 bias at the .05 level. This requires, first, that

5.8 $$P_{.5}(S) \geq .95,$$

and second, that

5.9 $$P_p(S) \leq .05$$

whenever p is outside of $.5 \pm .06$.

Denoting $\mathsf{S}_n = \mathsf{X}_1 + \cdots + \mathsf{X}_n$, write

$$P_p(S) = P_p\left(\frac{n}{2} - N \leq \mathsf{S}_n \leq \frac{n}{2} + N\right).$$

Under $p = .5$, S_n has mean $n/2$ and variance $n/4$. The above right- and left-hand bounds are $-2N/\sqrt{n}$ and $2N/\sqrt{n}$ σ-units away from the mean. Therefore (see p. 41)

$$P_{.5}(S) = P\left(\frac{-2N}{\sqrt{n}} \leq \mathsf{Z} \leq \frac{2N}{\sqrt{n}}\right),$$

so to satisfy 5.8 we need $2N/\sqrt{n} \geq 2.0$ or $N \geq \sqrt{n}$. For each n, taking $N = \sqrt{n}$ gives the smallest .05 level test, and therefore the test with the highest detection capability. Under $P_{.44}$, S_n has mean $.44n$ and standard deviation $.496\sqrt{n} \simeq .5\sqrt{n}$. Write the acceptance set as

$$\frac{n}{2} - \sqrt{n} \leq \mathsf{S}_n \leq \frac{n}{2} + \sqrt{n}.$$

The lower bound is $(n/2 - \sqrt{n} - .44n)/(.5\sqrt{n}) = (.12\sqrt{n} - 2)$ σ-units from the mean. The upper limit is $(.12\sqrt{n} + 2)$ σ-units from the mean. Hence

$$P_{.44}(S) \simeq P(.12\sqrt{n} - 2 \leq \mathsf{Z} \leq .12\sqrt{n} + 2).$$

From an $N(0, 1)$ table, the minimum n such that 5.9 is satisfied satisfies

$$-2 + .12\sqrt{n} \simeq 1.6,$$

or $\sqrt{n} \simeq 30$. Therefore $n \simeq 900$, so detecting a .06 bias at the .05 level requires quite a large sample size. Increasing the detection level to .1, we still need a sample size of about 600 to detect a bias of .06. But if we need only detect a .10 bias (say at .05), then the sample size required drops to 330, or to 210 at the .1 level. Still, it is disappointing not to be able to reliably detect a bias less than .10 with a sample size of about two or three hundred.

Problem 6 In the example of the first section, assume that the scores have a standard deviation of 10, and use an acceptance region of the form $|\hat{\mu} - 100| < \varepsilon$, where $\hat{\mu}$ is the sample mean. Select ε to get a test at level .05. Graph the operating characteristic. If $\hat{\mu} = 97.1$, do you accept or reject the hypothesis that $\mu^* = 100$? What is the detection capability of the test at the .05 level? How large a sample would you need to detect a 5-point difference ($|\mu - 100| \geq 5$) at the .1 level?

A Two Point Parameter Space

For the remainder of this chapter our chief concern will be how to construct good tests, that is, how to decide what points of Ω should be included in S. A given level α test is called *uniformly most powerful* if among all level α tests its operating characteristic $\phi(\theta)$ is as small or smaller than the operating characteristic of any other test *at every value of θ outside of Θ_1*.

Unfortunately, in most of the usual hypothesis testing situations, there is no uniformly most powerful test. But in a very simple situation, such a test does exist and its form gives some insight into how to construct good tests in more general situations.

Suppose that Θ contains only two parameter values, θ_1 and θ_2. That is, there are only two possible underlying probability distributions for the model. Here is an example:

Example b If a certain event occurs, an automatic radio transmitter is instructed to send a $+1$ signal over a short time interval. During this interval, the signal is sampled at the receiver 10 times. (The reason for the repeated sampling is that there is a fair amount of additive noise in the channel.) As a model, assume that the successive noise components are independent and normally distributed with mean 0 and variance 1. If there is a signal present, then the 10 received signals are sampled from a $N(1, 1)$ distribution. If there is no signal present, we receive the outcomes of 10 $N(0, 1)$ variables. Thus, the possible distributions for the data vector $\mathsf{X}_1, \ldots, \mathsf{X}_{10}$ are specified by saying that they are independent

$N(\mu, 1)$ variables, where μ equals 0 or 1. Given that we received $x_1, \ldots, x_{10}$, how do we decide whether there was a signal present or not?

The parameter space Θ consists of two values corresponding to $\mu = 0$ and $\mu = 1$. Let θ_1 correspond to $\mu = 1$, θ_2 correspond to $\mu = 0$. Take as hypothesis the presence of signal: $\theta = \theta_1$. Among all level α tests, the best or most powerful test has the property that

$$P_{\theta_1}(S) \geq 1 - \alpha$$

while $P_{\theta_2}(S)$ is as small as possible. Put another way, S should include all those points $\mathbf{x}$ whose probability is high under θ_1 and small under θ_2. This suggests the following procedure: Set some critical value c and include the point $\mathbf{x}$ in S only if the ratio of the densities satisfies

$$\frac{f_{\theta_1}(\mathbf{x})}{f_{\theta_2}(\mathbf{x})} \geq c.$$

Then determine c in terms of the level α by the requirement $P_{\theta_1}(S) = 1 - \alpha$.

Applying this procedure to our example,

$$f_{\theta_1}(\mathbf{x}) = \frac{1}{(2\pi)^5} \exp\left[-\tfrac{1}{2} \sum_1^{10} (x_k - 1)^2\right]$$

$$f_{\theta_2}(\mathbf{x}) = \frac{1}{(2\pi)^5} \exp\left[-\tfrac{1}{2} \sum_1^{10} x_k^2\right],$$

so that

$$\frac{f_{\theta_1}(\mathbf{x})}{f_{\theta_2}(\mathbf{x})} = \exp\,[-5] \exp\left[\sum_1^{10} x_k\right].$$

Then we accept if

$$\exp\,[-5] \exp\left[\sum_1^{10} x_k\right] \geq c.$$

Taking logs, this is equivalent to:

$$\sum_1^{10} x_k \geq 5 + \log c = c'.$$

Therefore, the rule is: Accept if the sum of the received signals is greater than a critical value, otherwise reject. The critical value c' is determined from

$$P_{\theta_1}\left[\sum_1^{10} \mathrm{X}_k \geq c'\right] = 1 - \alpha.$$

Note that

$$Z = \sum_{1}^{10} \frac{X_k - 1}{\sqrt{10}}$$

is an $N(0, 1)$ variable under θ_1. Hence the above equation can be written as

$$P\left(Z \geq \frac{c' - 10}{\sqrt{10}}\right) = 1 - \alpha.$$

Solve for c' by using an $N(0, 1)$ table. For instance, to get a .05 level test, use the fact that

$$P(Z \geq -1.65) = .95.$$

Then put $(c' - 10)/\sqrt{10} = -1.65$ and solve to get $c' = 4.78$. Also,

$$P_{\theta_2}(S) = P_{\theta_2}\left(\sum_{1}^{10} X_k \geq 4.78\right)$$

$$= P\left(Z \geq \frac{4.78}{\sqrt{10}}\right).$$

Using the $N(0, 1)$ table gives $P_{\theta_2}(S) = .066$.

The reasoning above seems quite sensible, but there are many other tests that also appear sensible. For instance, we could set some critical level c and accept the hypothesis if the largest signal received is larger than c. For this test,

$$S = \{\mathbf{x};\ \max(x_1, \ldots, x_{10}) \geq c\}.$$

To compare this test with the preceding one take this test and determine c by the requirement that S be a .05 level test. Computing, we get $P_{\theta_2}(S) = .40$. Surprisingly, the second test has error probability at θ_2 *over six times* as large for the same significance level as the first test. As with estimators, intuition is not enough. Seemingly reasonable tests may have considerably higher error probabilities than the best or nearly best tests.

Actually, our first approach works whenever $\Theta = \{\theta_1, \theta_2\}$ and produces the best test:

5.10 ***Neyman-Pearson Lemma*** *In a model where $\{P_\theta\}$ contains only two possible distributions, $P_{\theta_1}, P_{\theta_2}$, define a test region S as the set of all $\mathbf{x}$ such that*

$$\frac{P_{\theta_1}(\mathbf{x})}{P_{\theta_2}(\mathbf{x})} \geq c \qquad \textit{(discrete case)}$$

$$\frac{f_{\theta_1}(\mathbf{x})}{f_{\theta_2}(\mathbf{x})} \geq c \qquad \textit{(density case).}$$

Let $P_{\theta_1}(S) = 1 - \alpha$. Then S is the best level α test of the hypothesis $\theta = \theta_1$.

Best means that among all level α tests, the one given by the Neyman-Pearson lemma minimizes $P_{\theta_2}(S)$. The proof is simple. Consider the density case. Let S be the Neyman-Pearson test, and S' any other level α test. Define sets A, A', B as in Figure 5.2.

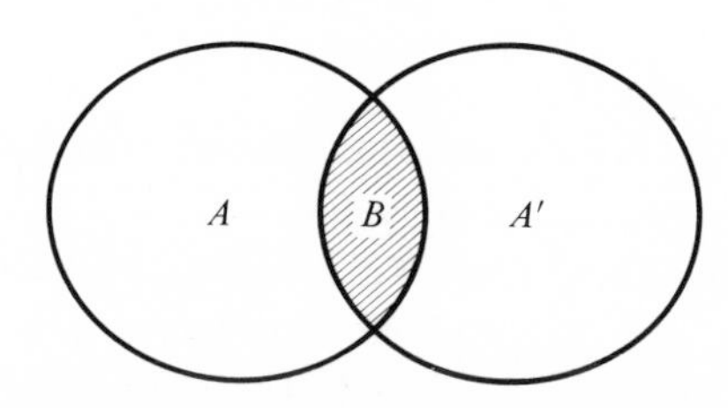

Figure 5.2 $B = A \cap A'$

To prove the lemma it is sufficient to show that

$$P_{\theta_2}(A') \geq P_{\theta_2}(A).$$

For θ either of θ_1, θ_2,

$$P_\theta(S) = P_\theta(B) + P_\theta(A)$$

$$P_\theta(S') = P_\theta(B) + P_\theta(A').$$

Since $P_{\theta_1}(S) = 1 - \alpha$ and S' is a level α test,

$$P_{\theta_1}(A') \geq P_{\theta_1}(A).$$

Since A' is outside S, if $\mathbf{x}$ is in A', then

$$f_{\theta_1}(\mathbf{x}) \leq c f_{\theta_2}(\mathbf{x}),$$

and integrating over A' gives

$$P_{\theta_1}(A') \leq c P_{\theta_2}(A').$$

Now A is in S, so if $\mathbf{x} \in A$, then

$$f_{\theta_1}(\mathbf{x}) \geq c f_{\theta_2}(\mathbf{x}).$$

Integrating over A gives

$$P_{\theta_1}(A) \geq c P_{\theta_2}(A).$$

Putting these inequalities together;

$$c P_{\theta_2}(A') \geq P_{\theta_1}(A') \geq P_{\theta_1}(A) \geq c P_{\theta_2}(A)$$

which completes the proof. The discrete case is similar.

As in our example, the Neyman-Pearson region is usually found by setting the level α and adjusting c so that $P_{\theta_1}(S) = 1 - \alpha$. But in the discrete case this usually cannot be done. For instance, if there are 50 possible values of $\mathbf{x}$, each having probability 1/50 under P_{θ_1}, how can we satisfy $P_{\theta_1}(S) = .95$? The practical thing to do is to select c so that $P_{\theta_1}(S)$ comes as close as possible to the desired value.

Problem 7 According to one theory, the distribution of a certain variable is double exponential with mean 0, variance 1. Another theory describes it as $N(0, 1)$. On the basis of n observations, construct the .05 level Neyman-Pearson test for the hypothesis that the first theory is valid. The evaluation of c is difficult! One way to do it is to assume n large enough so that the normal approximation can be applied to the appropriate sum of random variables.

Problem 8 In the example of the previous section, find the maximum likelihood estimator for θ. Compare this estimator with the Neyman-Pearson test of the hypothesis $\theta = \theta_1$. Is there any resemblance?

Problem 9 A common anthropological classification problem is this: A skull has been unearthed that can belong to only one of two early human types, say I and II. Skulls of both types have been subjected to a battery of measurements on 8 specified dimensions, and the distribution of these 8 variables has been approximated by a joint normal distribution. For type I, denote the means vector and moment matrix by μ_1 and $[\Gamma_1]$, and for II, by μ_2 and $[\Gamma_2]$. Find the Neyman-Pearson procedure for testing the hypothesis that the newly unearthed skull is type I at the .05 level. (Do not evaluate c.)

The Likelihood Ratio Tests

In the last section the reasoning that led us to the best test was that S should include all points $\mathbf{x}$ whose probability is high if the hypothesis holds and low if the hypothesis does not hold. When Θ has more than two points, it is no longer clear how to apply this reasoning. If, for instance, Θ_1 contains many values of θ, then $f_\theta(\mathbf{x})$ may be large for some of these values and small for others. Even so, there is a procedure that usually produces good tests and has nice asymptotic properties.

5.11 ***Definition*** *The likelihood ratio λ for the hypothesis Θ_1 is*

$$\lambda(\mathbf{x}) = \frac{\max_{\theta \in \Theta_1} P_\theta(\mathbf{x})}{\max_{\theta \in \Theta} P_\theta(\mathbf{x})} \qquad \textit{(discrete case)}$$

or

$$\lambda(\mathbf{x}) = \frac{\max_{\theta \in \Theta_1} f_\theta(\mathbf{x})}{\max_{\theta \in \Theta} f_\theta(\mathbf{x})} \qquad \textit{(density case).}$$

The likelihood ratio test for Θ_1 consists of fixing a critical ratio value c and accepting if $\lambda(\mathbf{x}) \geq c$.

The value of c is determined as follows: Pick the level α. As c is increased, the acceptance region

$$S = \{\mathbf{x};\ \ \lambda(\mathbf{x}) \geq c\}$$

gets smaller. Pick the largest value of c such that $P_\theta(S) \geq 1 - \alpha$ for all $\theta \in \Theta_1$. That is, use the smallest level α test of the given form.

Incidentally, why do we want S as small as possible within the restrictions that it gives a level α test? The likelihood ratio, as we defined it, is unsymmetric between Θ_1 and the alternate hypothesis consisting of the complement of Θ_1. This lack of symmetry is reflected in the fact that the likelihood ratio will not produce tests that achieve a large level of significance. If you want a test at a level higher than the likelihood ratio can produce, then either

apply the likelihood ratio test to the alternative hypothesis

or equivalently,

use the ratio

5.12
$$\lambda_1(\mathbf{x}) = \frac{\max_{\theta \in \Theta_1} P_\theta(\mathbf{x})}{\max_{\theta \notin \Theta_1} P_\theta(\mathbf{x})}$$

instead of $\lambda(\mathbf{x})$ to get test regions.

Either of these two procedures symmetrizes the likelihood ratio tests. In its symmetric form 5.12, you can see that we are considering the ratio of the maximum probability that $\mathbf{x}$ has under the hypothesis $\theta \in \Theta_1$ to its maximum probability under the alternative $\theta \notin \Theta_1$. When Θ consists of only two points $\{\theta_1, \theta_2\}$, the likelihood ratio test reduces to the Neyman-Pearson test. In this sense, the likelihood ratio concept is an attempt to generalize the idea that produces the best test in the two-point parameter space-situation.

The likelihood ratio method stands in the same honored position with respect to hypothesis testing as the maximum likelihood method in estimation. It usually gives good tests for both large and small sample sizes. If there is a test for a problem that is "best," then often it is the likelihood ratio test. For large sample sizes, in smooth problems, these tests are "asymptotically optimal." Furthermore, in a large class of problems the distribution of λ becomes simple for large sample sizes and makes the computation of c more routine.

The main reason for using the likelihood ratio as we originally defined it rather than in the symmetric form 5.12 is that it is usually easier to compute. The denominator is $\max_{\theta \in \Theta} P_\theta(\mathbf{x})$ or $\max_{\theta \in \Theta} f_\theta(\mathbf{x})$. The value of θ that maximizes $P_\theta(\mathbf{x})$ or $f_\theta(\mathbf{x})$ is the maximum likelihood estimate $\hat{\theta}$. Hence the denominator of $\lambda(\mathbf{x})$ has the simple form $P_{\hat{\theta}}(\mathbf{x})$ or $f_{\hat{\theta}}(\mathbf{x})$.

Notice also that $\lambda(\mathbf{x}) \leq 1$ for all $\mathbf{x}$, so $c \leq 1$; otherwise the test region $\lambda(\mathbf{x}) \geq c$ will contain no points. If the value of $\hat{\theta}$ falls in Θ_1, then the numerator equals the denominator, and $\lambda(\mathbf{x}) = 1$. Since $c \leq 1$, this gives the rule:

> *In using form (5.11) of the likelihood ratio test, always accept the hypothesis if* $\hat{\theta} \in \Theta_1$.

To compute S, it is very often convenient to work with

$$l(\mathbf{x}) = -2 \log \lambda(\mathbf{x}).$$

The inequality $\lambda(\mathbf{x}) \geq c$ can also be expressed as

$$-2 \log \lambda(\mathbf{x}) \leq -2 \log c,$$

or

$$-2 \log \lambda(\mathbf{x}) \leq d.$$

Hence an equivalent form of the likelihood ratio test is: Accept if

$$l(\mathbf{x}) \leq d, \tag{5.13}$$

where we use the minimum value of d that gives a level α test.

The likelihood ratio method is particularly successful in producing best tests when used on normally distributed samples. The next three sections contain three well known and important examples. These were selected not only because of their importance, but also because they are the classical models for problems that recur in later chapters, treated in a different way.

Problem 10 Let $x_1, \ldots, x_n$ be sampled from an $N(\mu, \sigma^2)$ distribution with known σ^2. For the hypothesis $\mu = \mu_0$, show that

a $l(\mathbf{x}) = \dfrac{n}{\sigma^2}(\hat{\mu} - \mu_0)^2$.

b The likelihood ratio test at level α accepts if

$$|\hat{\mu} - \mu_0| \leq \frac{\sigma z}{\sqrt{n}},$$

where z satisfies

$$P(|Z| \leq z) = 1 - \alpha.$$

Problem 11 Binary digits are transmitted over a communication channel as ± 1 signals. But the channel adds a noise drawn from an $N(0, 1)$ distribution, independently at each transmission. Let $\mathbf{s} = (s_1, \ldots, s_n)$ be any message of n -1's and $+1$'s, and let Θ be the set of all possible such messages.

a Given that the n received signals are $x_1, \ldots, x_n$, find the maximum likelihood estimate for $\mathbf{s}$.

b Use **a.** to find the form of the likelihood ratio test for the hypothesis $\mathbf{s} = \mathbf{s}^{(0)}$, where $\mathbf{s}^{(0)}$ is a fixed message.

(Ans. **a** $\hat{s}_k = x_k/|x_k|$. **b** Accept if $\sum_{\hat{s}_k \neq s_k^{(0)}} |x_k| < c'$.)

Problem 12 A communication channel takes the input and adds noise independently drawn from an $N(0, \sigma^2)$, distribution with known σ^2. After n successive outputs $x_1, \ldots, x_n$, the question is whether a signal is present or not. A signal is defined as n successive inputs $s_1, \ldots, s_n$ consisting of any real numbers not all 0.

a Find the form of the likelihood ratio test for the hypothesis of no signal ($\mathbf{s} = 0$).

b For $n = 20$, evaluate the constant c so as to get a .05 level test.

Problem 13 A roadway detector registers the passage time of every vehicle passing a given point. A set of data taken one morning is suspect. The detector may have malfunctioned in such a way that it registered the passage times of *every other* car. Knowing that one can safely assume that the times between passages of successive vehicles are independent and exponentially distributed, propose a test for the hypothesis that the detector is functioning properly. (Use (5.12).)

Problem 14 The distribution underlying the data $x_1, \ldots, x_n$ is suspected to be either normal or double exponential.

a Formulate an appropriate model and specify the parameter space Θ.

b Take as hypothesis that the underlying distribution is normal and use (5.12) to derive the form of the likelihood ratio test. (Simplify as much as you can.)

Problem 15 (Difficult!) Take $x_1, \ldots, x_n$ to be sampled from an $N(\mu, \sigma^2)$ distribution with both μ and σ^2 unknown.

a Find the form of the likelihood ratio test for the hypothesis $\sigma = \sigma_0$.

b How would you evaluate the critical constant for n small? For n large?

c Find the detectable difference at the .1 level for n large using the normal approximation.

Student's *t* Test

In Example **a.** of Section 1, we discussed the formulation: Test the hypothesis $\mu = 100$ when σ^2 is unknown. This is typical of a common class of problems which read: Let $x_1, \ldots, x_n$ be sampled from an $N(\mu, \sigma^2)$ distribution with σ^2 unknown and μ either ranging over the whole line or restricted to some interval. Then, test some hypothesis involving the value of μ alone.

For instance, in a paired comparison, experiment individuals (or rats, or plots of ground) are paired so that certain characteristics match. One of the two is given the treatment being investigated (drug, fertilizer, diet). After a predetermined period, the value of a symptomatic variable (weight gain, blood pressure) is measured for both individuals, and the score assigned to the pair is the difference of the pair of measurements. The classical model assumes that the differences $x_1, \ldots, x_n$ are sampled from an $N(\mu, \sigma^2)$ distribution with σ^2 unknown. The hypothesis that the treatment has no effect is the hypothesis $\mu = 0$. If the effect of the treatment is completely unknown, this would be reflected by taking the possible values of μ to range over $(-\infty, \infty)$. But in many situations it is virtually certain that the treatment will not have a negative effect. Then μ is taken to range over $[0, \infty)$, and the test becomes

$$\mu = 0 \qquad \textit{versus} \qquad \mu > 0$$

(that is, no effect as against the alternative of positive effect). Or, it could be that the effects of the treatment we want to test are negative or no effect against the alternative of positive effect. Then we test

$$\mu \leq 0 \qquad \textit{versus} \qquad \mu > 0.$$

Obviously, there are an infinity of different hypotheses we can formulate involving μ. The most frequently appearing problems are of the form

$$\mu = \mu_0 \qquad \textit{versus} \qquad \mu \neq \mu_0$$

$$\mu = \mu_0 \qquad \textit{versus} \qquad \mu > \mu_0 \qquad (\text{or } \mu < \mu_0)$$

$$\mu \leq \mu_0 \qquad \textit{versus} \qquad \mu > \mu_0 \qquad (\text{or} > \text{versus} \leq).$$

In the first problem the alternative includes all values of μ both to the right and left of μ_0. In the second and third the alternative includes only those values of μ to the right of μ_0.

5.14 ***Definition*** *If the alternative to a hypothesis consists of a single interval, call it a one-sided alternative. If the hypothesis is $\theta = \theta_0$ and the alternative consists of intervals on both sides of θ_0, call it a two-sided alternative.*

The same distinction holds for tests on any single coordinate in a multidimensional parametric family.

We start with the problem of testing $\mu = \mu_0$ versus $\mu \neq \mu_0$. The denominator of $\lambda(\mathbf{x})$ is

$$f_{\hat{\theta}}(\mathbf{x}) = \frac{1}{(2\pi\hat{\sigma}^2)^{n/2}} \exp\left[-\frac{1}{2\hat{\sigma}^2} \sum (x_k - \hat{\mu})^2\right]$$

$$= \frac{1}{(2\pi\hat{\sigma}^2)^{n/2}} \exp\left[-n/2\right].$$

The numerator of $\lambda(\mathbf{x})$ is found by substituting for σ^2 its maximum likelihood estimate $\hat{\sigma}^2$ when $\mu = \mu_0$. From the equation

$$\frac{\partial}{\partial\sigma} L_{(\mu_0,\sigma)}(\mathbf{x}) = 0$$

we get

$$\hat{\sigma}_0^2 = \frac{1}{n}\sum_1^n (x_k - \mu_0)^2.$$

Hence

$$\max_{\theta \in \Theta_1} f_\theta(\mathbf{x}) = f_{(\mu_0,\hat{\sigma}_0)}(\mathbf{x}) = \frac{1}{(2\pi\hat{\sigma}_0^2)^{m/2}} \exp\,[-n/2]$$

and

$$\lambda(\mathbf{x}) = \left(\frac{\hat{\sigma}^2}{\hat{\sigma}_0^2}\right)^{n/2}.$$

The acceptance region is

$$\left(\frac{\hat{\sigma}^2}{\hat{\sigma}_0^2}\right)^{n/2} \geq c.$$

Write this as

$$\frac{\hat{\sigma}_0^2}{\hat{\sigma}^2} \leq c^{-n/2}$$

or

$$\frac{\hat{\sigma}_0^2}{\hat{\sigma}^2} - 1 \leq c^{-n/2} - 1 = c'.$$

Now,

$$\begin{aligned}\frac{\hat{\sigma}_0^2}{\hat{\sigma}^2} - 1 &= \frac{\hat{\sigma}_0^2 - \hat{\sigma}^2}{\hat{\sigma}^2}\\ &= \frac{1}{n\hat{\sigma}^2}\left(\sum_{k=1}^n (x_k - \mu_0)^2 - \sum_{k=1}^n (x_k - \hat{\mu})^2\right)\\ &= \frac{(\hat{\mu} - \mu_0)^2}{\hat{\sigma}^2} = \frac{n}{n-1}\,\frac{(\hat{\mu} - \mu_0)^2}{\hat{s}^2},\end{aligned}$$

so the test region can be written as

$$\left|\frac{\sqrt{n}(\hat{\mu} - \mu_0)}{\hat{s}}\right| \leq \sqrt{c'(n-1)} = t.$$

For any state of nature (μ_0, σ) in the hypothesis set, the variable inside the absolute value sign has a Student's t distribution with $n - 1$ degrees of freedom (see Chapter 2, p. 41). Because of this, the test is called *Student's t test* of the hypothesis $\mu = \mu_0$ with σ^2 unknown. The above leads to an easy evaluation of t. Let S have a t_{n-1} distribution. From the table values of this distribution, find t such that

$$P(|S| < t) = 1 - \alpha.$$

The t test is exactly what we would expect on intuitive grounds. We can ignore the factor $\sqrt{n}$—it is only there to normalize the variable to a standard t distribution. The test says: Take the difference between $\hat{\mu}$ and μ_0. Divide this difference by an estimate of σ. If this ratio is too large, reject the hypothesis; otherwise, accept.

The operating characteristic of the t test depends only on $\Delta\mu/\sigma$, where $\Delta\mu = |\mu - \mu_0|$. The characteristic is graphed in Figure 5.3 for $\alpha = .05$ and various sample sizes.

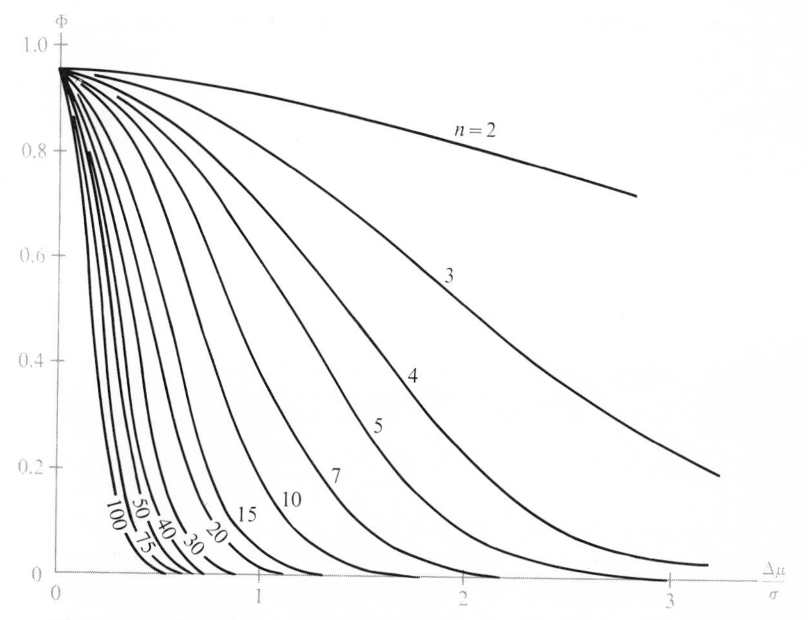

Figure 5.3 *Operating characteristic of the t test for various sample sizes* (C.D. Ferris, F. E. Grubbs, and C. L. Weaver, "Operating Characteristics for the Common Statistical Tests of Significance," *The Annals of Mathematical Statistics* 17 (1947):178–197)

For large sample size we can compute the operating characteristic and the detection capability by noting that

$$\frac{\sqrt{n}(\hat{\mu} - \mu)}{\hat{s}}$$

is approximately $N(0, 1)$ under $P_{(\mu,\sigma)}$, since $\hat{s} \simeq \sigma$. The minimum difference $\Delta\mu$ detectable at the .05 level is given by

$$\frac{\Delta\mu}{\sigma} = \frac{3.6}{\sqrt{n}}. \tag{5.15}$$

At the .1 level,

5.16 $$\frac{\Delta\mu}{\sigma} = \frac{2.9}{\sqrt{n}}.$$

For a sample size of 100, say, $\Delta\mu/\sigma \simeq .3$. Here, we cannot decide a priori, or before the experiment, what difference $\mu - \mu_0$ we want to be able to detect and then fix the sample size on that basis. Because the detectable difference is given in terms of its ratio to σ, and σ is assumed unknown. Usually, though, you have some idea of the order of magnitude of σ and select the sample size on this basis. Then compute $\hat{s}$ and use this as the value of σ in 5.15 or 5.16 to estimate the detection capability. If the first sample you selected does not give the desired discrimination, then on the basis of your estimated value of σ, recompute the sample size you need and take the additional observations.

The one-sided alternative illustrates a useful point. Take the hypothesis to be positive effect, $\mu > 0$, with the alternative $\mu \leq 0$. The denominator of $\lambda(\mathbf{x})$ is the same as in the two-sided test we treated above. If $\hat{\mu} > 0$, we automatically accept. So the only problem is $\hat{\mu} \leq 0$. Look at the density

$$f_{(\mu,\sigma)}(\mathbf{x}) = \frac{1}{(2\pi\sigma^2)^{n/2}} \exp\left[-\frac{1}{2\sigma^2}\sum_k (x_k - \mu)^2\right].$$

Write the exponent as

$$-\frac{1}{2\sigma^2}\sum_k (x_k - \hat{\mu} + \hat{\mu} - \mu)^2 = -\frac{1}{2\sigma^2}\sum_k (x_k - \hat{\mu})^2 - \frac{n}{2\sigma^2}(\hat{\mu} - \mu)^2.$$

To maximize this exponent with respect to μ under the restriction $\mu > 0$, reason that no matter what the value of σ we want $(\hat{\mu} - \mu)^2$ to be as small as possible. For $\hat{\mu} \leq 0$, this minimum is achieved by letting $\mu \to 0$. Then we get

$$\max_{\mu>0} f_{(\mu,\sigma)}(\mathbf{x}) = \frac{1}{(2\pi\sigma^2)^{n/2}} \exp\left[-\frac{1}{2\sigma^2}\sum_k x_k^2\right].$$

This is exactly the density we would get under the hypothesis $\mu = 0$. Thus, we get the same numerator and hence the same test as in the two-sided test of $\mu = 0$. Putting the two pieces together, we accept if

a $$\hat{\mu} > 0$$

b $$\hat{\mu} \leq 0 \quad \text{and} \quad \left|\frac{\sqrt{n}\hat{\mu}}{\hat{s}}\right| \leq t^+.$$ one-tailed

We combine **a.** and **b.** as: Accept if

$$\frac{\sqrt{n}\hat{\mu}}{\hat{s}} \geq -t^+.$$

The value of t^+ has to be adjusted to give a level α test. We want

$$P_{(\mu,\sigma)}\left(\frac{\sqrt{n}\hat{\mu}}{\hat{s}} \geq -t^+\right) \geq 1 - \alpha \tag{5.17}$$

for all $\mu > 0$ and all σ. For a fixed value of σ, it becomes least likely that $(\sqrt{n}\hat{\mu})/\hat{s}$ will have positive or only slightly negative values when μ is closest to 0. That is,

$$\min_{\mu>0} P_{(\mu,\sigma)}\left(\frac{\sqrt{n}\hat{\mu}}{\hat{s}} \geq -t^+\right) = P_{(0,\sigma)}\left(\frac{\sqrt{n}\hat{\mu}}{\hat{s}} \geq -t^+\right).$$

So 5.17 will be satisfied if

$$\min_{\sigma} P_{(0,\sigma)}\left(\frac{\sqrt{n}\hat{\mu}}{\hat{s}} \geq -t^+\right) = 1 - \alpha.$$

But for every value of $(0, \sigma)$, $\sqrt{n}\hat{\mu}/\hat{s}$ has a Student's t distribution with $n - 1$ degrees of freedom. Take S to have this distribution and determine t^+ from tables so that

$$P(\mathrm{T} \geq -t^+) = 1 - \alpha.$$

Now the test is completely specified.

Problem 16 Let $x_1, \ldots, x_n$ be sampled from an $N(\mu, \sigma^2)$ distribution with known σ^2. Find the .05 level likelihood ratio test for the hypothesis $\mu \geq 0$.

Problem 17 Assuming normally distributed test scores in the example of the first section, discuss how you would decide if the tribal children had lower test scores than American children

a assuming $\sigma^2 = 100$

b assuming σ^2 is unknown but $\hat{\sigma}^2 = 59.5$.

Problem 18 The increases in weight of 100 adult rats over a short period of time when put on a diet containing DDT are assumed sampled from an $N(\mu, \sigma^2)$ distribution. The experimenter wants to see if the presence of DDT causes any significant loss in weight. Formulate this as a hypothesis testing problem and set up a likelihood ratio test.

Problem 19 A good approximation for the variables of the preceding section is that under $P_{(\mu,\sigma)}$, for n large,

$$\sqrt{n}\,\frac{(\hat{\mu} - \mu_0)}{\hat{s}} \simeq \sqrt{n}\,\frac{(\hat{\mu} - \mu_0)}{\sigma}$$

$$= \sqrt{n}\,\frac{(\hat{\mu} - \mu)}{\sigma} + \sqrt{n}\,\frac{\Delta}{\sigma}$$

$$= \mathsf{Z} + \sqrt{n}\,\frac{\Delta}{\sigma},$$

where $\Delta = \mu - \mu_0$, and Z is an $N(0, 1)$ variable.

a Use this to find the minimum detectable $\Delta\mu$ of the test for $\mu = \mu_0$ at the .1 level.

b Find the minimum detectable $\Delta\mu$ for the one-sided t test of the hypothesis $\mu > 0$.

The ρ Test for Independence

Let $(x_1, y_1), \ldots, (x_n, y_n)$ be n samples drawn from a bivariate normal distribution, and take as parameter space $(\mu_1, \mu_2, \sigma_1, \sigma_2, \rho)$. Suppose we wish to test some hypothesis about the correlation coefficient ρ. The most common hypothesis is

independence: $\rho = 0$.

Others are

positive correlation: $\rho > 0$
negative correlation: $\rho < 0$.

Here is a very concrete example: Sixty consecutive one-minute counts are taken on both lanes of a two lane one-way freeway segment. Denote the count on the inner lane during the kth minute by $n_k^{(1)}$ and the count on the outer lane by $n_k^{(2)}$. Consider these as outcomes of $\mathbf{N}_k = (\mathbf{N}_k^{(1)}, \mathbf{N}_k^{(2)})$. Previous tests run on this and similar data make it plausible to assume that $\mathbf{N}_1, \ldots, \mathbf{N}_{60}$ are independent. The one-hour data period was selected so that the traffic flow remained fairly constant, and tests have confirmed that there seems to be no detectable trend in the size of the counts. Therefore, we assume that $\mathbf{N}_1, \ldots, \mathbf{N}_{60}$ are identically distributed. The question is whether the counts in adjacent lanes are independent; that is, are $\mathsf{N}_k^{(1)}$ and $\mathsf{N}_k^{(2)}$ independent? The counts, standardized by subtracting means and dividing by the standard deviations, are known from other considerations to be roughly approximated by an $N(0, 1)$ distribution. Therefore, as a first try, it is not unreasonable to assume that $\mathsf{N}_k^{(1)}$ and $\mathsf{N}_k^{(2)}$ have a joint normal distribution specified by the 5 parameters $(\mu_1, \mu_2, \sigma_1, \sigma_2, \rho)$, where $\mu_1 > 0$, $\mu_2 > 0$, $\sigma_1 > 0$, $\sigma_2 > 0$, and $-1 \leq \rho \leq 1$. Two variables having a joint normal distribution are independent if and only if $\rho = 0$. The hypothesis of independence becomes

$$\Theta_1 = \{(\mu_1, \mu_2, \sigma_1, \sigma_2, 0); \quad \mu_1 > 0, \mu_2 > 0, \sigma_1 > 0, \sigma_2 > 0\}.$$

The numerator of $\lambda(\mathbf{x}, \mathbf{y})$ is $f_{\hat{\theta}_0}(\mathbf{x}, \mathbf{y})$, where $\hat{\theta}_0$ is the maximum likelihood estimate of $(\mu_1, \mu_2, \sigma_1, \sigma_2)$ under the hypothesis of independence. The denominator is $f_{\hat{\theta}}(\mathbf{x}, \mathbf{y})$, where $\hat{\theta}$ is the maximum likelihood estimate of $(\mu_1, \mu_2, \sigma_1, \sigma_2, \rho)$. Computations give the result that in both numerator and denominator, maximum likelihood estimates of $\mu_1, \mu_2, \sigma_1^2, \sigma_2^2$ are the usual sample means and variances $\hat{\mu}_1, \hat{\mu}_2, \hat{\sigma}_1^2, \hat{\sigma}_2^2$. The maximum likelihood estimate of ρ is given by

$$\hat{\rho} = \frac{1}{n\hat{\sigma}_1\hat{\sigma}_2} \sum_k (x_k - \hat{\mu}_1)(y_k - \hat{\mu}_2).$$

Omitting the computations, when these estimates are substituted into the joint density we get

$$f_{\hat{\theta}_0}(\mathbf{x}, \mathbf{y}) = \frac{1}{(2\pi\hat{\sigma}_1\hat{\sigma}_2)^n} \exp\left[-\frac{n}{2}\right]$$

$$f_{\hat{\theta}}(\mathbf{x}, \mathbf{y}) = \frac{1}{(2\pi\hat{\sigma}_1\hat{\sigma}_2)^n} \cdot \frac{1}{(1 - \hat{\rho}^2)^{n/2}} \exp\left[-\frac{n}{2}\right].$$

The test is: Accept if

$$(1 - \hat{\rho}^2)^{n/2} \geq c$$

or

$$(1 - \hat{\rho}^2) \geq c^{2/n}.$$

That is,

$$\hat{\rho}^2 \leq 1 - c^{2/n}$$

or

$$|\hat{\rho}| \leq c'. \tag{5.18}$$

The likelihood ratio method therefore produces the most intuitive test one could think of for independence: Accept if $|\hat{\rho}|$ is small enough; otherwise reject.

We want to determine c' for a level α test and find the operating characteristic and the detection capability. For n large, this problem is simplified by using the fact that, since this is a smooth problem, $\hat{\rho}$ is an asymptotically normal and relatively unbiased estimate for ρ. But the most helpful simplification comes from the fact that the distribution of $\hat{\rho}$ depends only on ρ.

For n large (see Chapter 4, Problem 14)

$$V(\hat{\rho}) \simeq \frac{(1 - \rho^2)^2}{n}.$$

Therefore

$$Z = \sqrt{n}\,\frac{\hat{\rho} - \rho}{(1 - \rho^2)}$$

is approximately $N(0, 1)$ under $(\mu_1, \mu_2, \sigma_1, \sigma_2, \rho)$. Let z satisfy

$$P(|Z| \leq z) = 1 - \alpha.$$

Then the critical level c' in 5.18 is $c' = z/\sqrt{n}$. The minimum detectable correlation for a .05 level test of independence is $\Delta\rho = (3.6)/\sqrt{n}$ and at the .1 level, $\Delta\rho = (2.9)/\sqrt{n}$.

For a sample size of 100, $\Delta\rho \simeq .3$. Thus with a sample size of 100, a correlation coefficient less than $\pm.3$ cannot be reliably detected at the .1 level. But how dependent are X and Y if they have a correlation coefficient of .3? Here is one way to judge this: Suppose you predict that the value of some random variable X will be x_0. Then the expected squared error between X and your predicted value is $E(X - x_0)^2$. The value of x_0 that minimizes this is $x_0 = EX$, and the minimum expected squared error is $\sigma^2 = E(X - EX)^2$. The deviation σ is a measure of the scatter of the values of X around the best predicted value of X.

Now suppose X and Y have a joint distribution and we wish to first observe the value y of Y and then use y to predict a value of X. The best predictor of X based on Y is that function $g^*(Y)$ of Y which minimizes $E[X - g(Y)]^2$ (see PSP, pp. 145–146). The square root of $E[X - g^*(Y)]^2$, the root-mean-square distance of X from its predicted value, is a measure of the scatter of the values of X around its best predicted values based on Y. This scatter compared with the scatter of X around its best predicted value EX when no knowledge of Y is assumed gives a measure of the dependence between the variables.

For bivariate normal variables X, Y (see PSP, p. 300)

$$g^*(Y) = \frac{\sigma_1}{\sigma_2}\rho(Y - \mu_2) + \mu_1$$

and

$$E[X - g^*(Y)]^2 = \sigma_1^2(1 - \rho^2).$$

Hence for correlation ρ, the scatter is diminished by a factor of $\sqrt{1 - \rho^2}$. For $\rho = .3$,

$$\sqrt{1 - \rho^2} \simeq .95.$$

Therefore, by the measure we are using, a correlation coefficient of .3 does not indicate a very strong dependence between variables.

The normality of $\hat{\rho}$ sets in slowly, but the transformed variable

$$\hat{w} = \tfrac{1}{2}\log\left(\frac{1 + \hat{\rho}}{1 - \hat{\rho}}\right)$$

is an asymptotically relatively unbiased estimator for

$$w = \tfrac{1}{2}\log\left(\frac{1 + \rho}{1 - \rho}\right)$$

which converges to normality much more rapidly, and can be taken as normal for $n \geq 30$. Its squared error loss is approximately equal to $1/n$ for all ρ. By transforming to w and $\hat{w}$ we can use the normal approximation at considerably smaller sample sizes.

Problem 20 **a** Criticize the use of the ρ test in analyzing the traffic count data for dependence.

b Defend its use.

The *F* Test

One of the most familiar questions in statistics is: Do these two or more populations have the same underlying distribution or are the values of some of the populations systematically larger than others? The classical formulation of this for three populations is: Let

$$\begin{array}{l} x_1, \ldots, x_n \\ y_1, \ldots, y_m \\ z_1, \ldots, z_l \end{array}$$

be independent samples from three normal populations $N(\mu_1, \sigma_1^2)$, $N(\mu_2, \sigma_2^2)$, $N(\mu_3, \sigma_3^2)$. Since we are primarily interested in differences in location, we make the simplifying assumption that the unknown variances are the same:

$$\sigma_1^2 = \sigma_2^2 = \sigma_3^2 = \sigma^2.$$

(Of course, this has to have some foundation in reality. If you know that one variance is much larger than another, this model is unsuitable.) Then the hypothesis is

$$\mu_1 = \mu_2 = \mu_3.$$

The parameter space is $\Theta = (\mu_1, \mu_2, \mu_3, \sigma)$; the maximum likelihood estimates are the sample means $\hat{\mu}_1, \hat{\mu}_2, \hat{\mu}_3$ and

$$\hat{\sigma}^2 = \frac{1}{n + m + l}\left(\sum_k (x_k - \hat{\mu}_1)^2 + \sum_k (y_k - \hat{\mu}_2)^2 + \sum_k (z_k - \hat{\mu}_3)^2\right).$$

Under any state in the hypothesis space

$$\Theta_1 = \{(\mu, \mu, \mu, \sigma), -\infty < \mu < \infty, \sigma^2 > 0\}$$

the three samples are drawn from the same distribution $N(\mu, \sigma^2)$, so we can consider them as one sample of length $n + m + l$. The maximum likelihood estimates are

$$\hat{\mu} = \frac{1}{n + m + l}\left(\sum_k x_k + \sum_k y_k + \sum_k z_k\right)$$

and

$$\hat{\sigma}_0^2 = \frac{1}{n + m + l}\left(\sum_k (x_k - \hat{\mu})^2 + \sum_k (y_k - \hat{\mu})^2 + \sum_k (z_k - \hat{\mu})^2\right).$$

Substituting into the density gives

$$f_{\hat{\theta}}(\mathbf{x}, \mathbf{y}, \mathbf{z}) = \frac{1}{(2\pi\hat{\sigma}^2)^{(n+m+l)/2}} e^{-(n+m+l)/2},$$

and the maximum over Θ_1 is

$$f_{\hat{\theta}_0}(\mathbf{x}, \mathbf{y}, \mathbf{z}) = \frac{1}{(2\pi\hat{\sigma}_0^2)^{(n+m+l)/2}} e^{-(n+m+l)/2}.$$

The likelihood ratio test accepts if

$$\left(\frac{\hat{\sigma}^2}{\hat{\sigma}_0^2}\right)^{(n+m+l)/2} \geq c$$

or if

$$\frac{\hat{\sigma}_0^2}{\hat{\sigma}^2} \leq c^{-(n+m+l)/2}.$$

That is, if

$$\frac{\hat{\sigma}_0^2 - \hat{\sigma}^2}{\hat{\sigma}^2} \leq c'. \tag{5.19}$$

Now

$$\sum_k (x_k - \hat{\mu})^2 = \sum_k (x_k - \hat{\mu}_1 + \hat{\mu}_1 - \hat{\mu})^2$$

$$= \sum (x_k - \hat{\mu}_1)^2 + n(\hat{\mu}_1 - \hat{\mu})^2,$$

because the cross product term

$$2 \sum_k (\hat{\mu}_1 - \hat{\mu})(x_k - \hat{\mu}_1) = 2(\hat{\mu}_1 - \hat{\mu}) \sum_k (x_k - \hat{\mu}_1) = 0.$$

Thus

$$\sum_k (x_k - \hat{\mu})^2 - \sum_k (x_k - \hat{\mu}_1)^2 = n(\hat{\mu}_1 - \hat{\mu})^2.$$

Repeating this computation on the other samples, the left side of 5.19 becomes (canceling $1/(n + m + l)$ from bottom and top),

$$\frac{n(\hat{\mu}_1 - \hat{\mu})^2 + m(\hat{\mu}_2 - \hat{\mu})^2 + l(\hat{\mu}_3 - \hat{\mu})^2}{\Sigma_k(x_k - \hat{\mu}_1)^2 + \Sigma_k(y_k - \hat{\mu}_2)^2 + \Sigma_k(z_k - \hat{\mu}_3)^2}. \tag{5.20}$$

Standardizing the denominator by dividing it by $n + m + l - 3$ gives $\hat{s}^2$, the unbiased estimate of the common variance σ^2. The numerator divided by σ^2 has a χ^2 distribution with 2 degrees of freedom, and we divide it by two. In its final form, the test accepts the hypothesis if

$$\frac{1}{2} \frac{n(\hat{\mu}_1 - \hat{\mu})^2 + m(\hat{\mu}_2 - \hat{\mu})^2 + l(\hat{\mu}_3 - \hat{\mu})^2}{\hat{s}^2} \leq f, \tag{5.21}$$

where f is a constant. If $\mu_1 = \mu_2 = \mu_3$, the variable on the left side of 5.21 has an $F_{2,n+m+l-3}$ distribution (see p. 42). Once we have tables of the $F_{2,n+m+l-3}$ distribution, then we can determine f by the usual requirement that the test has significance level α. If $n + m + l$ is large, then $\hat{s}^2 \simeq \sigma^2$ and the variable in 5.21 has the distribution of a χ_2^2 variable divided by two.

F test

Look at the form of this test. You evaluate the overall mean $\hat{\mu}$ of all samples. Compute the squared differences between the individual sample means and the overall sample mean and sum these as weighted by the sample size. Then divide by an estimate of the common variance and reject if this ratio is too large. The detection capability of this test illustrates the idea of a minimum detection distance in a multidimensional problem. For any means vector $\boldsymbol{\mu} = (\mu_1, \mu_2, \mu_3)$, define a distance $d(\boldsymbol{\mu})$

from the hypothesis space $\mu_1 = \mu_2 = \mu_3$ this way: Define the overall mean weighted according to sample size as

$$\mu = \frac{1}{n + m + l}(n\mu_1 + m\mu_2 + l\mu_3),$$

and take

$$d(\boldsymbol{\mu}) = n(\mu_1 - \mu)^2 + m(\mu_2 - \mu)^2 + l(\mu_3 - \mu)^2.$$

Then we can detect $\boldsymbol{\mu}$ at level α if

$$d(\boldsymbol{\mu}) \geq k\sigma^2, \tag{5.22}$$

where k is a constant determined by α and the sample size. If n, m, l are large, then for $\alpha = .05$, $k \simeq 15$. For example, if $n = m = l = 100$, then we can detect $\boldsymbol{\mu}$ at the .05 level if

$$(\mu_1 - \mu)^2 + (\mu_2 - \mu)^2 + (\mu_3 - \mu)^2 \geq .15\sigma^2.$$

Problem 21 Given three samples from $N(\mu_1, \sigma^2)$, $N(\mu_2, \sigma^2)$, $N(\mu_3, \sigma^2)$ distributions, here are two ways to test for equality of the three means:

a Use the first two samples to test $\mu_1 = \mu_2$. Then use the second and third samples to test $\mu_2 = \mu_3$.

b Use all three samples and the F test to test $\mu_1 = \mu_2 = \mu_3$.

c What might be the advantages and disadvantages of each method?

d Find the likelihood ratio test for the hypothesis $\mu_1 = \mu_2$ using the first two samples.

Analyzing the Differences

If a hypothesis is rejected, then the important question is: Why was it rejected? Sometimes the answer is obvious. But if Θ is multidimensional and if Θ_1 is also multidimensional, the situation may be unclear. If we reject, our conclusion is that θ^* is outside of Θ_1. But which coordinate or coordinates of θ^* are causing the trouble? Is θ^* deduced to be outside Θ_1 because, say, the estimates for its second and fifth cordinates are too large? In other words, what particular feature of the data has led to rejection? This second phase of the problem, which involves estimation and confidence regions, can reveal more about the character of the phenomenon being measured than the hypothesis testing phase.

For example, suppose we have independent samples of size $n_1, n_2, \ldots, n_K$ from J normal populations $N(\mu_1, \sigma^2), \ldots, N(\mu_J, \sigma^2)$. The means

$\mu_1, \ldots, \mu_J$ and the common variance σ^2 are unknown. The hypothesis is that the means are all equal,

$$\mu_1 = \mu_2 = \cdots = \mu_J.$$

This is a straightforward generalization of the three sample problem of the last section. The likelihood ratio test is similar. Let $\hat{\mu}_j$, $\hat{\sigma}_j^2$ be the sample mean and variance for the jth sample. An unbiased estimate σ^2 is given by

$$\hat{s}^2 = \frac{1}{n - J} \sum_1^J n_j \hat{\sigma}_j^2,$$

where $n = n_1 + \cdots + n_J$. Denote the combined sample mean by $\hat{\mu}$. Then reject the hypothesis if

$$\frac{1}{J - 1} \cdot \frac{\sum_1^J n_j(\hat{\mu}_j - \hat{\mu})^2}{\hat{s}^2} \geq f.$$

The ratio on the left has an $F_{J-1,n-J}$ distribution. This distribution is standard; there are tables for n of moderate size. For n large, $\hat{s}^2 \simeq \sigma^2$ and the ratio has the distribution of a χ_{J-1}^2 variable divided by $J - 1$.

If this test rejects the hypothesis, then we conclude that the means are not all equal, and the question becomes: Where is the inequality? Is it that $\mu_3 \neq \mu_1$ or perhaps that $\mu_5 \neq \mu_2$? Furthermore, if $\mu_3 \neq \mu_1$, how large is the difference? To answer the above questions, it would be convenient and useful to compute confidence intervals for the differences $\mu_j - \mu_k$. But to systematically explore these differences we want simultaneous confidence intervals. On the other hand, for J any size at all, there are so many differences that the procedure for constructing simultaneous confidence intervals (Chapter 4, p. 107) leads to intervals that are much too large.

Another procedure, due to Tukey, is more efficient and gives considerably shorter intervals. We assume that the J sample sizes are equal, say $n_j = m$ for all j. Let $\mathsf{u}_j = \hat{\mu}_j - \mu_j$. The u_j variables have an $N(0, \sigma^2/m)$ distribution. The *range* of these variables is defined as

$$\max_j \mathsf{u}_j - \min_j \mathsf{u}_j.$$

An unbiased estimate of the variance of each of the u_j variables is given by $\hat{s}^2/m$. Now divide the range by the estimate $\hat{s}/\sqrt{m}$ of the standard deviation of the u_j. This division essentially cancels the value of σ, and the resulting variable

$$\mathsf{R} = \frac{\max_j \mathsf{u}_j - \min_j \mathsf{u}_j}{\hat{s}/\sqrt{m}}$$

has a standard distribution not depending on the values of any of the parameters. This distribution is called the *Studentized range*, with sample size J and $n - J$ degrees of freedom. (See Table 6) Define the value r by

5.23
$$P(\mathsf{R} \le r) = \gamma.$$

5.24 ***Theorem*** *With probability γ, all differences $\mu_k - \mu_j$ simultaneously satisfy*

$$\mu_k - \mu_j \in \hat{\mu}_k - \hat{\mu}_j \pm r \frac{\hat{\mathsf{s}}}{\sqrt{m}}.$$

The proof is interesting and simple. From 5.23, with probability γ

$$\max \mathsf{u}_j - \min \mathsf{u}_j \le r \frac{\hat{\mathsf{s}}}{\sqrt{m}}.$$

For all k and j this inequality is equivalent to

$$|\mathsf{u}_k - \mathsf{u}_j| \le r \frac{\hat{\mathsf{s}}}{\sqrt{m}}$$

or

$$-r \frac{\hat{\mathsf{s}}}{\sqrt{m}} \le \mathsf{u}_k - \mathsf{u}_j \le r \frac{\hat{\mathsf{s}}}{\sqrt{m}}.$$

Substituting $\mathsf{u}_k = \hat{\mu}_k - \mu_k$, $\mathsf{u}_j = \hat{\mu}_j - \mu_j$ into these inequalities gives the result of the theorem.

Problem 22 Consider the three sets of observations

1	2	3
2.79	3.36	3.59
3.31	1.09	4.64
1.29	2.37	3.63
3.08	0.25	4.53
4.29	1.69	3.01
5.14	1.55	4.49
1.74	1.27	4.93
4.61	0.71	4.77
1.83	0.00	6.32
3.93	3.41	2.71

a Show that the F test rejects equality of means at the .05 level.

b Find simultaneous confidence intervals for all differences of means at the 90% level using the Studentized range.

The Asymptotic Distribution of Likelihood Ratio Tests

Often the most difficult part of setting up a level α likelihood ratio test is the determination of the critical constant; that is, either the maximum value of c such that for all $\theta \in \Theta_1$,

$$P_\theta(\boldsymbol{\lambda} \geq c) \geq 1 - \alpha$$

or the minimum value of d such that

$$P_\theta(\boldsymbol{l} \leq d) \geq 1 - \alpha$$

for all $\theta \in \Theta_1$. Here $\boldsymbol{\lambda} = \lambda(\mathbf{x})$ and $\boldsymbol{l} = -2 \log \boldsymbol{\lambda}$.

If Θ_1 consists of a single point θ_0, then usually d can be determined by the use of the following theorem.

5.25 ***Theorem*** *If Θ is a subset of N-dimensional space, if the hypothesis point θ_0 is not a boundary point of Θ, and if the problem is smooth, then under P_{θ_0}, for large sample sizes, $\boldsymbol{l}$ has approximately a χ^2 distribution with N degrees of freedom.*

When the conditions of this theorem hold, a table of the χ_N^2 distribution can be used to determine the value of d such that

$$P_{\theta_0}(\boldsymbol{l} \leq d) = 1 - \alpha.$$

The situation in which Θ_1 consists of one point is not encountered very often. More frequently, Θ_1 will consist of a line, plane, or hyperplane (see definition below) through the parameter space. For example, if the underlying distributions are $N(\mu, \sigma^2)$, both μ and σ^2 unknown, then the hypothesis set $\mu = \mu_0$ is a straight line through the 2-dimensional parameter space of points (μ, σ), $\sigma > 0$. The hypothesis set corresponding to independence in the bivariate normal distribution consists of the 4-dimensional hyperplane of all points such that $\rho = 0$ running through the 5-dimensional space of points $(\mu_1, \sigma_1, \mu_2, \sigma_2, \rho)$ such that $\sigma_1 \geq 0$, $\sigma_2 \geq 0$, $-1 \leq \rho \leq 1$.

In general, let Θ be an N-dimensional subset of Euclidean space. An $(N - M)$-dimensional hyperplane in this space is the set of all points $\theta = (\theta_1, \ldots, \theta_N)$ satisfying some system of M linearly independent equations

$$\begin{aligned}
a_{11}\theta_1 + a_{12}\theta_2 + \cdots + a_{1N}\theta_N &= b_1 \\
&\vdots \\
a_{M1}\theta_1 + a_{N2}\theta_2 + \cdots + a_{MN}\theta_N &= b_M
\end{aligned}$$

for specified constants a_{ij}, $i = 1, \ldots, M, j = 1, \ldots, N$.

5.26 ***Theorem*** *Let Θ_1 consist of all points on an $(N - M)$-dimensional hyperplane in Θ. If the problem is smooth, then for any parameter value in Θ_1 not on the boundary of Θ,* $\boldsymbol{l}$ *has a χ^2 distribution (approximately) with M degrees of freedom for large sample size.*

Problem 23 Let $x_1, \ldots, x_n$, n large, be drawn from a double exponential family with known scale parameter β and unknown location parameter α. What is the likelihood ratio test of the hypothesis $\alpha = 0$? Evaluate the critical constant for level .05. (Assume the problem is smooth.)

Problem 24 Your data will consist of 1000 rolls of a single die; you are to decide if the die is fair. Find the .05 level likelihood ratio test for the hypothesis of fairness. (Evaluate the critical constant.)

Problem 25 Assume that 5 different companies are manufacturing the same component. Samples of size $n \gg 1$ are selected from the runs of each manufacturer and tested until failure. Assuming exponential failure time distributions, set up the likelihood ratio test at level .05 for the hypothesis that the 5 underlying failure distributions are the same. (Ans. Accept if

$$2n\left(5 \log \hat{\beta} - \sum_{1}^{5} \log \hat{\beta}_j\right) \leq 9.5,$$

where $\hat{\beta}_j$ is the average of the failure times for the jth sample and $\hat{\beta}$ is the average failure time over all samples.)

Problem 26 To which of the problems on pages 142, 143, 144 can Theorems 5.25 or 5.26 be applied, and why?

How Does Likelihood Ratio Work?

The likelihood ratio tests can be difficult to compute. But like the maximum likelihood estimates, the computations are mechanical. The drawback thus far is that although we have carried out the computations in a number of examples, we still have no real understanding of what the likelihood ratio test regions generally look like.

Consider the following approach to testing the hypothesis $\theta = \theta_0$: Take a good estimate $\hat{\theta}$ of θ. If $\hat{\theta}$ is too far from θ_0, reject; otherwise accept. For example, to test if a coin is fair, compute an estimate $\hat{p}$ of p. Then accept if

$$|\hat{p} - \tfrac{1}{2}| \leq \varepsilon;$$

otherwise reject. The maximum distance ε is fixed by the requirement that the test be at a predetermined level α. If we use the maximum likelihood method to estimate, the above test is the same as the likelihood ratio test (see Problem 27).

More generally, suppose the hypothesis is $\theta \in \Theta_1$. Our proposed procedure is this: Take a good estimate $\hat{\theta}$ of θ. Let Θ_2 be some set of parameter values containing Θ_1 plus a small belt of parameter values around the boundary of Θ_1. *Accept* the hypothesis if $\hat{\theta} \in \Theta_2$, and *reject* if $\hat{\theta} \notin \Theta_2$. The reasoning behind this is clear: If the true state of nature is in Θ_1, then the values of $\hat{\theta}$ should be in Θ_2 with high probability. If the state of nature is well outside of Θ_2, then $\hat{\theta}$ should be outside of Θ_2 with high probability. If the state of nature lies outside of Θ_1 but not too far outside of Θ_2, then $\hat{\theta}$ can fall both inside and outside of Θ_2 with nonnegligible probability. Setting the level of the test at α usually determines how much larger Θ_2 must be than Θ_1.

If Θ is 1-dimensional and if the hypothesis is $\theta = \theta_0$, then this procedure accepts if

$$\hat{\theta} \in [\theta_0 - \varepsilon_1, \theta_0 + \varepsilon_2].$$

If the hypothesis is that θ lies in the interval $[\theta_1, \theta_2]$, then the acceptance region has the form

$$\theta \in [\hat{\theta}_1 - \varepsilon_1, \theta_2 + \varepsilon_2].$$

In testing a coin for absence of bias or in testing $\mu = \mu_0$ in normally distributed samples with known variance, the likelihood ratio test leads to the same acceptance region as the simplified procedure based on the maximum likelihood estimator. But in testing whether the location parameter α is α_0 in a double exponential family with known scale parameter β, the likelihood ratio test becomes: Accept if

$$\sum_k |x_k - \alpha_0| - \sum_k |x_k - \hat{\nu}| \leq c,$$

where $\hat{\nu}$ is the sample median. On the other hand, the simplified procedure, using $\hat{\nu}$ as an estimate of α, accepts if

$$\hat{\nu} \in [\alpha_0 - \varepsilon_1, \alpha_0 + \varepsilon_2].$$

These two acceptance regions seem different. Yet for large sample size we have

5.27 **Theorem** *In a smooth problem the likelihood ratio test is asymptotically equivalent to a test which accepts the hypothesis $\theta \in \Theta_1$ if the maximum likelihood estimate $\hat{\theta}$ is in a set Θ_2 containing Θ_1.*

Here we take the phrase *asymptotically equivalent* to mean that for every $\theta \in \Theta$, the probability under P_θ that $\mathbf{x}$ falls into one test region but not the other goes to 0 rapidly as the sample size increases.

Looking at Theorem 5.27, one is tempted to ask why we bother with likelihood ratio at all. Why not use the maximum likelihood estimate $\hat{\theta}$, find Θ_2, and then test accordingly? The problem is in finding Θ_2. If Θ is 1-dimensional and we are testing $\theta = \theta_0$, it is clear that Θ_2 should be of the form $[\theta_0 - \varepsilon_1, \theta_0 + \varepsilon_2]$. But in the multidimensional case, we really know little about Θ_2. For example, look at the three-sample F test we constructed in the section on the F test for testing the hypothesis $\mu_1 = \mu_2 = \mu_3$. Here the parameter space is 4-dimensional; Θ is the set of all points $(\mu_1, \mu_2, \mu_3, \sigma)$ and Θ_1 consists of the 2-dimensional plane of points (μ, μ, μ, σ). The maximum likelihood estimates are the usual sample means $\hat{\mu}_1, \hat{\mu}_2, \hat{\mu}_3$, and

$$\hat{\sigma}^2 = \frac{1}{n + m + l}(n\hat{\sigma}_1^2 + m\hat{\sigma}_2^2 + l\hat{\sigma}_3^2)$$

where $\hat{\sigma}_1^2, \hat{\sigma}_2^2, \hat{\sigma}_3^2$ are the usual sample variances. Now all we know is that we want Θ_2 to be some region enclosing the plane Θ_1 such that

$$P_\theta[(\hat{\mu}_1, \hat{\mu}_2, \hat{\mu}_3, \hat{\sigma}) \in \Theta_2] \geq 1 - \alpha$$

for all $\theta \in \Theta_1$. There are two difficulties. First, this condition does not uniquely determine Θ_2. Second, to compute a Θ_2 that will do the job is considerably more difficult than just writing out and simplifying the likelihood ratio.

Theorem 5.27 can be useful if Θ is 1-dimensional, or if we are testing a hypothesis that involves only one coordinate of θ. When Θ is 1-dimensional, the most frequently appearing alternatives are of the two-sided form

$$\theta \neq \theta_0$$

or the one-sided

$$\theta \leq \theta_0 \qquad \text{or} \qquad \theta \geq \theta_0.$$

Assume that the sample size is large and that $\hat{\theta}$ is asymptotically normal and relatively unbiased, with variance $\sigma^2(\theta)$ under P_θ. Then, under P_{θ_0},

$\hat{\theta} - \theta_0$ is normal with mean 0 and standard deviation $\sigma(\theta_0)$. If z is determined from an $N(0, 1)$ table by

$$P|Z| \leq z) = 1 - \alpha,$$

a level α test based on $\hat{\theta}$ accepts if

5.28
$$\theta_0 - z\sigma(\theta_0) \leq \hat{\theta} \leq \theta_0 + z\sigma(\theta_0).$$

The detection distance for this test can be computed by using the fact that $\phi(\theta)$, the probability under P_θ that

$$\theta_0 - z\sigma(\theta_0) \leq \hat{\theta} \leq \theta_0 + z\sigma(\theta_0)$$

can be approximated as the probability that

5.29
$$\frac{\theta_0 - z\sigma(\theta_0) - \theta}{\sigma(\theta)} \leq Z \leq \frac{\theta_0 + z\sigma(\theta_0) - \theta}{\sigma(\theta)}.$$

Since $\phi(\theta)$ goes to 0 rapidly as θ moves away from θ_0, we do our calculations for θ near θ_0, where we assume $\sigma(\theta) \simeq \sigma(\theta_0)$. Then 5.29 can be written as

5.30
$$\frac{\theta - \theta_0}{\sigma(\theta_0)} - z \leq Z \leq \frac{\theta - \theta_0}{\sigma(\theta_0)} + z.$$

Substituting

$$\theta = \theta_0 + z_0\sigma(\theta_0)$$

into 5.30 gives

5.31 ***Proposition*** *The minimum difference $\Delta\theta$ from θ_0 that can be detected by the above test is*

$$\Delta\theta = z_0\sigma(\theta_0),$$

where z_0 is computed from the equations

$$P(|Z| \leq z) = 1 - \alpha$$

$$P(z_0 - z \leq Z \leq z_0 + z) = \alpha.$$

For given α, z_0 can be found using $N(0, 1)$ tables. (See the table below.)

α	.10	.05	.01
z	1.65	1.96	2.57
z_0	2.93	3.61	4.90

It is somewhat surprising to see how large the differences must be before they are detectable. For example, suppose we want detection at the 10% level ($\alpha = .1$), with the sample size $n = 100$. If we are estimating the parameter β in an exponential family by maximum likelihood, then the standard deviation is $\beta/\sqrt{n}$, and $\Delta\beta/\beta_0 \simeq .29$. This means that *we cannot detect less than almost a* 30% *change in the parameter at the* .1 *level with a sample size of* 100.

The test for the hypothesis $\theta \geq \theta_0$ versus $\theta < \theta_0$ is of the form: Accept if

$$\hat{\theta} \geq \theta_0 - \varepsilon,$$

where ε is determined by

$$P_\theta(\hat{\theta} \geq \theta_0 - \varepsilon) \geq 1 - \alpha$$

for all $\theta \geq \theta_0$. Unless $\sigma(\theta)$ changes very erratically this condition will be hardest to satisfy at $\theta = \theta_0$. Therefore, ε must satisfy only the single condition

$$P_{\theta_0}(\hat{\theta} \geq \theta_0 - \varepsilon) = 1 - \alpha.$$

Define z^+ by

$$P(Z \leq z^+) = 1 - \alpha.$$

Then the level α test for $\theta \geq \theta_0$ accepts if

$$\hat{\theta} \geq \theta_0 - z^+\sigma(\theta_0).$$

Using the same approximations as in the two-sided test, the minimum distance of θ below θ_0 that can be detected at level α is given by

$$\Delta\theta = z_0^+\sigma(\theta_0)$$

where z_0^+ is computed from

$$P(Z \geq z_0^+ - z^+) = \alpha.$$

Some values of z_0^+ are tabled below. Notice that $z_0^+ = 2z^+$.

α	.10	.05	.01
z^+	1.28	1.64	2.33
z_0^+	2.56	3.29	4.65

Now, suppose that $\theta = (\theta_1, \ldots, \theta_k)$ and we want to test some hypothesis depending on one coordinate only, for instance $\theta_1 = 0$ or

$$2 \leq \theta_2 \leq 5.$$

This can be handled similarly to the 1-dimensional situation. There our essential tool was an estimator $\hat{\boldsymbol{\theta}}$ such that $(\hat{\boldsymbol{\theta}} - \theta)/\sigma(\theta)$ has (approximately) an $N(0, 1)$ distribution under P_θ. In the multidimensional case, assume that $\hat{\theta} = (\hat{\theta}_1, \ldots, \hat{\theta}_k)$ is a consistent estimate for θ such that $\hat{\boldsymbol{\theta}}_1$ is asymptotically normal and relatively unbiased.

In some important examples, the variance of $\hat{\theta}_1$ depends only on θ_1, and not on the other coordinates. Then the problem becomes 1-dimensional. To test any hypothesis about θ_1, we ignore the coordinates $\theta_2, \ldots, \theta_k$ and just use the fact that $\hat{\boldsymbol{\theta}}_1$ is an asymptotically normal and relatively unbiased estimator with variance $\sigma_1^2(\theta_1)$.

But in general, $\hat{\boldsymbol{\theta}}_1$ will have a variance $\sigma_1^2(\theta)$ that depends on other coordinates besides θ. For large sample size, we use the approximation

$$\sigma_1(\theta) \simeq \sigma_1(\hat{\theta}).$$

Then, using the same z as above, we have that a level α test for $\theta_1 = \theta_1^{(0)}$ accepts if

$$\theta_1^{(0)} - z\sigma_1(\hat{\theta}) \leq \hat{\theta}_1 \leq \theta_1^{(0)} + z\sigma_1(\hat{\theta}). \tag{5.32}$$

The analogous test using z^+ holds in the one-sided case.

Furthermore, for the same z_0 as defined in the last section, the detectable distance at level α is

$$\Delta\theta_1 = z_0\sigma_1(\hat{\theta}).$$

Now $\sigma_1(\hat{\boldsymbol{\theta}})$ is random, and actually we should take $\Delta\theta = z_0\sigma_1(\theta)$, where θ is the underlying state. But after the data are in, using $z_0\sigma_1(\hat{\theta})$ provides a more useful evaluation, since θ is unknown.

Problem 27 If n_1 is the number of heads in n tosses of a coin, and $\hat{p} = n_1/n$, show that the likelihood ratio test that the coin is unbiased reduces to: Accept if $|\hat{p} - \frac{1}{2}| \leq \varepsilon$.

Problem 28 If $x_1, \ldots, x_n$ are sampled from a Poisson distribution, show that the likelihood ratio test for $\lambda = \lambda_0$ reduces to: Accept if $\hat{\lambda} \in [\lambda_0 - \varepsilon_1, \lambda_0 + \varepsilon_2]$.

Problem 29 Given the model of Problem 11, what simplified procedure might you use for testing the hypothesis

$$\mathbf{s} = \mathbf{s}^{(0)}?$$

Problem 30 Let $x_1, \ldots, x_{100}$ be sampled from an $N(\mu, 1)$ distribution, $-\infty < \mu < \infty$. Find an acceptance region at the .05 level for the hypothesis $\mu \geq 0$ and graph its operating characteristic. If $\mu = -.10$, how frequently do you incorrectly accept the hypothesis? What if $\mu \leq -.35$?

Problem 31 A telephone dialing system is designed to be usable under conditions such that with probability greater than .99, the number of incoming calls per second does not exceed 3. One hundred observations are made of the number of calls per second during the peak period. Construct a test at the .01 level for the hypothesis that the system is usable. (It is known that the Poisson distribution is a good approximation to the number of incoming calls per unit time. Assume also that the numbers of incoming telephone calls in different one-second intervals are independent and identically distributed. Use the Poisson Table 5.)

Problem 32 Measurements are to be made on the mass of what is suspected to be a new fundamental particle. The known particle closest in mass to the suspected new particle has mass m_0. From past experience it is known that the standard deviation of the measuring procedure for particles with mass near m_0 is $.07m_0$. Find the sample size necessary to test the hypothesis that the suspected new particle is identical to the known particle that will detect at the .01 level if the true mass of the suspected new particle differs from m_0 by more than 1%.

Problem 33 For the example of the first section, set up a test of the hypothesis that $\sigma = 10$, at the .05 level, based on the unbiased sample variance. If the unbiased sample variance computed from the actual data is 59.5, do you accept or reject? (Use the normal approximation.)

Problem 34 The hypothesis concerning the failure time of a component is: The probability of failure during 100 hours of use is less than .02. Given that a large sample of size n will be tested until failure, specify a test at the .025 level. (Assume an exponential distribution.)

Problem 35 Over the last 60 years, total monthly rainfalls at a certain location have been recorded. Assume that the rainfalls recorded over 60 consecutive Februarys can be considered independent and uniformly distributed over some range $[0, l]$. The hypothesis we want to test is: The probability of a February rainfall greater than 6.1 inches is less than .01. The observed maximum over the 60-year period was 5.9. Decide whether to accept or reject, using a test at the .05 level. (See Chapter 4, Problem 15.)

Problem 36 A number of rats are run through a maze which looks like this:

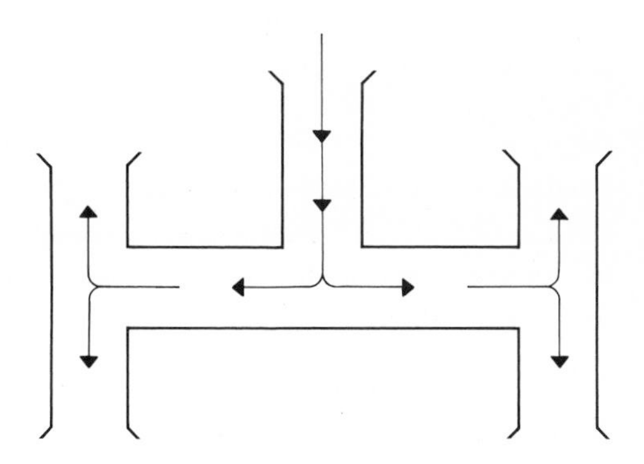

The purpose of the experiment is to find out whether the first direction of turn (right or left) in the maze has any influence on the direction of the second turn. A simple model is constructed as follows: Assume that on each trial, the probability that a rat initially turns right or left is $\frac{1}{2}$. If the rat turned right on the first turn, he has probabilities p_1, q_1 of turning right and left respectively on the second; that is,

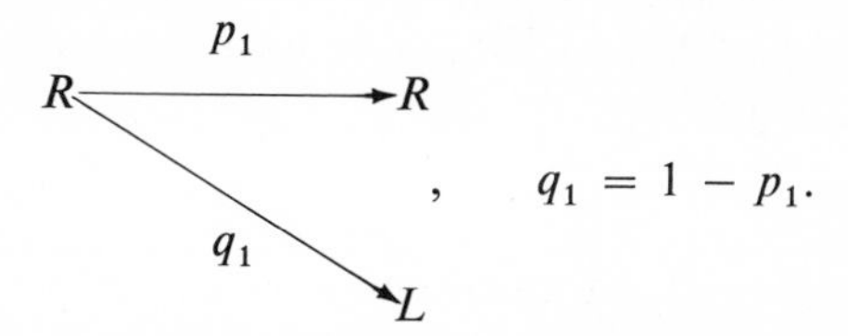

But if he turned left on the first turn, the situation is symmetrically opposite:

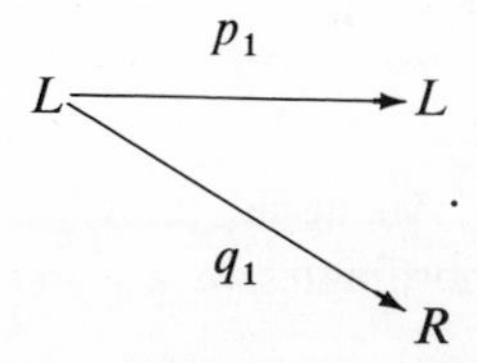

Assuming there are 75 trials in all, and that these are independent and identically distributed, construct a test for the hypothesis that the first direction of turn has no influence on the second.

Problem 37 A hypothesis was formulated as: The probability that the sex of a newly born infant is female is greater than .5. The data were that of 15,632 recorded births in one city in one year, 7809 were female. A level .05 test was constructed and the hypothesis accepted. What difference in birth probability from .5 is detectable at the .05 level? What conclusion would be likely if you took as hypothesis: The probability of a male infant is greater than .5, and constructed a .05 level test? Is there a contradiction here? Discuss!

Problem 38 The half-minute counts of cars passing a point on a certain freeway are independent and approximately normal with $\sigma \simeq 6$ and $\mu \simeq 30$. How quickly could you detect a 10% change in μ at the .1 level, assuming σ remains constant? (Ans. 17 minutes.)

Problem 39 Particles are arriving at a counter in a Poisson process with parameter λ. At some time t_0, the parameter value may have changed by 5%. What is, approximately, the minimum number of additional particle arrivals we have to record in order to detect this change at the .1 level? (Ans. $3.4 \cdot 10^3$.)

Problem 40 A large city has 57% of its voters registered as Democrats. Recently a poll selected 1600 voters at random in this city and asked them how they would register if they had to do it that day. Only 54.5% responded "Democratic." Does this indicate a trend? Why?

Problem 41 A distribution is known to be double exponential with scale parameter 1. Construct a level .05 test of the hypothesis that the location parameter is zero. What is the minimum .1 level detectable difference?

Problem 42 An aerial photographic system has as final outputs the (x, y) coordinates in a standard geographic coordinate system of various landscape features. The specifications are that the system give locations of points to within 3 feet of their true location with probability at least .95. To check the system a point with known location is photographed and read out a total of 75 times. Assume that the errors Δx, Δy in the x- and y-coordinates are independent $N(0, \sigma^2)$ variables with unknown σ and set up a test to see if specifications are being met. (Specify the level yourself.)

Problem 43 A manufacturer is faced with the problem of proving that in 100 hours of use, certain components have a failure probability of less than .001. He and the customer both feel secure about assuming an exponential failure distribution. The problem he faces is this: The probability of failure in 100 hours is

$$p = 1 - e^{-100/\beta}.$$

Solving for β and assuming $p \ll 1$ so that we can use $\log(1 - p) \simeq -p$, gives $\beta \simeq 100/p$. For p on the order of .001, β is on the order of 10^5 hours, about 4000 days or over 10 years. To test until failure a batch of these components having an average failure time of about 10 years is out of the question. As an alternative, the manufacturer proposes testing 2000 components over a 4000-hour period and keeping track of failure times during this period. Discuss the various aspects of this problem and design a test based on this data. (One approach is to assume that the total number of failures in 4000 hours is $\ll 2000$. Break the 4000-hour period up into 40 disjoint 100-hour periods. Show that the numbers of failures in the 100-hour periods should be independent Poisson variables. Reformulate the hypothesis into a hypothesis on the size of the Poisson parameter.) Discuss the minimum .05 level detectable difference in the 100-hour failure probability. Do you protect your client adequately if you take as hypothesis: The failure probability in 100 hours is $\leq .001$?

Problem 44 Design an experiment to test whether a certain subject has the ESP ability to detect the color (red or black) of a playing card which he cannot see. Discuss how to insure independence, what sample size you would require, and other design factors you might insist on.

Problem 45 Show that the minimum detectable proportional change in the standard deviation σ is approximately half of the minimum detectable proportional change in the variance for the same sample size. (See Chapter 3, Problem 10.)

Problem 46 In a sample of size 50 drawn from a double exponential distribution, what percentage difference in the scale parameter is detectable at the .1 level?

Problem 47 A test that assigns a number on an "authoritarian scale" has been criticized on the grounds that it is, more or less, another form of an intelligence test. How would you set up an experiment to check this allegation and how would you handle the resulting data?

Problem 48 A method for determining the pH of solutions is claimed by the manufacturer to be accurate to within $\pm .02$. Testing the device on 10 solutions of known pH gives errors as follows:

$$-.00663, \quad .01194, \quad .00714, \quad .00312, \quad -.02173,$$
$$-.00003, \quad .00913, \quad .00488, \quad -.00465, \quad -.01016.$$

Assuming normality and that the accuracy specifications are the $\pm 2\sigma$ values, what can you conclude from the data about the manufacturer's claim? Use the exact χ^2 distribution and tables.

Problem 49 Given a location-scale family, how would you go about constructing an acceptance region for some hypothesis on the location parameter? Illustrate using the double exponential family. Find the minimum detectable difference in location at the .1 level.

Confidence Intervals Revisited

Some of this material on hypothesis testing is reminiscent of our previous discussion of confidence intervals. In fact, there is a simple-minded approach to hypothesis testing based on confidence intervals: we want to test the hypothesis $\theta = \theta_0$ at level α. Take $\hat{\boldsymbol{\theta}}$ to be an estimator of θ with $100\,(1 - \alpha)\%$ confidence intervals $J(\hat{\boldsymbol{\theta}})$. For the given data compute $\hat{\theta}$.

5.33 *If θ_0 is in $J(\hat{\theta})$ accept the hypothesis; otherwise, reject.*

This is an intuitively appealing procedure. Does it work—that is, does it give us a level α test? Is $P_{\theta_0}(S) \geq 1 - \alpha$? The answer is quick—

$$P_{\theta_0}(S) = P_{\theta_0}[\theta_0 \in J(\hat{\theta})]$$

and the right-hand side is greater than or equal to $1 - \alpha$ by the definition of confidence intervals.

Here is a more explicit form of the acceptance region: Let

$$J(\hat{\theta}) = [a(\hat{\theta}), b(\hat{\theta})];$$

then $\theta_0 \in J(\hat{\theta})$ is expressed by

$$a(\hat{\theta}) \leq \theta_0 \leq b(\hat{\theta}).$$

If we solve these for $\hat{\theta}$ in terms of θ_0, we usually get some interval

$$A(\theta_0) \leq \hat{\theta} \leq B(\theta_0)$$

as an equivalent definition of the acceptance region.

Example c Take $x_1, \ldots, x_n$ to be drawn from an exponential distribution with scale parameter β. Estimate β by the sample mean $\hat{\beta}$. For moderate to large n, 95% confidence intervals are given by

$$J(\hat{\beta}) = \hat{\beta}\left(1 \pm \frac{2.0}{\sqrt{n}}\right).$$

Then a .05 level test for $\beta = \beta_0$ is

$$\textit{accept if } \beta_0 \in J(\hat{\beta}).$$

This is equivalent to

$$\hat{\beta}\left(1 - \frac{2.0}{\sqrt{n}}\right) \leq \beta_0 \leq \hat{\beta}\left(1 + \frac{2.0}{\sqrt{n}}\right).$$

Solving the left-hand inequality gives

$$\hat{\beta} \leq \beta_0\left(1 - \frac{2.0}{\sqrt{n}}\right)^{-1} \simeq \beta_0\left(1 + \frac{2.0}{\sqrt{n}}\right),$$

and the right-hand side yields

$$\hat{\beta} \geq \beta_0\left(1 + \frac{2.0}{\sqrt{n}}\right)^{-1} \simeq \beta_0\left(1 - \frac{2.0}{\sqrt{n}}\right).$$

Together, these give the acceptance region

$$\beta_0\left(1 - \frac{2.0}{\sqrt{n}}\right) \leq \hat{\beta} \leq \beta_0\left(1 + \frac{2.0}{\sqrt{n}}\right).$$

This confidence-interval approach works even when θ is multidimensional, say $\theta = (\theta_1, \ldots, \theta_k)$. Take $\hat{\theta}$ to be an estimate for θ, and $J_1(\hat{\theta})$ to be a $100(1 - \alpha)\%$ confidence interval for θ_1. To test the hypothesis $\theta_1 = \theta_1^{(0)}$, use the rule:

5.34 *accept if* $\theta_1^{(0)} \in J_1(\hat{\theta})$; *reject otherwise.*

Use the same reasoning as above to see that this gives a level α test. We illustrate by an example.

Example d If $x_1, \ldots, x_n$ are drawn from an $N(\mu, \sigma^2)$ distribution with both μ and σ^2 unknown, then $100(1 - \alpha)\%$ confidence intervals for μ are given by $\hat{\mu} \pm t\hat{s}/\sqrt{n}$, (see 4.24, p. 119) where t is obtained from a table of Student's t distribution with $n - 1$ degrees of freedom, $\hat{\mu}$ is the sample mean, and $\hat{s}$ the unbiased sample variance. To test the hypothesis $\mu = \mu_0$ at level α, this procedure says: Accept only if

$$\mu_0 \in \left[\hat{\mu} - t\frac{\hat{s}}{\sqrt{n}}, \hat{\mu} + t\frac{\hat{s}}{\sqrt{n}}\right].$$

Write this as

$$\hat{\mu} - t\frac{\hat{s}}{\sqrt{n}} \leq \mu_0 \leq \hat{\mu} + t\frac{\hat{s}}{\sqrt{n}},$$

or

$$\left|\sqrt{n}\,\frac{\hat{\mu} - \mu_0}{\hat{s}}\right| \leq t.$$

Of course, this is the same acceptance region we came up with on page 146.

Testing $\theta_1 = \theta_1^{(0)}$ in the multidimensional parameter case reduces, for large sample size, to 5.32. We could have gotten this same result by using the confidence-interval approach and the fact that $100(1 - \alpha)\%$ confidence intervals for θ_1 are given by $\hat{\theta}_1 \pm z\sigma(\hat{\theta})$. The relationship between confidence intervals and tests of a hypothesis expressed in 5.33 and 5.34 works both ways. In this section we derived acceptance regions from known confidence intervals. In the future, we will make use of the other direction, using known tests of a hypothesis to get confidence intervals. Here is how this procedure goes: Let $S(\theta_0)$ be the acceptance region for testing the hypothesis $\theta = \theta_0$ at level α. This means that for every θ_0, $S(\theta_0)$ is a set of data vectors such that

$$P_{\theta_0}[\mathbf{X} \in S(\theta_0)] \geq 1 - \alpha.$$

Now define

$$R(\mathbf{x}) = \{\theta; \mathbf{x} \in S(\theta)\}.$$

That is, $R(\mathbf{x})$ is a region in the parameter space Θ consisting of all parameter values whose acceptance regions include the data vector $\mathbf{x}$. Then

5.35 ***Theorem*** *$R(\mathbf{x})$ is a $100(1 - \alpha)\%$ confidence region for θ.*

To show that

$$P_\theta[\theta \in R(\mathbf{X})] \geq 1 - \alpha, \qquad \text{for all} \qquad \theta \in \Theta,$$

notice that $\theta \in R(\mathbf{x})$ is equivalent to $\mathbf{x} \in S(\theta)$. Hence

$$P_\theta[\theta \in R(\mathbf{X})] = P_\theta[\mathbf{X} \in S(\theta)] \geq 1 - \alpha, \qquad \text{for all} \qquad \theta \in \Theta.$$

This works exactly the same in the multidimensional case, yielding confidence intervals for θ_1 from tests $S(\theta_1^{(0)})$ at level α of the hypothesis $\theta_1 = \theta_1^{(0)}$.

Problem 51 The following numbers are drawn from a distribution uniform on $[0, l]$:

$$.020, \quad .248, \quad .731, \quad .767, \quad .896,$$
$$.009, \quad .101, \quad .991, \quad .696, \quad .331.$$

We have reason to believe that $l = 1.10$. Construct a test for this hypothesis starting from the confidence intervals for l.

Problem 52 Let $\{P_\theta(dx)\}$, $\theta \in \Theta$, be some multidimensional parameter family of continuous distributions, each of which is symmetric around the point $x = \theta_1$. A simple way to test the hypothesis $\theta_1 = \theta_1^{(0)}$ is to let N_n be the number of values in $\mathbf{x} = (x_1, \ldots, x_n)$ that are greater than $\theta_1^{(0)}$, and accept the hypothesis if

$$\left|N_n - \frac{n}{2}\right| \leq k.$$

It is not hard to get a level α test. If the distribution is symmetric around $\theta_1^{(0)}$, then $\mathsf{X}_k \geq \theta_1^{(0)}$ with probability $\frac{1}{2}$ and N_n is a binomial variable corresponding to the number of successes in n tosses of a fair coin. Hence, under symmetry about $\theta_1^{(0)}$, $\mathsf{N}_n - n/2$ is approximately normal with mean 0 and variance $n/4$, so that $(2\mathsf{N}_n - n)/\sqrt{n}$ is nearly $N(0, 1)$. Take z such that $P(|\mathsf{Z}| \leq z) = \alpha$; then the test S given by

$$\textit{accept if} \quad \left|N_n - \frac{n}{2}\right| \leq \frac{z\sqrt{n}}{2}$$

is at level α. Use this result to get $100\alpha\%$ confidence intervals for θ_1.

(Ans. Let $x_{(1)} \leq \cdots \leq x_{(n)}$ be the order statistics for the sample, and let $k = z\sqrt{n}/2$. Then $100\alpha\%$ intervals are given by

$$J(\mathbf{x}) = (x_{n/2-k}, x_{n/2+k}).$$

Thus, we get $J(\mathbf{x})$ as the interval formed by the kth reading below the median to the kth reading above the median. The basic estimator for θ_1 we are using here thus turns out to be the median.)

Summary

This chapter gives an introduction to hypothesis testing in parametric models. The important things to remember are:

a A hypothesis is a statement that the true state of nature lies in a specified subset Θ_1 of the parameter space.

b A test of a hypothesis is defined by a set S in the space Ω of all possible data vectors. The test results in acceptance if the data vector falls in S.

c The operating characteristic of the test is given by the acceptance probability $\phi(\theta) = P_\theta(S)$.

d The test is at level α if $P_\theta(S) \geq 1 - \alpha$ for all $\theta \in \Theta_1$.

e A test S can detect a parameter value θ outside of Θ_1 at level α if it is a level α test and if $P_\theta(S) \leq \alpha$.

Constructing Tests

a The basic principle is that S should contain all $\mathbf{x}$ with high probability when $\theta \in \Theta_1$, as compared to its probability for $\theta \notin \Theta_1$.

b When Θ has only two points the likelihood ratio test is a best test.

c The likelihood ratio tests are usually good, and if there is a test that is "best" in some standard statistical sense, then often it is the likelihood ratio test. In normally distributed samples, likelihood ratio leads to three "best" tests—Student's t test, the ρ test, and the F test.

d The large sample distribution of -2 (log of the likelihood ratio) has a simple standard form in a smooth problem when Θ_1 is a linear subspace of Θ.

e The likelihood ratio test is asymptotically equivalent to

$$\textit{accept if } \hat{\theta} \in \Theta_2,$$

where $\hat{\theta}$ is the maximum likelihood estimate and Θ_2 is some range (depending on n) of parameter values that includes Θ_1.

f The set Θ_2 is difficult to determine directly when Θ and Θ_1 are multidimensional. When Θ is one-dimensional or the hypothesis is a condition on one coordinate, if $\hat{\boldsymbol{\theta}}$ is asymptotically normal and relatively unbiased, and if the sample size is large, Θ_2 is easily determined.

g Given that $J(\hat{\theta})$ is a $100(1 - \alpha)\%$ confidence interval for θ, the procedure which accepts the hypothesis $\theta = \theta_0$ when $\theta_0 \in J(\hat{\theta})$ is a level α test. Conversely, for every θ_0, given a level α test of $\theta = \theta_0$, we can construct $100(1 - \alpha)\%$ confidence intervals for θ. This extends to intervals and tests on one or several coordinates of a multidimensional parameter problem.

6 Fitting the Underlying Distribution

Introduction

Suppose we have some data $x_1, \ldots, x_n$ which we assume are independently drawn from the same underlying distribution. Before we specify a parametric model, we usually should look at the data and see if they could have come from one of the family of distributions we are intending to use in our model. There are two general methods for doing this. The first approach is through the histogram, which essentially gives an estimate of the density of the underlying distribution. The second is through the sample distribution function, which gives an estimate of the underlying cumulative distribution function.

Example a Here are 50 failure times:

262.8,	1.0,	36.4,	4.0,	59.4,	35.3,	70.5,	22.6,	3.7,	5.8,
32.1,	0.5,	17.4,	77.6,	46.7,	182.4,	76.7,	3.5,	13.4,	29.7,
6.1,	15.1,	110.5,	45.9,	31.7,	22.4,	27.8,	10.0,	33.0,	26.7,
8.0,	6.8,	63.0,	70.9,	30.0,	12.2,	29.6,	3.3,	32.2,	12.3,
128.2,	24.6,	7.0,	39.8,	71.1,	19.4,	5.4,	4.4,	54.4,	24.8.

We first construct a histogram for this data using the intervals [0, 20), [20, 40), [40, 60), [60, 80), [80, 100), [100, ∞). Letting the height of each column be the number of data points in the given interval, and grouping all values over 100 into the 100–120 interval, we get Figure 6.1.

We can compare this histogram with the exponential density in different ways. The sample mean for the data is 41.1. If the data come from an exponential distribution, then the maximum likelihood estimate for β is 41.1, and our best estimate for the density is

$$f(x) = \frac{1}{41.1} e^{-x/41.1}.$$

Standardize the histogram so that the total area under the rectangles is 1; that is, divide all heights by 1000. This standardized graph is the graph of a density function, since the area under the curve is 1. Now superimpose the graph of the exponential density with $\beta = 41.1$ on this graph (see Figure 6.2), and compare the two.

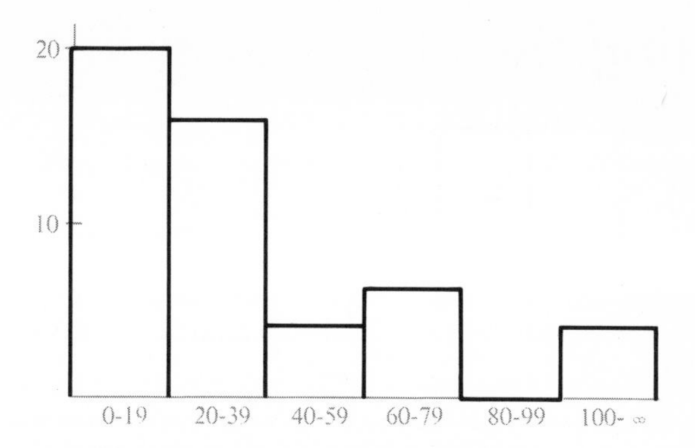

Figure 6.1 *Histogram of failure times in Example* a.

Another method of comparison is the following: For an exponential distribution with $\beta = 41.1$, the probabilities that a sample point will fall into the intervals used in the histogram are

$$
\begin{aligned}
P([0, 20)) &= .385 \\
P([20, 40)) &= .236 \\
P([40, 60)) &= .145 \\
P([60, 80)) &= .089 \\
P([80, 100)) &= .055 \\
P([100, \infty)) &= .088.
\end{aligned}
\tag{6.1}
$$

Form the proportions of the histogram by dividing each height in the histogram of Figure 6.1 by $n = 50$. The new height is the proportion of sample points falling in the corresponding interval and should be an estimate for the corresponding probability in 6.1. In Figure 6.3 we plot the probabilities 6.1 and the proportions histogram.

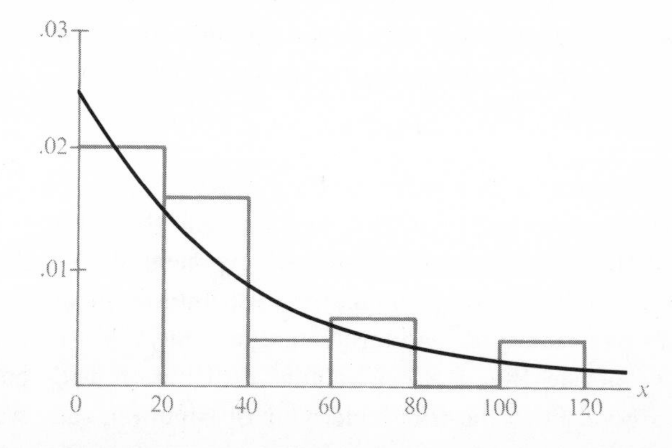

Figure 6.2 *Exponential density with* $\beta = 41.1$ *compared to the standardized histogram of the failure times in Example* a.

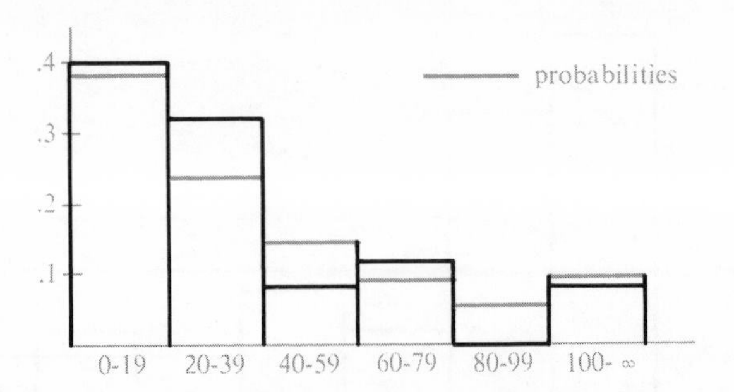

***Figure 6.3** Probabilities computed from an exponential distribution with β = 41.1—proportion of failure times falling into the interval*

Visual comparisons are haphazard. We need some systematic way of deciding whether the given histogram could have been generated by an exponential distribution; that is, we need some way of measuring how far apart the two graphs in Figure 6.3 are.

Other questions that must be answered are: How were the intervals used in constructing the histogram chosen? Why were intervals of width 20 selected? How sensitive will our answer be to the selection of intervals? A method that bypasses the interval selection problem uses the sample distribution function

6.2 $$\hat{F}(x) = \frac{1}{n}(\text{no. of } \{x_1, \ldots, x_n\} < x);$$

$\hat{F}(x)$ is the proportion of sample points in the interval $(-\infty, x)$. For every x, $\hat{F}(x)$ should be a good estimate of $F(x)$, the underlying cumulative distribution function. Hence comparing $\hat{F}(x)$ and a proposed candidate $F(x)$ for the underlying cumulative distribution should give some idea of their agreement or disagreement. Graphed in Figure 6.4 is the sample distribution function for the 50 data points above, and superimposed as a dotted line is the cumulative distribution function

$$F(x) = 1 - e^{-x/41.1}$$

for the exponential distribution with $\beta = 41.1$.

Now the question is: Are these two significantly different? For example, suppose the underlying distribution is exponential with $\beta \simeq 41.1$. How far from the underlying cumulative distribution function $F(x)$ is $\hat{F}(x)$ liable to be because of random fluctuations only?

If we decide that the exponential distribution with parameter 41.1 is not a good fit to the true underlying distribution, then we must decide why. Is it possible that 41.1 is a poor estimate of the parameter? Or perhaps no exponential will give a good fit. If the latter is true, then our

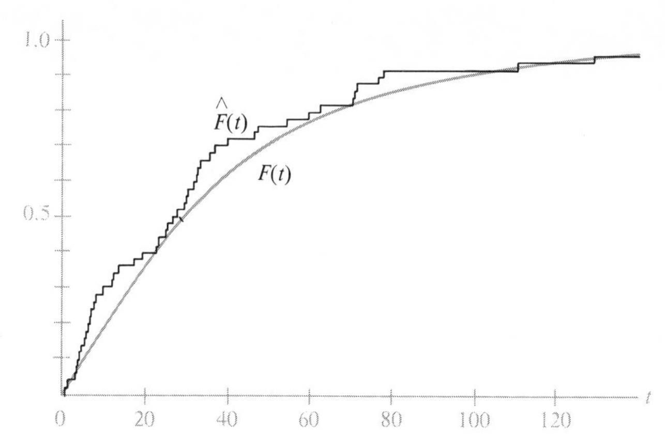

***Figure 6.4** Sample distribution function $\hat{F}(t)$ for the failure time data, and the cumulative distribution function $F(t)$ for the exponential distribution with $\beta = 41.1$*

analysis shifts to finding the significant aspects in which the underlying distribution is not exponential. For example, the distribution might have too long a tail to be exponential, or it might assign too little probability to the interval 60–80.

Example b Here are 10, assumed independent, measurements of the percent by weight of an element in a certain compound:

$$\begin{array}{ccccc} 54.641, & 54.246, & 54.156, & 54.623, & 54.874, \\ 54.774, & 54.504, & 54.468, & 54.029, & 54.926. \end{array}$$

Is it reasonable to assume that they come from some normal distribution? Constructing a sample histogram for only 10 data points hardly makes sense. The sample distribution function is graphed in Figure 6.5, with the graph of the cumulative distribution function of a normal distribution with μ and σ^2 given by the sample mean and unbiased sample variance superimposed.

Furthermore, with only 10 points can one deduce that a normal distribution gives a better fit than almost any other reasonable distribution? For instance, suppose we try to see how a double exponential will do. The sample median is any number between 54.504 and 54.623. By our earlier convention we take the halfway point $\hat{\nu} = 54.56$. The estimate of scale in this family by maximum likelihood is

$$\hat{\beta} = \frac{1}{n} \sum_{1}^{n} |x_k - \hat{\nu}|,$$

and for our sample we get $\hat{\beta} = .24$. The cumulative distribution function for a double exponential with $\alpha = 54.56$, $\beta = .24$ is also shown in Figure 6.5.

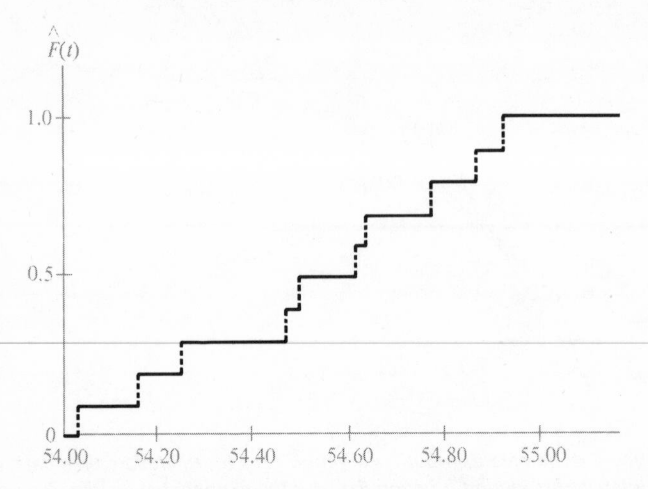

***Figure 6.5** Sample distribution function of the weight measurements in Example* b.

Finally, is it possible that the data points came from a distribution uniform on $[m - (l/2), m + (l/2)]$? The maximum likelihood estimates are

$$\hat{m} = \frac{\max(x_1, \ldots, x_n) + \min(x_1, \ldots, x_n)}{2}$$

$$\hat{l} = \max(x_1, \ldots, x_n) - \min(x_1, \ldots, x_n).$$

For our sample, $\hat{m} = 54.48$ and $\hat{l} = .91$. To get an unbiased estimate, take $(n + 1)/n \cdot \hat{l} = 1.0$ as the estimate of the scale l. The graph of the uniform cumulative distribution function with $m = 54.48$ and $l = 1.0$ is also shown in Figure 6.5. Looking at Figure 6.5, we can see that the sample distribution function differs much more from any of the three given cumulatives than they differ between themselves.

Assuming that the underlying distribution is normal, computing exact 90% confidence intervals for μ using the t distribution gives $54.53 \pm .18$. If the distribution is assumed uniform, then an exact 90% confidence procedure for m gives $54.48 \pm .15$.

These are two different, but not inconsistent, possible locations for the midpoint of the distribution. We could get many other possible locations by assuming other underlying distributions. But can we show, on the basis of the data, that the normal distribution gives a better fit than the uniform or double exponential? And if we can't do it with a sample of size 10, how large a sample size do we need? This last question brings us back to the idea of detectability.

Example c Sixty pairs of one-minute counts recorded in adjacent lanes of a freeway are listed below:

lane 1 (inner)	24	30	32	29	37	31	29	40	22	32	26	23	22	35	31	36	34
lane 2 (outer)	13	16	19	12	17	19	18	19	18	13	20	20	17	18	16	19	18

1	20	30	31	26	24	29	20	35	33	24	33	44	22	33	35	20	18	44	26	44	24
2	18	20	22	20	17	19	16	16	18	19	18	22	24	19	24	13	19	26	17	18	15

1	22	37	35	26	35	38	34	26	31	36	39	25	15	38	32	33	37	23	25	34	26	25
2	17	19	15	22	20	22	19	16	15	16	23	19	12	22	20	17	22	19	14	17	18	15

On the basis of some theoretical considerations, it is assumed that these pairs were drawn from a bivariate normal distribution. Assuming successive pairs are independent samples, how can we use the data to decide if normality is a reasonable assumption?

For a bivariate sample, there is nothing like the histogram or sample distribution function that can be easily graphed and examined. Of course, we can define a 2-dimensional histogram by breaking up the plane into rectangles or squares $R_1, \ldots, R_n$ and counting the data pairs (x_1, y_1) that fall into each of the rectangles.

One simple approach is to make a histogram of the counts in the first lane only and another histogram of the counts in the second lane. The maximum likelihood estimates of the 5 parameters in a bivariate normal distribution are computed from the pairs of counts and yield $\mu_1 = 29.8$, $\sigma_1^2 = 57$, $\mu_2 = 18.2$, $\sigma_2^2 = 11$, $\rho = .44$. Therefore, the first histogram should be close to an $N(29.8, 57)$ distribution and the second close to an $N(18.2, 11)$ distribution. If one or the other is not adequately fitted by a normal distribution, then certainly the joint distribution cannot be bivariate normal. If both histograms are close to normal, we have strong but not conclusive evidence that the joint distribution is close to bivariate normal.

Another approach is this: Define the 2-dimensional sample distribution function $\hat{F}(x, y)$ as the proportion of the pairs whose first coordinate is less than x and whose second coordinate is less than y. This should give an estimate for the cumulative joint distribution function of the underlying distribution. Hence some measure of the discrepancy between $\hat{F}(x, y)$ and the joint distribution function for a bivariate normal with the parameters as estimated above should give us a quantitative idea of the goodness of fit. While similar in spirit to the 1-dimensional approach using the sample distribution function, this is much more complicated and involves a long and tedious computation. The bivariate problem seems inherently more difficult than the univariate.

The Histogram and a Touch Estimate of the Density

Suppose you have data $x_1, \ldots, x_n$ drawn independently from some underlying distribution. Often what you want first is a rough picture of the distribution. If the distribution is discrete, then there is an obvious way to do this. Suppose the data take only the integer values $0, 1, 2, \ldots$. Let $\hat{n}_j$ be the number of times that we get outcome j. Equivalently, $\hat{n}_j$ is the number of $x_1, \ldots, x_n$ equal to j. Then $\hat{p}_j = \hat{n}_j/n$ is the maximum likelihood estimate for p_j, the probability of outcome j. To get an idea of what the distribution with probabilities $p_1, p_2, \ldots$ looks like, simply plot a line graph of $\hat{n}_j$ as a function of j, or normalize these and plot $\hat{n}_j/n$. For instance, counting particles over 1000 intervals of one-minute duration gave the data:

Number of times j particles occurred

j	0	1	2	3	4	5	6	7	8
$\hat{n}_j$	381	360	173	65	13	15	2	0	1

The graph of $\hat{p}_j$ is seen in Figure 6.6.

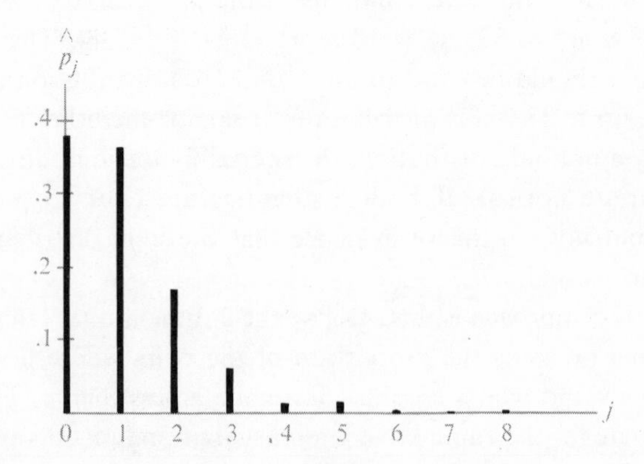

***Figure 6.6** Estimates for the probability of j particles*

Suppose we are trying to estimate p_j for only one j, say $j = 3$. Then we have the biased coin situation. Each outcome is either 3 or not 3 with probabilities p_3, $1 - p_3$, respectively. The estimate for p_3 is $\hat{p}_3 = \hat{n}_3/n$, with squared error loss

$$E(\hat{p}_3 - p_3)^2 = \frac{p_3(1 - p_3)}{n}.$$

Let $J(\hat{p}_3)$ be a $100\gamma\%$ confidence interval for p_3. For $\hat{n}_3 \gg 1$, we know that we can take

$$J(\hat{p}_3) = \hat{p}_3 \pm z\sqrt{\frac{\hat{p}_3(1 - \hat{p}_3)}{n}}, \tag{6.3}$$

where z is determined from an $N(0, 1)$ table. Assume that $\sqrt{1 - \hat{p}_3} \simeq 1$; then we can write

$$J(\hat{p}_3) = \hat{p}_3\left(1 \pm \frac{z}{\sqrt{\hat{p}_3 n}}\right) = \hat{p}_3\left(1 \pm \frac{z}{\sqrt{\hat{n}_3}}\right). \tag{6.4}$$

Even if $\hat{n}_3$ is not large, 6.4 gives answers within an order of magnitude to the question of how much the underlying probabilities $p_1, p_2, \ldots$ can differ from $\hat{p}_1, \hat{p}_2, \ldots$. Expression 6.4 gives a simple prescription—the confidence interval around $\hat{p}_j$ for any j is $\pm 100z/\sqrt{\hat{n}_j}\%$ of $\hat{p}_j$. Taking $z = 2$ for 95% intervals, these limits for the above particle data are

j	$J(\hat{p}_j)$
0	.381(1 ± .10)
1	.360(1 ± .11)
2	.173(1 ± .15)
3	.065(1 ± .25)
4	.013(1 ± .55)
≥ 5	.008(1 ± .79).

(We grouped all outcomes $j \geq 5$.) The accuracy of the last two confidence intervals is very suspect because $\hat{n}_4$ and $\hat{n}_5$ are too small for the normal approximation. Notice how rapidly the percent error goes up as $\hat{p}_j$ goes down. The smaller the probability, the harder it is to estimate it. On the other hand, 10% error in .381 is $\simeq .04$, but 80% error in .008 is $\simeq .006$. The smaller the estimated probability, the smaller the confidence interval around it. Now we plot these intervals in a line graph similar to Figure 6.6 (see Figure 6.7).

Actually, this is a situation in which simultaneous confidence intervals for $p_1, \ldots, p_5$ are more useful than individual intervals.

From Chapter 4, page 107, we know that if there are a total of J outcomes and we construct $100\gamma\%$ intervals around each $\hat{p}_j$, then the

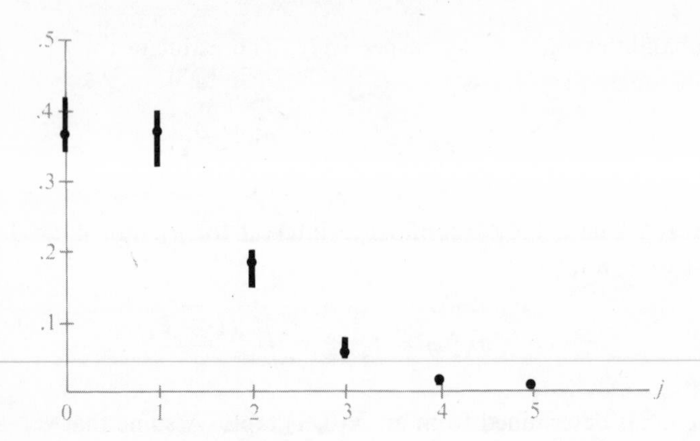

Figure 6.7 *95% confidence intervals for the probability of j particles; large dots are the Poisson probabilities for $\lambda = 1.00$*

probability is at least $1 - (1 - \gamma)J$ that $p_j \in J(\hat{p}_j)$ for every j. If we select z above so as to get 99% intervals, then the probability that every interval contains the underlying probability for that outcome is at least .94.

If we suspect that the Poisson distribution makes a good model for the data, we estimate λ by forming the sample mean. This computation gives $\hat{\lambda} = 1.00$, and the corresponding probabilities are

j	0	1	2	3	4	≥ 5
p_j	.368	.368	.184	.061	.015	.004.

These are plotted as large dots in Figure 6.7. They all fall into the corresponding confidence interval, and for this reason we conclude that the Poisson is a possible candidate for the underlying distribution.

We can set up a similar procedure for data from a continuous distribution by making the data discrete. Divide the range of the variable into disjoint intervals (or "cells") $I_1, \ldots, I_J$. If the range of the variable is infinite, these intervals cannot all be the same length; at least one and possibly two intervals must be infinite. It is usually convenient, but not necessary, to take all finite intervals to be of the same length. Then the histogram is plotted by letting the number of the data points $\hat{n}_j$ in the jth interval be the height of the column over that interval. The selection of cells converts this continuous distribution into a multinomial distribution with J outcomes and probabilities $p_j = P(I_j)$, $j = 1, \ldots, J$, where $P(dx)$ is the underlying continuous distribution. Again, $\hat{n}_j/n = \hat{p}_j$ is a good estimate for $P(I_j)$. Let $f(x)$ denote the density of $P(dx)$ and assume $f(x)$ is continuous. By the Mean Value theorem for integrals,

$$P(I_j) = \int_{I_j} f(x)\,dx = f(x_j)\|I_j\|, \qquad \text{for some } x_j \in I_j,$$

where $\|I_j\|$ denotes the length of I_j. So $\hat{n}_j/n\|I_j\|$ is an estimate of $f(x)$ for some x in I_j. With intervals of equal length, $\hat{n}_j$ is proportional to the estimate, and in this case the histogram should look something like the graph of $f(x)$. But if the lengths are not equal, we can get a picture of $f(x)$ by plotting $\hat{n}_j/\|I_j\|$ as the height over the jth interval. This gives a graph whose heights are roughly proportional to $f(x)$. By plotting

$$\hat{f}_j = \frac{\hat{n}_j}{n\|I_j\|}$$

for $x \in I_j$, we get a histogram that is a direct estimate of $f(x)$.

For n_j large enough, confidence intervals for $P(I_j)$ are the same as in the discrete case. Let $\hat{p}_j = \hat{n}_j/n$; then for $\hat{n}_j$ large enough, the inequality

6.5 $$\hat{p}_j\left(1 - \frac{z}{\sqrt{\hat{n}_j}}\right) \leq P(I_j) \leq \hat{p}_j\left(1 + \frac{z}{\sqrt{\hat{n}_j}}\right)$$

holds with probability at least γ, where z is computed from $N(0, 1)$ tables. But since

$$\|I_j\| \min_{x \in I_j} f(x) \leq P(I_j) \leq \|I_j\| \max_{x \in I_j} f(x),$$

6.5 implies that

6.6

a $$\min_{x \in I_j} f(x) \leq \hat{f}_j\left(1 + \frac{z}{\sqrt{\hat{n}_j}}\right)$$

b $$\max_{x \in I_j} f(x) \geq \hat{f}_j\left(1 - \frac{z}{\sqrt{\hat{n}_j}}\right).$$

Therefore both of the above must hold with probability at least γ. If 6.6 holds, then for $x \in I_j$, $f(x)$ cannot always lie above the value $\hat{f}_j(1 + z/\sqrt{\hat{n}_j})$, because then a. would be violated. Similarly, it cannot always lie below $\hat{f}_j(1 - z/\sqrt{\hat{n}_j})$ without violating b. Therefore, if we draw the bar above the interval I_j that stretches from height $\hat{f}_j(1 - z/\sqrt{\hat{n}_j})$ to height $\hat{f}_j(1 + z/\sqrt{\hat{n}_j})$, then the graph of $f(x)$ must touch this bar at least once in the interval. It does not have to remain inside the bar for all $x \in I_j$; all it has to do is touch at least once (see Figure 6.8).

Now, above every interval $I_j, j = 1, \ldots, J$, draw these $100\gamma\%$ confidence bars. The graph of the underlying density touches each of these touch confidence bars with probability at least γ. By the same argument used in the discrete case, it touches all of them with probability at least $1 - (1 - \gamma)J$.

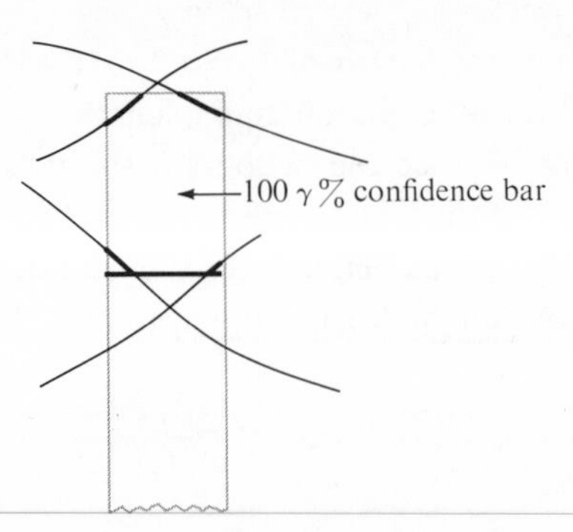

***Figure 6.8** Possible ways that the graph of the density f(x) can touch the 100γ% confidence bar*

How many cells $I_1, \ldots, I_J$ should we use; how large should they be? If there are too few cells, information is lost; if there are too many cells, information is not effectively conveyed. For instance, suppose we have a sample size of 1000 and use 4 cells $(-\infty, -2]$, $(-2, 0]$, $(0, 2]$, $(2, \infty)$ with corresponding $\hat{n}_1 = 16$, $\hat{n}_2 = 468$, $\hat{n}_3 = 513$, $\hat{n}_4 = 3$. Obviously, the histogram constructed using these cells gives very little information about the shape of $f(x)$. On the other hand, if we have a cell for which $\hat{n}_j \leq 10$, then the height of the touch confidence bar is of about the same order as the histogram height. If too many of the cells have frequencies $\hat{n}_j \leq 10$, then we have constructed a histogram where the "noise" or fluctuation level is the same order of magnitude as the meaningful "signal," and we get very little useful information out of it.

There is a secondary reason for using cells large enough so that $\hat{n}_j \geq 10$ for all j, or at least $\hat{n}_j \geq 5$: then the central limit approximations to the confidence intervals for $P(I_j)$ are fairly good, and we can use the simple expressions given above for the touch confidence bars. On the other hand, if there are cells such that n_j is large, say $n_j \geq 100$, then the possibility of getting more detailed information about $f(x)$ by subdividing these cells should be considered.

Problem 1 For the purposes of constructing a histogram, data are grouped into the intervals $(-\infty, -4)$, $[-4, -3)$, $[-3, -2)$, $[-2, -1)$, $[-1, 0)$, $[0, 1)$, $[1, 2)$, $[2, \infty)$. Say there are 3 points in $(-\infty, -4)$ and 5 in $[2, \infty)$. What would you do with these points when plotting the histogram? What procedures would you use and why? What would be the height of the histogram for $x < -4$ or $x \geq 2$?

Problem 2 The following table gives the frequency distribution

Cell	< 0	[0, 1)	[1, 2)	[2, 3)	[3, 4)
$\hat{n}_j$	1	8	26	58	99

Cell	[4, 5)	[5, 6)	[6, 7)	[7, 8)	[8, 9)
$\hat{n}_j$	135	153	149	126	95

Cell	[9, 10)	[10, 11)	[11, 12)	[12, 13)	≥ 13
$\hat{n}_j$	65	40	23	12	6

of certain sample data when grouped into the given cells. Assume that the sample consists of independent outcomes from the same distribution. Graph the $\hat{f}_j$ histogram and the 99% touch confidence bars. Compute an approximate sample mean $\hat{\mu}$ and sample variance $\hat{\sigma}^2$ from the grouped data. Plot the density of the $N(\hat{\mu}, \hat{\sigma}^2)$ distribution on the same graph. What do you conclude?

The χ^2 Test

Confidence intervals and touch bars are exploratory tools, and are not suited to give a single overall measure of how well an assumed distribution or family of distributions fits the histogram data. The χ^2 (chi-squared) test is tailored for this purpose.

Suppose the data $x_1, \ldots, x_n$ consist of independent repetitions of an experiment with J possible outcomes $1, \ldots, J$. As the simplest goodness of fit problem, suppose we think that the underlying distribution is $\mathbf{p}^{(0)} = (p_1^{(0)}, \ldots, p_J^{(0)})$; for example, suppose we think that all outcomes are equally likely. How can we use the data to check our assumption?

This may be called a goodness of fit problem, but it is in fact a hypothesis testing problem. The question is: Does $\mathbf{p} = \mathbf{p}^{(0)}$? To formalize the model, take Θ to be the set of all possible probabilities for a J-outcome experiment. That is,

$$\Theta = \left\{\mathbf{p};\ p_1 \geq 0, \ldots, p_J \geq 0, \quad \sum_1^J p_j = 1\right\},$$

and the set of underlying probabilities is given by

$$P_{\mathbf{p}}(x_1, \ldots, x_n) = p_1^{\hat{n}_1} \cdots p_J^{\hat{n}_J},$$

where $\hat{n}_j$ is the number of times that the jth outcome appears in the sequence $x_1, \ldots, x_n$.

We know, in principle, how to construct a test of the hypothesis $\mathbf{p} = \mathbf{p}^{(0)}$. We take a good estimate $\hat{\mathbf{p}}$ of $\mathbf{p}$, and let $D(\hat{\mathbf{p}}, \mathbf{p}^{(0)})$ be some measure of the distance between $\hat{\mathbf{p}}$ and $\mathbf{p}^{(0)}$. Reject the hypothesis if D is too large; otherwise, accept. Recall that for the multinomial distribution the maximum likelihood estimate for $\mathbf{p}$ is

$$\hat{p}_j = \frac{\hat{n}_j}{n}, \qquad j = 1, \ldots, J.$$

A convenient measure of the distance between $\hat{\mathbf{p}}$ and $\mathbf{p}^{(0)}$ is a weighted sum of squares

6.7
$$D = \sum_1^J \lambda_j (\hat{p}_j - p_j^{(0)})^2.$$

Now we want to find the distribution of D under $\mathbf{p}^{(0)}$. Our aim is to fix the level α and find δ such that

$$P_{\mathbf{p}^{(0)}}(\mathsf{D} \leq \delta) = 1 - \alpha.$$

Then accept the hypothesis if $D \leq \delta$; otherwise, reject.

Fix j and think of each repetition of the experiment as producing outcome j or outcome *not* j. This reduces the problem to a biased coin problem with probability p_j of heads and $q_j = 1 - p_j$ of tails. In this context, we know that $\hat{p}_j$ is an asymptotically relatively unbiased and normal estimate for p_j with variance $p_j q_j / n$. Hence, if $\mathbf{p}^{(0)}$ is the underlying probability distribution, then

$$\mathsf{X}_j = \sqrt{n}\, \frac{\hat{\mathsf{p}}_j - p_j^{(0)}}{\sqrt{p_j^{(0)} q_j^{(0)}}}$$

is approximately $N(0, 1)$. Naturally, we want to select the constants λ_j to make the distribution of D as simple as possible. If we take

$$\lambda_j = \frac{n}{p_j^{(0)} q_j^{(0)}},$$

then

$$\mathsf{D} = \sum_1^J \mathsf{X}_j^2.$$

If $\mathsf{X}_1, \ldots, \mathsf{X}_J$ were independent, then this selection would give D a χ_J^2 distribution. Unfortunately, they are not independent, since the identity

$$\hat{\mathsf{n}}_1 + \cdots + \hat{\mathsf{n}}_J = n$$

implies that $X_1, \ldots, X_J$ are linearly dependent. However, by an argument similar to that used in Chapter 3, Problem 24, we can show that if we take

$$\lambda_j = \frac{n}{p_j^{(0)}},$$

then the resulting D has a χ^2 distribution with $J - 1$ degrees of freedom. We formalize this important result [drop the superscript (0)]:

6.8 ***Theorem*** *Under the probability distribution $(p_1, \ldots, p_J)$, for large n*

$$D = \sum_1^J \frac{n(\hat{p}_j - p_j)^2}{p_j} = \sum_1^J \frac{(\hat{n}_j - np_j)^2}{np_j}$$

has approximately a χ^2_{J-1} distribution, and the test based on D is called the χ^2 test.

The proof is interesting and not difficult. The outline is given in Problem 6 below. The usual nomenclature is to call the $\hat{n}_j$ the observed frequencies and $n_j = np_j$ the expected frequencies. If p_j is the underlying probability of outcome j, then clearly $E\hat{n}_j = n_j$. Now write D as

6.9
$$D = \sum_1^J \frac{(\hat{n}_j - n_j)^2}{n_j}.$$

Example d A group of students observed a roulette wheel continuously for two days for a total of 4000 games. They recorded the following frequencies for the 36 red and black numbers and for zero and double zero;

number	1	2	3	4	5	6	7	8	9	10
red	121	89	123	112	130	130	113	118	95	88
black	107	92	89	119	117	131	94	106	108	91

number	11	12	13	14	15	16	17	18	0	00
red	113	125	107	90	97	93	98	105	109	111
black	97	100	102	81	92	108	97	102		

The problem is to decide whether there is any detectable bias in this wheel. Take absence of bias as the hypothesis. Then by (6.9)

$$D = \sum_1^{38} \frac{(\hat{n}_j - 105)^2}{105},$$

where $105 = 4000/38 = n_j$. Substituting the observed frequencies gives $D = 61$. This value should be compared with a test computed from a χ^2_{37} distribution. That is, take δ so that for C having a χ^2_{37} distribution,

$$P(C \leq \delta) = 1 - \alpha.$$

However, the χ^2 table only goes up to 30 degrees of freedom, for the following reason: Let $X_1, \ldots, X_k$ be independent $N(0, 1)$ variables. Then the χ^2_m sum with m degrees of freedom

$$C = \sum_1^m X_k^2$$

has mean m and variance $2m$. But C is a sum of independent identically distributed random variables. The standardized sum

$$\frac{C - m}{\sqrt{2m}}$$

has for $m \geq 30$ a distribution that is adequately approximated by $N(0, 1)$.
Fix z so that for X an $N(0, 1)$ variable, $P(Z \leq z) = 1 - \alpha$. Then

$$P(C_m \leq m + z\sqrt{2m}) = 1 - \alpha.$$

Therefore, the acceptance region for D when there are J outcomes, $J \geq 30$, is

$$D \leq (J - 1) + z\sqrt{2(J - 1)}. \tag{6.10}$$

At the 5% level, $z = 1.65$. For the roulette wheel $J = 38$, and the acceptance region is

$$D \leq 37 + 1.65\sqrt{2 \cdot 37} = 51.2.$$

We reject the hypothesis of no bias at the 5% level. Now the question is: Where is the bias? What should be done is to construct the line graph of the $\hat{p}_j$. This is assigned as Problem 3.

The way we have gone about setting up this test was somewhat arbitrary. First, we decreed that we would take the distance between $\mathbf{p}$ and $\mathbf{p}^{(0)}$ as a weighted sum of squares. Then we selected the weights "by convenience" to give D a simple distribution. It is comforting to know that if we start with the likelihood ratio test for the hypothesis $\mathbf{p} = \mathbf{p}^{(0)}$ and make some reasonable approximations, we get the χ^2 test. Furthermore the fact that under the null hypothesis D has a χ^2_{J-1} distribution is actually a consequence of Theorem 5.25 on the large sample size distribution of l. In this light, the χ^2 test has a firmer, less arbitrary foundation. (See Problem 8.)

Problem 3 Draw a line graph of $\hat{p}_j$ for the roulette wheel data of Example **d.** above. Noticing that $\sqrt{\hat{n}_j}$ varies from 9 to about 11.5, take $1/\sqrt{\hat{n}_j} \simeq .1$ and draw in the 95% confidence intervals around every $\hat{p}_j$. What can you conclude?

Problem 4 The first 608 digits in the decimal expansion of π have the following frequencies:

0	1	2	3	4	5	6	7	8	9
60	62	67	68	64	56	62	44	58	67.

How do the data square with the assumption that each digit is equally likely? (Assume independence!)

Problem 5 In a famous experiment, Gregor Mendel observed the shape and color of peas that resulted from certain crossbreedings. The results of one such experiment were

round yellow	315
round green	108
wrinkled yellow	101
wrinkled green	32.

According to his theory, the frequencies should be in the ratio 9:3:3:1. What would he have concluded from a χ^2 test?

Problem 6 To test a coin for bias, toss it n times and let $\hat{n}_1$, $\hat{n}_2$ be the number of heads and tails, respectively. To analyze the results we have the method given in the last chapter: Take $\hat{p}_1 = \hat{n}_1/n$; accept the hypothesis of unbiasedness if $|\hat{p}_1 - \frac{1}{2}| \leq \gamma$ and reject otherwise, where γ is selected to give a level α test. But we also have the method of this section: Let

$$D = \frac{(\hat{n}_1 - n/2)^2}{n/2} + \frac{(\hat{n}_2 - n/2)^2}{n/2}$$

and reject if $D \geq \delta$, where δ is fixed so as to give a level α test. What is the relationship between these two tests?

Problem 7 (Difficult!) To show that the limiting distribution of D is χ^2 with $J - 1$ degrees of freedom, define

$$\mathsf{X}_m^{(j)} = \begin{cases} 1, & \text{if the outcome of the } m\text{th experiment is } j \\ 0, & \text{if the outcome of the } m\text{th experiment is not } j. \end{cases}$$

Then the vector variables

$$\mathbf{X}_1 = (\mathsf{X}_1^{(1)}, \ldots, \mathsf{X}_1^{(J)}), \ldots, \mathbf{X}_m = (\mathsf{X}_n^{(n)}, \ldots, \mathsf{X}_n^{(J)})$$

are independent, and

$$\hat{\mathsf{n}}_j = \mathbf{X}_1^{(j)} + \cdots + \mathbf{X}_n^{(j)}, \qquad j = 1, \ldots, J.$$

Notice that

$$E\mathsf{X}_m^{(j)} = p_j, \qquad V(\mathsf{X}_m^{(j)}) = p_j(1 - p_j).$$

Define $\mathbf{Y} = (\mathsf{Y}_1, \ldots, \mathsf{Y}_J)$ by

$$\mathsf{Y}_j = \frac{\hat{\mathsf{n}}_j - np_j}{\sqrt{n}}, \qquad j = 1, \ldots, J$$

or

$$\mathbf{Y} = \frac{1}{\sqrt{n}} \sum_1^n (\mathbf{X}_k - \boldsymbol{\mu}).$$

Since $\mathbf{Y}$ is a sum of small independent vector variables, by the multidimensional central limit theorem $\mathsf{Y}_1, \ldots, \mathsf{Y}_J$ have a joint normal distribution. The covariance matrix is given by

$$\begin{aligned} E\mathsf{Y}_i\mathsf{Y}_j &= E(\mathsf{X}_1^{(i)} - p_i)(\mathsf{X}_1^{(j)} - p_j) \\ &= \begin{cases} -p_i p_j, & i \neq j \\ p_j(1 - p_j), & i = j. \end{cases} \end{aligned}$$

Now define a new set of variables $\mathsf{U}_1, \ldots, \mathsf{U}_J$ by

$$\mathsf{U}_j = \frac{\mathsf{Y}_j}{\sqrt{p_j}}.$$

Select a set of orthogonal unit vectors $\mathbf{e}_1, \ldots, \mathbf{e}_J$ by taking

$$\mathbf{e}_J = (\sqrt{p_1}, \ldots, \sqrt{p_J})$$

and $\mathbf{e}_1, \ldots, \mathbf{e}_{J-1}$ to be any other unit vectors orthogonal to $\mathbf{e}_J$ and to each other. Define $\mathsf{Z}_1, \ldots, \mathsf{Z}_J$ by

$$\mathsf{Z}_j = \mathbf{e}_j \cdot \mathbf{U}.$$

Show that

a
$$\mathsf{Z}_J = 0,$$

b
$$E\mathsf{Z}_i\mathsf{Z}_j = \begin{cases} 0, & i \neq j \\ 1, & i = j, \end{cases}$$

and use **a.** and **b.** to show that 6.8 holds.

Problem 8 **a** For the likelihood ratio test for the hypothesis $\mathbf{p} = \mathbf{p}^{(0)}$, show that

$$l(\mathbf{x}) = -2 \sum_{j=1}^{J} \hat{n}_j \log\left(\frac{p_j^{(0)}}{\hat{p}_j}\right).$$

b If $\hat{\mathbf{p}} = (\hat{p}_1, \ldots, \hat{p}_J)$ is not too far from $\mathbf{p}^{(0)}$, write

$$p_j^{(0)} = \hat{p}_j + (p_j^{(0)} - \hat{p}_j)$$

and use the partial Taylor expansion $\log(1 + x) \simeq x - x^2/2$ to show that

$$l(\mathbf{x}) \simeq \sum_{1}^{J} \frac{n(\hat{p}_j - p_j^{(0)})^2}{\hat{p}_j};$$

or, using $\hat{p}_j \simeq p_j^{(0)}$ in the denominator,

$$l(\mathbf{x}) = \sum_{1}^{J} \frac{n(\hat{p}_j - p_j^{(0)})^2}{p_j^{(0)}}.$$

Notice that the right-hand side is exactly the D function of the χ^2 test.

c Discuss why a consequence of **b.** is that the acceptance regions given by the likelihood ratio test and by the χ^2 test will be nearly the same for large sample size.

d How can Theorem 5.25 now be used to conclude that D has a χ^2_{J-1} distribution under $P_{\mathbf{p}^{(0)}}$?

Maximum Likelihood Estimates and χ^2 Tests

The usual question in analyzing data is not whether the data can be fit by one specified distribution, but rather, if they can be fit by one of a class of distributions. For example, could the data have come from *some* Poisson distribution or *some* normal distribution?

Suppose we have counted particles over 1000 intervals of one-minute duration, and the numbers of times that 0, 1, 2, 3, 4, or 5 or more particles occurred are as follows:

Number of Times k Particles Occurred

k	0	1	2	3	4	≥ 5
$\hat{n}_k$	381	360	173	65	13	8

The question is: Could this data have been generated by sampling from a Poisson distribution? For any λ, suppose we compute the deviation $D(\lambda)$ between the given data and the Poisson probabilities

$$p_0(\lambda), p_1(\lambda), \ldots, p_5(\lambda),$$

where

$$p_k(\lambda) = \frac{\lambda^k}{k!} e^{-\lambda}, \qquad k = 0, 1, 2, 3, 4,$$

and

$$p_5(\lambda) = 1 - p_0(\lambda) - \cdots - p_4(\lambda).$$

Thus,

$$D(\lambda) = \sum_{k=0}^{5} \frac{[\hat{n}_k - np_k(\lambda)]^2}{np_k(\lambda)}, \qquad n = 1000.$$

To see if there is a good fit to *some* Poisson distribution, we want to look at the minimum value of $D(\lambda)$ over all $\lambda \geq 0$. That is, we want to compare the data with that Poisson distribution that fits best. So we look for the value of λ which minimizes $D(\lambda)$. Expanding the square in the numerator of $D(\lambda)$, we get

$$D(\lambda) = \sum_k \frac{\hat{n}_k^2}{np_k(\lambda)} - 2 \sum_k \hat{n}_k + \sum_k np_k(\lambda).$$

Since $\sum_k \hat{n}_k = n$ and $\sum_k p_k(\lambda) = 1$,

$$D(\lambda) = \sum_k \frac{\hat{n}_k^2}{np_k(\lambda)} - n.$$

The minimizing λ is a solution of

$$\frac{dD(\lambda)}{d\lambda} = -\sum_k \frac{\hat{n}_k^2}{np_k^2(\lambda)} \frac{d}{d\lambda} [p_k(\lambda)] = 0.$$

Write this as

6.11
$$\sum_k \frac{\hat{n}_k^2}{np_k(\lambda)} \frac{d}{d\lambda} [\log p_k(\lambda)] = 0.$$

If the underlying distribution is Poisson, $\hat{n}_k/n \simeq p_k(\lambda^*)$, where λ^* is the true value of λ. Assuming that the λ that gives the best fit, i.e. minimizes $D(\lambda)$, is close to λ^*, then, everywhere in the vicinity of the minimizing λ,

$$\frac{\hat{n}_k}{np_k(\lambda)} \simeq 1.$$

Therefore, 6.11 can be approximated in this vicinity by

6.12
$$\sum_k \hat{n}_k \frac{d}{d\lambda} \log p_k(\lambda) = 0.$$

Of course, we want to be able to show that the solution of the approximate equation 6.12 is close to the minimizing value of λ. This turns out to be true.

Now recognize 6.12 as the equation $dL/d\lambda = 0$, where L is the log of the probability

$$p_0(\lambda)^{\hat{n}_0} p_1(\lambda)^{\hat{n}_1} \cdots p_5(\lambda)^{\hat{n}_5}$$

of getting $\hat{n}_0$ outcomes 0, $\hat{n}_1$ outcomes 1, . . . , $\hat{n}_5$ outcomes 5 or more. We have therefore reached the gratifying conclusion that *the λ which minimizes $D(\lambda)$ is the maximum likelihood estimator for λ* (approximately, for large n).

The maximum likelihood estimator for λ in this case is not quite the same as the maximum likelihood estimator for the Poisson parameter, because we have grouped together all particle counts of 5 or more into one category. Ordinarily, the maximum likelihood estimator for the Poisson parameter is the number of particles counted per unit of time. If $\hat{n}$ is the total particle count, we would estimate $\hat{\lambda} = \hat{n}/n$. If we retain the original counts $\hat{n}_0, \hat{n}_1, \ldots$ instead of taking $\hat{n}_5$ to be the total number of times that we observed 5 or more particles, then the total particle count is

$$1 \cdot (\text{No. of times 1 particle occurred})$$
$$+ \; 2 \cdot (\text{No. of times 2 particles occurred}) + \cdots$$

or

$$\hat{n} = 1 \cdot \hat{n}_1 + 2 \cdot \hat{n}_2 + \cdots.$$

However, if we put

$$\hat{n} = 1 \cdot \hat{n}_1 + 2 \cdot \hat{n}_2 + \cdots + 5 \cdot \hat{n}_5,$$

then since $\hat{n}_5$ may include one occurrence when we had, for example, an 8-particle count, $\hat{n}$ underestimates the total particle count. Thus $\hat{n}/n$ will underestimate λ. However, ignoring this, the maximum likelihood estimate of λ is .993; if we solve the exact maximum likelihood equations for λ, we get .994. With the latter value of λ we get

$$p_0 = .369, \qquad p_1 = .367, \qquad p_2 = .184,$$
$$p_3 = .061, \qquad p_4 = .015, \qquad p_5 = .004.$$

So D is

$$\frac{12^2}{369} + \frac{7^2}{367} + \frac{11^2}{187} + \frac{4^2}{61} + \frac{2^2}{15} + \frac{4^2}{4} = 5.72.$$

Using $\hat{\lambda}$, the expression for D is

$$D(\hat{\lambda}) = \sum_{k=0}^{5} \left(\frac{\hat{n}_k - np_k(\hat{\lambda})}{\sqrt{np_k(\hat{\lambda})}} \right)^2.$$

The $p_k(\hat{\lambda})$ are now functions of the observations, not fixed constant probabilities, and thus they are random variables. So there is no reason to believe that $D(\hat{\lambda})$ is χ_5^2. In fact, if we use the best estimates

$$\hat{p}_k = \frac{\hat{n}_k}{n}, \qquad k = 0, \ldots, 5,$$

for the probabilities p_k selected from the unrestricted 5-dimensional space

$$(p_0, p_1, \ldots, p_5), \qquad p_0 + p_1 + \cdots + p_5 = 1,$$

then

$$D(\hat{p}_1, \ldots, \hat{p}_5) = \sum_{k=0}^{5} \frac{(\hat{n}_k - n\hat{p}_k)^2}{\hat{n}p_k} \equiv 0.$$

It is the following remarkable result that makes the use of χ^2 so universal.

6.13 ***Theorem*** *Let $\hat{n}_1, \ldots, \hat{n}_J$ be the number of outcomes $1, \ldots, J$ in n independent identical repetitions of an experiment with J outcomes. If the true underlying probability $(p_1^{(0)}, \ldots, p_J^{(0)})$ is one of the set*

$$\{[p_1(\theta), \ldots, p_J(\theta)]\},$$

where θ is an M-dimensional parameter $\theta = (\theta_1, \ldots, \theta_M)$, then for n large, the value of

$$D(\theta) = \sum_{j=1}^{J} \frac{[\hat{n}_j - np_j(\theta)]^2}{np_j(\theta)},$$

minimized over all θ in the M-dimensional parameter space Θ has a χ_{J-M-1}^2 distribution.

Since maximum likelihood estimates give the minimizing parameter values (approximately), another form of Theorem 6.13 is: *Let $\hat{\theta}$ be the maximum likelihood estimator for θ; then for n large*

$$D(\hat{\theta}) = \sum_{j=1}^{J} \frac{[\hat{n}_j - np_j(\hat{\theta})]^2}{np_j(\hat{\theta})}$$

has a χ_{J-M-1}^2 distribution. We drop a degree of freedom for every real-valued parameter we estimate. In our Poisson example, $J = 6$. We estimate the 1-dimensional parameter λ. So $D(\hat{\lambda})$ should have a χ_4^2 distribution if the underlying distribution is Poisson. Comparing the computed value with Table 2, we certainly would not reject this hypothesis.

Application of the χ^2 Test to Goodness of Fit

The χ^2 test can and should be used as a first or second test of whether the data can be reasonably well fit by some member of a family $\{P_\theta\}$, $\theta \in \Theta$ of distributions. For instance, the χ^2 test should be used to answer such questions as: Is the data exponentially distributed? Is it normally distributed? The χ^2 test is a histogram-based test. To see how it is applied, we go back to the first example of this chapter.

Example a (Revisited) The 50 failure times given in this example have an estimated mean $\hat{\beta} = 41.1$. The histogram was plotted for the following intervals:

$$\begin{matrix} I_1 & I_2 & I_3 & I_4 & I_5 & I_6 \\ [0, 20) & [20, 40) & [40, 60) & [60, 80) & [80, 100) & [100, \infty). \end{matrix}$$

For these intervals, we have the following data frequencies and expected frequencies:

	1	2	3	4	5	6
$\hat{n}_j$	20	16	4	6	0	4
n_j	19.2	11.8	7.2	4.5	2.7	4.4

The computed value of D is 6.2. Compare this with $\delta = 9.5$ which gives the 95% acceptance region for χ_4^2. Incidentally, why are there 4 degrees of freedom instead of 5?

So far, there is no reason to reject the hypothesis of an exponential distribution. But on the other hand, fitting a lognormal distribution to the histogram using $\mu = 3.3$ and $\sigma = 1.1$ gives the expected frequencies

$$20, 12, 6, 3, 2.2, 6.3$$

and a D value of 7.5. Again, we cannot reject at the 95% confidence level. Yet the two densities are quite dissimilar (see Figure 6.9).

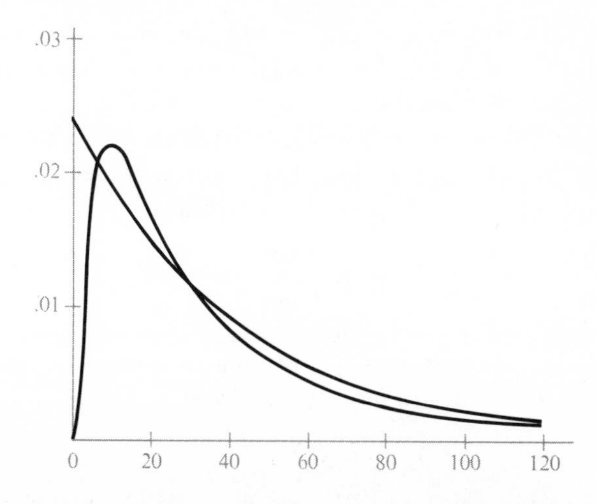

***Figure 6.9** Densities of the exponential and lognormal distributions as fitted to the failure time data*

Example c (Revisited) In Example **c.** of the first section, how can we use a χ^2 test to see if the counts in the first lane, say, are approximately normal? To draw a histogram, use intervals of length 3:

$$I_1 = [0, 20.5], I_2 = [20.5, 23.5), \ldots, I_8 = [38.5, \infty).$$

The histogram looks like Figure 6.10.

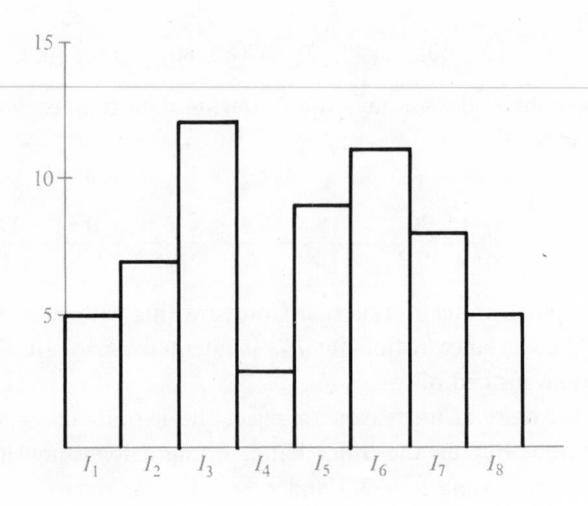

***Figure 6.10** Histogram of lane count data*

The resemblance of this histogram to a normal density seems rather distant. But what does the χ^2 test say? The sample mean and variance are $\hat{\mu} = 29.8$, $\hat{\sigma}^2 = 57$. Computing χ^2 by using the probabilities from an $N(29.8, 57)$ distribution gives $D = 10.5$. The rejection point for χ_5^2 at the .1 level is 9.2; at the .05 level it is 11.1. Normality cannot be comfortably rejected, but it is marginal.

The general procedure for using χ^2 to test goodness of fit of some family $\{P_\theta\}$, $\theta \in \Theta$, seems quite clear. Use the data to get an estimate $\hat{\theta}$ of θ and divide the range of the variable into cells $I_1, \ldots, I_J$. *Then*

$$\text{6.14} \qquad D(\hat{\theta}) = \sum_{j=1}^{J} \frac{(\hat{n}_j - n_j)^2}{n_j}$$

continuous case

has a χ^2_{J-M-1} distribution, where M is the dimension of Θ and the expected frequency n_j is computed by using $P_{\hat{\theta}}$, that is,

$$n_j = nP_{\hat{\theta}}(I_j), \qquad j = 1, \ldots, J.$$

But the procedure is not resolved yet. First, there is the important question of how to choose the cells $I_1, \ldots, I_J$ in order to get informative and accurate results. The first consideration is accuracy. In applications, the sample size n is rarely infinite, and the χ^2 distribution is only an approximation to the actual distribution of D. If the approximation is poor, then the acceptance region selected does not have the desired probability. More precisely, the χ^2 procedure is to compute δ from a χ^2_{J-M-1} table such that $P(\mathsf{C} \leq \delta) = 1 - \alpha$, then use the acceptance region $\mathsf{D} \leq \delta$ for the hypothesis that the underlying distribution is in $\{P_\theta\}$, $\theta \in \Theta$. But if the distribution of D is not closely approximated by χ^2_{J-M-1}, then $P(\mathsf{D} \leq \delta)$ is not, in general, equal to $1 - \alpha$.

What can make the approximation poor? The assumption underlying the χ^2 approximation is, in part, that each of the variables $(\hat{\mathsf{n}}_j - n_j)/\sqrt{nj}$ is normal. By the central limit theorem, this is approximately true if n_j is large. The trouble is with those outcomes or cells that have small expected frequencies. Most sources in the field quote the requirement that for a good χ^2 approximation, either $n_j \geq 5$ for all j, or $n_j \geq 10$ for all j. The latter condition seems overly cautious. Furthermore, we can afford one or two outcomes or cells with n_j as small as 1 if J is not too small and the other n_j are at least 5.

The best procedure is to define the cells so that the n_j are all equal. This can be done by estimating θ, fixing J, and defining the cells by the equations

$$P_{\hat{\theta}}(I_1) = \frac{1}{J},\ P_{\hat{\theta}}(I_2) = \frac{1}{J}, \ldots,\ P_{\hat{\theta}}(I_J) = \frac{1}{J}.$$

We call this an *equiprobable* partition. However, there are good reasons for not using this type of partition, chiefly convenience and simplicity. Computing the cells for an equiprobable partition is an additional computation which does not seem warranted in terms of returns versus the time involved, especially since a histogram may already have been plotted and observed cell frequencies computed. Of course, if the class of underlying distributions is discrete, then generally no equiprobable partition is possible.

Assume that cells have already been selected. For instance, suppose that they are all equal in length except possibly for the end cells. If some of the n_j are so small as to make the approximation to χ^2 suspect, then we resort to pooling outcomes, grouping two or more of the outcomes or cells together to produce n_j sufficiently large. For instance, suppose $n = 64$ and we have decided to use eight cells where the cell probabilities under $P_{\hat{\theta}}$ are

$$\tfrac{1}{4}, \tfrac{1}{4}, \tfrac{1}{4}, \tfrac{1}{8}, \tfrac{1}{16}, \tfrac{1}{32}, \tfrac{1}{64}, \tfrac{1}{64}.$$

The corresponding n_j are

$$16, 16, 16, 8, 4, 2, 1, 1.$$

The last three or four of these are too low. We can pool in various ways: by grouping the last three cells into a combined cell with probability 1/16 and $n_j = 4$; more conservatively, by grouping the last four together to get $n_j = 8$; or by grouping the 4th, 7th, and 8th cells and the 5th and 6th cells, getting the expected frequencies 16, 16, 16, 10, 6.

You can see that the decision on which cells to pool is somewhat arbitrary. One guiding principle is to do only the amount of grouping necessary to get a good approximation to χ^2. Why not take just a few cells and make the expected frequencies all fairly large? The answer is that if you pool too much, or take too few cells, information is lost. For instance, suppose we have a sample of 1000, and want to test goodness of fit to an $N(0, 1)$ distribution. Suppose we start with 20 cells, say, but decide to pool until we have only the two cells $I_1 = (-\infty, 0)$, $I_2 = [0, \infty)$. The approximation to χ_1^2 is very good; the expected frequencies are both 500. But while the accuracy is good, our conclusion may be very weak. If we reject the hypothesis $P = N(0, 1)$, there is no problem. But if we accept, we really don't know if the underlying distribution is anything like $N(0, 1)$. Any other distribution that assigns probabilities of .5, .5 to I_1, I_2 is just as acceptable on the basis of these two cells as an $N(0, 1)$.

For equiprobable partitions, the number of cells which gives maximum sensitivity to departures from the hypothesized distribution has expected frequencies of about 12 per cell if $n = 200$, 20 per cell for $n = 400$, 30 per cell for $n = 1000$. If there are cells in the partition with $n_j \geq 50$, then the sensitivity of the test can be appreciably improved by subdividing these intervals.

A second point is that in order to apply Theorem 6.13, which states that a degree of freedom is dropped for every parameter estimated by maximum likelihood, the estimation should be done from the *grouped and not the ungrouped data.* For example, with our failure data, checking goodness of fit to an exponential, we use cells $I_1, I_2, I_3, I_4, I_5, I_6$. To apply 6.13, let P_β be the exponential distribution with parameter β and define

$$p_j(\beta) = P_\beta(I_j), \qquad j = 1, \ldots, 6.$$

For the observed frequencies 20, 16, 4, 6, 0, 4, the maximum likelihood estimate of β is given by

$$\frac{\partial}{\partial \beta} L(\beta) = 0,$$

where

$$\begin{aligned} L(\beta) &= \log P(\hat{\mathsf{n}}_1 = 20, \ \hat{\mathsf{n}}_2 = 16, \ldots, \ \hat{\mathsf{n}}_6 = 4) \\ &= 20 \log p_1(\beta) + \cdots + 4 \log p_6(\beta). \end{aligned}$$

Solving gives $\hat{\beta} = 36.7$ as compared to the estimate $\hat{\beta} = 41.1$ obtained from the ungrouped data. The two different estimates will give different values of n_j and different distributions for D. To apply Theorem 6.13, we *should* estimate parameters from the grouped data, but it is usually more convenient and sensible to estimate them from the ungrouped data. If we follow the latter course, then the distribution of D is intermediate between χ^2_{J-M-1} and χ^2_{J-1}. For instance, in the exponential example, if we estimate β from the grouped data, then the approximate distribution is χ^2_4. But the distribution of D using the estimate $\hat{\beta}$ from the ungrouped data is approximately

$$\mathsf{Z}_1^2 + \mathsf{Z}_2^2 + \mathsf{Z}_3^2 + \mathsf{Z}_4^2 + 0.1\mathsf{Z}_5^2,$$

where $\mathsf{Z}_1, \ldots, \mathsf{Z}_5$ are independent $N(0, 1)$ variables. If the number of cells is large and each cell small in length, then we get about the same estimates using either the grouped or the ungrouped data, and we should subtract a degree of freedom for each parameter estimated. But if J is small, the ungrouped estimates may give computed values of D that are too large and lead to incorrect rejection. If you accept using the χ^2_{J-M-1} distribution or reject using the χ^2_{J-1} distribution, you are safe. This fact makes the acceptance of the normality of the traffic counts a bit less marginal.

When data is grouped, it is clear that information is lost. It should not be surprising that estimates based on grouped data are generally not as accurate as estimates based on ungrouped data. More precisely, it is not difficult to show that in a smooth 1-dimensional problem the information content $I(\theta)$ about θ is generally decreased when computed from the probabilities corresponding to the grouped data. Therefore, the maximum likelihood estimate of θ based on the grouped data has asymptotically a larger squared error risk than the maximum likelihood estimate based on the ungrouped data.

Problem 9 For the data of Problem 2, run a χ^2 test to check the fit of a normal distribution.

Problem 10 Table 10 consists of random numbers generated from a uniform distribution on [0, 1]. Take any 200 of these numbers, select cells, construct a histogram, and run a χ^2 test to check goodness of fit to a uniform distribution on [0, 1].

Problem 11 The χ^2 test can be easily applied to goodness of fit to bivariate or multivariate distributions. Let $\mathbf{x}_1, \ldots, \mathbf{x}_n$ be outcomes of independent identically distributed N-dimensional random vectors $\mathbf{X}_1, \ldots, \mathbf{X}_n$. The problem is to test goodness of fit to some family $\{P_\theta(d\mathbf{x})\}$, $\theta \in \Theta$, of N-variate distributions. Estimate θ by maximum likelihood. Partition N-dimensional space into J cells $R_1, \ldots, R_J$, and let $\hat{n}_j$ be the number of $\mathbf{x}_1, \ldots, \mathbf{x}_n$ falling into R_j. Now simply compute D as in the 1-dimensional case with $n_j = nP_\theta(R_j)$, and approximate its distribution by χ^2_{J-M-1}, where M is the dimension of Θ. Apply this approach to the pairs of counts given in Example **c.** of the first section. Select the cells yourself. Estimate the 5 parameters of the bivariate normal, and check goodness of fit to the bivariate normal.

do

Problem 12 A fish ladder has 7 steps. Over a period of time, the following numbers of fish were observed successfully negotiating the jth step:

Step	1	2	3	4	5	6	7
Number of fish	1263	1090	1031	881	765	693	597

A rough deterministic model that has been proposed for a fish ladder takes the number of fish that successfully negotiate the jth step to be a constant fraction r, $0 < r < 1$, of the fish that negotiate the $(j - 1)$th step. Translate this into a probabilistic model and check goodness of fit to the above data.

Problem 13 Try fitting the second lane count data (Example **c.** of the first section) by a Poisson distribution, and check the fit with a χ^2 test.

Detection Sensitivity of the χ^2 Test

If the χ^2 test rejects a fit, you are on safe ground. You know that your hypothesized distribution is not a good fit to the data. But if you accept, can you then confidently assert that the underlying distribution is the hypothesized one? For instance, for moderate sample sizes, as we will show, the χ^2 test accepts as normal almost every distribution with a symmetric, single maximum, density. How well can the χ^2 test discriminate between various possible underlying distributions? In more familiar terms, what is its detection capability?

There is a simple minimum detectable distance for the χ^2 test. Our hypothesis is that the underlying distribution is

$$\mathbf{p}^{(0)} = (p_1^{(0)}, \ldots, p_J^{(0)}).$$

Set up the χ^2 test for the hypothesis $p^{(0)}$ at level α, and define a measure of distance from $\mathbf{p}^{(0)}$. Then a value $D_{\min}$ of this distance is called the *minimum detectable distance* at level α if the underlying distributions that can be detected at level α consist of all those $\mathbf{p}$ whose distance from $\mathbf{p}^{(0)}$ is at least $D_{\min}$. In other words, even if the true state of nature is $\mathbf{p} \neq \mathbf{p}^{(0)}$, if the distance from $\mathbf{p}^{(0)}$ to $\mathbf{p}$ is less than the minimum detectable distance $D_{\min}$, we still accept the hypothesis $\mathbf{p}^{(0)}$ on the basis of the above with probability at least α. And the converse is true if the distance from $\mathbf{p}^{(0)}$ to $\mathbf{p}$ is greater than $D_{\min}$.

For instance, suppose we want to test goodness of fit to an $N(0, 1)$ distribution. We set up cells $I_1, \ldots, I_J$, compute cell probabilities $p_j^{(0)} = P_{(0,1)}(I_j)$, $j = 1, \ldots, J$, and accept or reject on the basis of $D^{(0)}$. Say our level is .1. Suppose that the true underlying distribution is a double exponential with $\alpha = 0$, $\beta = 1$, and assign the probabilities $\mathbf{p} = (p_1, \ldots, p_j)$ to the cells. If the distance from $\mathbf{p}^{(0)}$ to $\mathbf{p}$ is less than $D_{\min}$, then although the underlying distribution is double exponential (0, 1), at least 10% of the time we accept the hypothesis of normality.

Therefore, a minimum detectable distance would help us in knowing what we really accept when the χ^2 test says accept. Denote $p_j - p_j^{(0)} = \Delta p_j$, and define a distance measure $D(\mathbf{p}^{(0)}, \mathbf{p})$ from $\mathbf{p}^{(0)}$ to $\mathbf{p}$ by

6.15
$$D(\mathbf{p}^{(0)}, \mathbf{p}) = \sum_1^J \frac{(\Delta p_j)^2}{p_j^{(0)}}.$$

Note that this distance is not symmetric in $\mathbf{p}^{(0)}$ and $\mathbf{p}$.

6.16 ***Theorem*** *For n sufficiently large and $K = J - 1$, the minimum detection distance is given by*

$$D_{\min} = \gamma_K \cdot \frac{\sqrt{K}}{n},$$

where γ_K is nearly constant for a given level α.

For instance, for $\alpha = .10$ and K between 4 and 400, $\gamma_K \simeq 6$. For $\alpha = .05$ and $4 \leq K \leq 400$, $\gamma_K \simeq 8$.

To illustrate Theorem 6.16, suppose we propose to test goodness of fit to an $N(0, 1)$ distribution using the 10 cells illustrated below:

	−1.6		−1.2		−.8		−.4		0		.4		.8		1.2		1.6	
I_1		I_2		I_3		I_4		I_5		I_6		I_7		I_8		I_9		I_{10}

The probabilities are symmetric around 0. Starting with I_6 and going to the right, they are .155, .133, .097, .060, .055. The question is: How large a sample size is needed to detect a uniform distribution with mean zero and variance one? The uniform with variance one is uniform on $[-\sqrt{3}, \sqrt{3}]$. For this underlying distribution, the cell probabilities are, starting with I_6 and going to the right, .115, .115, .115, .115, .040. Therefore

$$D(\mathbf{p}^{(0)}, \mathbf{p}) = 2\left[\frac{(.040)^2}{.155} + \frac{(.018)^2}{.133} + \frac{(.018)^2}{.097} + \frac{(.055)^2}{.060} + \frac{(.015)^2}{.055}\right]$$

$$= .14.$$

Here, $J = 10$, $K = 9$, and $D_{\text{min}} = 18/n$ at the .1 level. Hence, to get $D \geq D_{\text{min}}$ we need

$$\frac{18}{n} \leq .14$$

or $n \geq 129$. Surprisingly enough, if the underlying distribution is uniform on $[-\sqrt{3}, \sqrt{3}]$, we still accept normality at least 10% of the time unless the sample size is well over 100. Repeating this computation with a double exponential, $\beta = .88$, we find that to detect at the 10% level the sample size must be larger than about 340. For a Cauchy, we begin to detect the heavy tail at about $n = 100$.

The upshot of this is that for $n \leq 100$, the χ^2 test will frequently accept as normal, distributions that differ as drastically from the normal as the uniform and the Cauchy. For $n \geq 100$, the differences in tail behavior between these latter distributions and the normal begin to show up. But you have to go up to a sample size of about 300 to detect the difference between the normal distribution and some symmetric distribution like the double exponential, whose tails taper off smoothly but rapidly. In this sense, for moderate sample size, many distributions appear normal.

Notice from 6.15 that the χ^2 test picks up changes in the small $p_j^{(0)}$ much more easily than in the large $p_j^{(0)}$. For instance, a change of .05 in $p_j^{(0)} = .1$ gives a contribution to $D(\mathbf{p}^{(0)}, \mathbf{p})$ of $(.05)^2/.1 = .025$. But the same change of .05 in $p_j^{(0)} = .4$ gives the contribution $(.05)^2/.4 = .006$.

Because of this sensitivity to change in the smallest of the p_j, $j = 1, \ldots, J$, you will find that frequently when the χ^2 test rejects, the major contributions to D come from those few outcomes with the smallest probabilities. At the same time, the remainder of the outcomes give a nominal contribution to D and seem to be fit quite reasonably by the hypothesized distribution. In this sense, the χ^2 test is sensitive to the tail behavior of the distribution.

Problem 14 A six-sided die is to be tested for bias. The hypothesis is that the outcomes are equally likely. What sample size would be necessary to detect a bias as pronounced as

$$\mathbf{p} = (\tfrac{1}{5}, \tfrac{1}{5}, \tfrac{1}{5}, \tfrac{1}{5}, \tfrac{1}{10}, \tfrac{1}{10})?$$

(Choose the level α yourself.)

Problem 15 Here is an example due to W. G. Cochran.* The hypothesis is that 100 observations follow a Poisson distribution with parameter 1. But suppose the true underlying distribution is given by

$$p_j = \frac{(j + 1)a^j}{b^{j+2}}, \qquad j = 0, 1, \ldots,$$

where $a = .5$, $b = 1.5$. This gives the two sets of expected frequencies

j	Poisson	Other
0	36.8	44.4
1	36.8	29.6
2	18.4	14.8
3	6.1	6.6
≥ 4	1.9	4.5

Is the difference detectable at the .1 level? What happens if the last two groups are pooled—is the difference still detectable at the .1 level?

Problem 16 For an experiment with 9 outcomes, the hypothesis is that each outcome is equally likely. The alternatives of interest are that the higher numbered outcomes have higher probabilities, say that

$$p_j = a + bj, \qquad j = 1, \ldots, 9.$$

With sample size n, what is the minimum value of b that could be detected at the .1 level using a χ^2 test? (Ans. $b_{\min} \simeq .18/\sqrt{n}$.)

* "Some methods for strengthening the common χ^2 tests," W. G. Cochran, *Biometrics*, Dec. 1954, pp. 417–451. Reprinted by permission of the publisher.

do

Problem 17 Two different theories lead to the following probabilities for the number of particles resulting from a certain two-particle collision:

Particles	1	2	3	4	5	6	7
Probabilities I	0	.31	.17	.27	.09	.14	.02
Probabilities II	0	.25	.16	.34	.08	.15	.02

460 = answer

What order of sample size would be necessary for reliable discrimination between these two sets of probabilities using a χ^2 test on either one?

Problem 18 For a double exponential with mean zero, fix β by the requirement that

$$P([-1, 1]) = .67.$$

Compute the probabilities of the ten cells set up in the preceding section to test $N(0, 1)$ fit, and compute the sample size needed to detect the double exponential distribution at the 10% level, with the $N(0, 1)$ distribution as the hypothesis.

Problem 19 Repeat Problem 18 for a Cauchy distribution with $\alpha = 0$. Where do the major contributions to D come from?

Problem 20 A *contaminated* $N(0, 1)$ distribution has the form

$$P(I) = pP_0(I) + qP_1(I), \qquad p \geq 0, q \geq 0, p + q = 1,$$

where $P_0(I)$ is an $N(0, 1)$ distribution and $P_1(I)$ is an $N(0, \sigma^2)$ distribution, with $\sigma^2 > 1$. Essentially, this means that we draw from $N(0, 1)$ a fraction p of the time and from $N(0, \sigma^2)$ a fraction q of the time. For $\sigma^2 = 9$, repeat Problem 18, and find the minimum detectable proportion of contamination q as a function of n ($\alpha = .1$). What value of q can be detected for $n = 100, 400$?

(Ans.

$$q_{\min} = 2.7/\sqrt{n};$$

for

$$n = 100, \; q_{\min} = .27;$$

for

$$n = 400, \; q_{\min} = .13.)$$

The Sample Distribution Function

We have used one simple but powerful idea as the key to constructing histograms, estimating densities, and setting up the χ^2 test: *For any underlying distribution, the proportion $\hat{P}(I)$ of the sample values that fall into an interval I is a good estimate of $P(I)$.* We can look at it this way: For each of $X_1, \ldots, X_n$, count it as a success if it falls in I; otherwise, count it a failure. But this is exactly the biased coin-tossing situation with $P(I)$ the underlying probability of success, and we know that in this situation, the proportion of successes $\hat{P}(I)$ is a good way of estimating the probability of success.

So far, our use of this idea has involved partitioning the outcomes into cells $I_1, \ldots, I_J$ and using $\hat{P}(I_1), \ldots, \hat{P}(I_J)$ as estimates of $P(I_1), \ldots, P(I_J)$. There is another approach which avoids the troublesome and arbitrary selection of cells. Recall that for any distribution P, the cumulative distribution function is defined by

$$F(x) = P[(-\infty, x)];$$

that is, $F(x)$ is the probability that an outcome will be less than x. Define $\hat{F}(x)$ as the proportion of the sample values that are less than x:

$$\hat{F}(x) = \frac{1}{n}(\# \text{ of } \{x_1, \ldots, x_n\} < x). \tag{6.17}$$

Using the key idea stated above, we argue that for every value of x, $\hat{F}(x)$ is a good estimate of $F(x)$. Now look at $\hat{F}(x)$ as a function of x.

6.18 ***Definition*** *The function $\hat{F}(x)$ defined for all x by* 6.17 *is called the sample cumulative distribution function.*

The sample cumulative distribution function $\hat{F}(x)$ ought to look like the underlying cumulative distribution function $F(x)$, just as the histogram ought to look like the underlying density $f(x)$. Now we consider $\hat{F}(x)$ as our estimate of $F(x)$ and look at its computation and properties.

Arranging the data to construct a histogram and run a χ^2 test involved a simple tally of the number of outcomes falling into each cell. Plotting $\hat{F}(x)$ involves the more tedious job of ordering the sample. Recall that if the data are $x_1, \ldots, x_n$, then the ordered sample is $x_{(1)} \leq x_{(2)} \leq \cdots \leq x_{(n)}$, where $x_{(1)}$ is the smallest of $x_1, \ldots, x_n$, $x_{(2)}$ the next smallest, and so on. For a small sample with $n \lesssim 100$, ordering by inspection is not too onerous. For large samples, machine programs for ordering the data are standard and easy to run. If you plan to do extensive work with the data, it is always a good idea to begin by ordering the sample. It is much easier to decide how big to take histogram or χ^2 cells by looking at the ordered sample than by looking at the raw data. The ordered data $x_{(1)}, \ldots, x_{(n)}$ are also called the *order statistics*.

Once your sample is ordered, $\hat{F}(x)$ is easy to plot for small samples. It jumps by the amount $1/n$ at each of the points $x_{(1)}, \ldots, x_{(n)}$, is zero to the left of $x_{(1)}$, and is constant otherwise. To plot, start at the left of $x_{(1)}$ and go up an amount $1/n$ every time you come to one of the points $x_{(1)}, \ldots, x_{(n)}$. Figure 6.4 is the graph of $\hat{F}(x)$ for the 50 failure times of Example **a.** of the first section. If your sample size is large, then the above plotting strategy becomes very time consuming, and it is better to plot $\hat{F}(x)$ at closely spaced values of x, and evaluate $\hat{F}(x)$ by seeing how many of the sample points are less than x. Since the machine program will usually number the ordered sample, this method is relatively painless.

Unfortunately, the graph of the sample distribution function does not give as clear a picture of the shape of the distribution as the histogram. However, once the data have been ordered and numbered, the order statistics can be used as followed to construct a histogram and χ^2 test that are improvements over those gotten by using the fixed cell method. The density $f(x)$ is the derivative of $F(x)$, so in order to estimate $f(x)$, take an approximate derivative of $\hat{F}(x)$. In the interval $[x_{(j)}, x_{(k)})$, $\hat{F}(x)$ jumps $k - j$ times and therefore increases by $(k - j)/n$. Hence an estimate for its derivative over this interval is

$$\frac{1}{n} \cdot \frac{k - j}{x_{(k)} - x_{(j)}}.$$

Now take m so that $n \simeq mJ + 1$; m will function as the number of points in every cell, and J as the number of cells. Let the cells be given by

$$I_1 = [x_{(1)}, x_{(m+1)}), I_2 = [x_{(m+1)}, x_{(2m+1)}), \ldots, I_J = [x_{(m(J-1)+1)}, x_{(mJ+1)}).$$

In the jth cell, the estimate for $f(x)$ is

$$\hat{f}_j = \frac{m}{n\|I_j\|}, \tag{6.19}$$

and these are the heights that are computed and plotted. If the last cell does not have exactly m points, say it has $m - 1$ or $m + 1$ or so on, then estimate $\hat{f}_J$ by using the correct number of points in the numerator of 6.19. Notice that in this method of constructing the histogram, the cells are determined from the sample. These cells have the important advantage that there are an equal number of points in each cell. Because of this we get a neat expression for the heights of the $100\gamma\%$ touch bars for $f(x)$. Determine z from an $N(0, 1)$ distribution: $P(-z \leq Z \leq z) = \gamma$. Then the heights of the bars are given, for $m \geq 5$, say, by

$$\hat{f}_j\left(1 \pm \frac{z}{\sqrt{m}}\right). \tag{6.20}$$

Thus, the touch bar heights are constant multiples of the histogram heights.

To set up the χ^2 test, note that each cell has the constant proportion m/n of the sample points, hence $\hat{P}(I_j) = m/n$. If P is the hypothesized distribution, compute the expression

$$D = \frac{n}{m}\sum_{j=1}^{J}\left(\frac{m}{n} - P(I_j)\right)^2.$$

This converges to χ^2_{J-1} for n large and is a better approximation to χ^2_{J-1} for moderate values of n, than the fixed cell χ^2 test. The approximation is fairly good if $m \geq 5$.

If the underlying distribution is discrete, it is possible that some values in the ordered sample repeat:

$$x_{(j)} = x_{(j+1)} = \cdots = x_{(k)}.$$

The above method of histogram construction then does not work. However, the distribution does not have a density.

Problem 21 Order the first 50 entries in the table of random $N(0, 1)$ outcomes and plot $\hat{F}(x)$. Plot $F(x)$ for an $N(0, 1)$ on the same graph.

Problem 22 Plot $\hat{F}(x)$ for the Lane 1 counts in Example **c.** of the first section. Plot $F(x)$ for $N(\hat{\mu}, \hat{\sigma}^2)$ on the same graph.

Problem 23 Order the failure data of Example **a.** of the first section, and construct a variable cell histogram using $m = 5$. Indicate the heights of the 90% touch bars. Compute D for these variable cells using $\hat{\beta} = 41.1$.

Confidence Bands

Estimates should rarely be used without an attached accuracy statement. In the last section $\hat{F}(x)$ was put forward as an estimate of the underlying cumulative distribution function. Now we want to explore how accurate an estimate it is. But to do this, we need some measure of how far apart two distribution functions $F(x)$ and $G(x)$ are. A useful measure is:

6.21 ***Definition*** *Define a distance $d(F, G)$ between two cumulative distribution functions by*

$$d(F, G) = \max |F(x) - G(x)|,$$

where the maximum is taken over all values of x in $(-\infty, +\infty)$.

To understand the geometric meaning of this, take either distribution function, say $F(x)$, graph it, and draw a band around this graph which extends β units above and below the graph (see Figure 6.11).

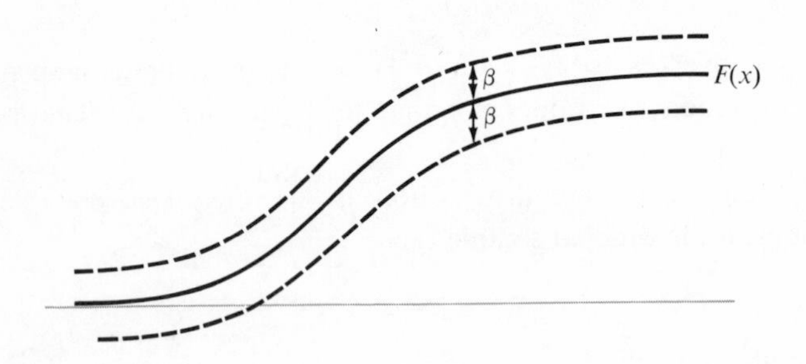

Figure 6.11

Then if $\beta \geq d$, the graph of $G(x)$ is completely contained in the band. If $\beta < d$, then $G(x)$ passes outside the band at least once. In other words, the band of width $2d$ around $F(x)$ is the narrowest band that completely contains the graph of $G(x)$.

Look at the distance between $\hat{F}(x)$ and the underlying cumulative distribution function $F(x)$. Since $\hat{F}(x)$ depends on the outcome of the random data vector $\mathbf{X}$, this distance is a random variable. Denote it by

$$d(\mathbf{X}) = \max_x |\hat{\mathsf{F}}(x) - F(x)|,$$

where for each x, $\hat{\mathsf{F}}(x)$ is the random variable defined as the proportion of the $\mathsf{X}_1, \ldots, \mathsf{X}_n$ less than x. The kind of confidence statement we are looking for would go something like this: Find a δ such that for any underlying distribution,

$$P(d(\mathbf{X}) \leq \delta) \geq \gamma. \tag{6.22}$$

If 6.22 holds, then no matter what the underlying cumulative distribution function is, with probability at least γ it is contained in the band of width 2δ around the sample distribution function.

Looking at this another way, we want an infinite set of simultaneous confidence intervals. That is, we want to find δ such that

$$F(x) \in (\hat{F}(x) - \delta, \hat{F}(x) + \delta)$$

holds true for all x simultaneously with probability at least γ.

The distribution of $d(\mathbf{X})$ depends on the distribution of $\mathbf{X} = (X_1, \ldots, X_n)$. Since the common underlying distribution of $X_1, \ldots, X_n$ can range over all 1-dimensional distributions, finding a δ that satisfies 6.22 would seem to be a formidable task. The fact that rescues us is this surprising result:

6.23 ***Theorem*** *The distribution of $d(\mathbf{X})$ is the same for all continuous underlying distributions.*

This means: Let $X_1, \ldots, X_n$ be independent with a common distribution $P_1(dx)$ which does not assign positive probability to any single point. Now using this distribution, compute the distribution of $d(\mathbf{X})$. Repeat this procedure using any other distribution $P_2(dx)$ for $X_1, \ldots, X_n$. If $P_2(dx)$ also has no discrete part, then, according to Theorem 6.22, we get exactly the same distribution for $d(\mathbf{X})$ as when we used $P_1(dx)$.

The proof of Theorem 6.23 is illuminating. Let $F(x)$ be the underlying cumulative distribution function of the variables $X_1, \ldots, X_n$. If $P_1(dx)$ does not assign positive probability to single points, then $F(x)$ is a continuous function. To make the proof easier, assume also that $F(x)$ is strictly increasing, or that $x_2 > x_1$ implies $F(x_2) > F(x_1)$. The proof if we assume only that $F(x)$ is continuous is similar, but requires more care. Make the change of variables

$$y = F(x)$$
$$Y_k = F(X_k), \qquad k = 1, \ldots, n.$$

Since $F(x)$ is smoothly increasing, the inequality $X_k < x$ is equivalent to $F(X_k) < F(x)$ or $Y_k < y$. Therefore

$$\hat{F}(x) = \frac{\#\text{ of } X_k < x}{n} = \frac{\#\text{ of } Y_k < y}{n} = \hat{G}(y).$$

Furthermore, the distribution function of Y_k is

$$P(Y_k < y) = P(X_k < x) = F(x) = y.$$

So Y_k has a uniform distribution with cumulative distribution function $G(y) = y$. Now write

$$|\hat{F}(x) - F(x)| = |\hat{G}(y) - y|.$$

As x ranges from $-\infty$ to $+\infty$, y ranges from 0 to 1, so

6.24 $$\max_{x} |\hat{F}(x) - F(x)| = \max_{0 \leq y \leq 1} |\hat{G}(y) - y|.$$

The expression on the right is the distance $d(\mathbf{Y})$ between the sample distribution function of the $Y_1, \ldots, Y_n$ and their underlying cumulative distribution function. Now 6.24 states that $d(\mathbf{X}) = d(\mathbf{Y})$. That is, by our substitutions $d(\mathbf{X})$ has been shown equal to $d(\mathbf{Y})$, where the variables $Y_1, \ldots, Y_n$ are uniform on $[0, 1]$.

Let $X_1, \ldots, X_n$ have uniform distributions on $[0, 1]$. For any γ, compute δ such that

$$P(d(\mathbf{X}) \leq \delta) = \gamma.$$

Then *for* $X_1, \ldots, X_n$ *having any continuous distribution,*

6.25 $$P(d(\mathbf{X}) \leq \delta) = \gamma.$$

The number δ depends only on n and γ. Here is a short table:

Table 6.1

n	$\gamma = .90$	$\gamma = .95$
10	.37	.41
20	.26	.29
30	.22	.24
large	$1.22/\sqrt{n}$	$1.36/\sqrt{n}$

Here is the way to form $100\gamma\%$ confidence bands under the assumption that the distribution of the variables is continuous: Use the data $x_1, \ldots, x_n$ to compute the sample distribution function $\hat{F}(x)$. Find the δ corresponding to the given n and γ. Then with probability γ, the underlying cumulative distribution function is in the $\pm\delta$ band around $\hat{F}(x)$.

Table 6.1 gives some idea of how well we can estimate the distribution from the data. For $\gamma = .9$, say, $\delta \simeq 1.22/\sqrt{n}$ for n large. For $n = 100$, $\delta = .12$, so $F(x)$ is within $\pm.12$ of $\hat{F}(x)$ for all x with probability .9. This is a fairly wide band. In fact, for some interval I, $P(I)$ could differ from the estimate $\hat{P}(I)$ by as much as .24 and $F(x)$ would still be in the $\pm.12$ band. But to be in error in an estimate of $P(I)$ by as much as .24 is a considerable mistake.

The problem is that the width of the band decreases slowly—only as fast as $1/\sqrt{n}$. To cut the width of the band in two, we have to quadruple sample size. To get a $\pm.06$ band we need a sample size of 400.

Problem 24 Draw 90% confidence bands around the graphs of $\hat{F}(x)$ in Problems 21 and 22.

The Kolmogorov-Smirnov Test for Goodness of Fit

We can use the results of the previous section to set up a goodness of fit test based on $\hat{F}(x)$. Suppose $F(x)$ is the hypothesized distribution.

6.26 ***Definition*** *The Kolmogorov-Smirnov test consists of: Accept the hypothesis if*

$$\max_x |\hat{F}(x) - F(x)| \leq \delta,$$

where δ is adjusted to give a level α test in accordance with Table 6.1.

But be warned—this test is applicable *only for continuous underlying distributions*. If the possible underlying distributions are discrete, go back to χ^2. But if the underlying distribution is continuous, then we know from Theorem 6.23 that the procedure of 6.26 gives a level α test.

The advantage of the Kolmogorov–Smirnov test over the χ^2 test is that it gets away from the necessity of partitioning into cells. Because the data are not grouped, information is not lost, and the Kolmogorov–Smirnov test is sometimes more sensitive than the χ^2 test. What happens if the hypothesis to be tested is that the distribution is one of a family $\{P_\theta\}$, $\theta \in \Theta$? As before, the reasonable procedure is to estimate θ by $\hat{\theta}$ and run the Kolmogorov–Smirnov test on the cumulative distribution function of $P_{\hat{\theta}}$. The effect that this has on the level of the test is not well known. The evidence we have is that the effect is not very important. For moderate to large sample size, it is probably safe to ignore the fact that θ was estimated.

To apply this test intelligently we need to know its detection capabilities. Unfortunately, the set of those distributions which can be detected at level α by the Kolmogorov–Smirnov test cannot be neatly characterized by a minimum detectable distance.

We do know the following:

6.27 ***Proposition*** *If*

$$\max_x |F(x) - G(x)| \lesssim \begin{cases} .65/\sqrt{n}, & \alpha = .1 \\ .84/\sqrt{n}, & \alpha = .05, \end{cases}$$

then we cannot detect at level α.

Another way of putting this proposition is that if

$$n \leq \frac{\gamma}{[d(F, G)]^2}, \qquad \gamma = .42, .71, \tag{6.28}$$

then detection at the .1, .05 levels, respectively, is impossible. Notice that the proposition does not say that if $n > \gamma/[d(F, G)]^2$, detection is possible. Depending on the specific forms of F and G, the sample size needed to detect may be 50–100% larger than the bound given in 6.28. In general, if the maximum difference of F and G occurs near the tails of the distribution, then it is easier to detect, and the required sample size is close to the bound given in 6.28.

Using 6.27 and $\alpha = .1$, we cannot detect the difference between an $N(0, 1)$ distribution and a distribution uniform on $[-\sqrt{3}, \sqrt{3}]$ with a sample size of less than 130. To detect at .1 between an $N(0, 1)$ and a double exponential distribution with $\beta = .88$ and mean zero takes a sample size of at least 500. The important point is the same: If you do not reject normality or exponentiality or whatever the hypothesized family is, and if your sample size is not large, then the conclusion that the underlying distribution is normal or exponential or whatever is suspect. When you "accept" normality, say, you also "accept" every distribution whose cumulative distribution function lies within the band around the $N(0, 1)$ cumulative whose width is specified by 6.27.

Problem 25 **a** Using the 50 observed failure times of Example **a.** of the first section, check goodness of fit to the exponential distribution with $\hat{\beta} = 41.1$ by the Kolmogorov–Smirnov test.

b Check goodness of fit of these data to a gamma distribution with $\alpha = 1$, $\beta = 38$. How do you explain your result?

Problem 26 For the data of Problem 21, check goodness of fit to a uniform distribution on $[-\sqrt{3}, \sqrt{3}]$.

Problem 27 By using calculus techniques to find maxima, find the distances between

a an $N(0, 1)$ and a uniform on $[-\sqrt{3}, \sqrt{3}]$,

b an $N(0, 1)$ and a double exponential with $\alpha = 0$, $\beta = .88$.

Use these distances to find minimum sample sizes for detection at the .1 level.

Problem 28 Let X have an $N(0, \sigma^2)$ distribution; then $Y = |X|$ is said to have a truncated normal distribution. What minimum sample size would be required to detect the difference between a truncated normal and an exponential with the same mean?

Problem 29 For a contaminated normal distribution, the density is given by

$$f(x) = pf_0(x) + qf_1(x),$$

where $f_0(x)$ is the $N(0, 1)$ density and $f_1(x)$ an $N(0, \sigma^2)$ density with $\sigma > 1$. For $\sigma = 3$, find the bound on the sample size below which the contaminated distribution cannot be detected from an $N(0, 1)$ distribution by the Kolmogorov-Smirnov test. (Ans. $n \geq 7.2/q^2$.)

The Search for Rejection

Rejection is much more informative than acceptance, unless the sample size is enormous. If, for example, you reject normality as an underlying distribution for a certain data vector **x**, interesting avenues of exploration are opened up: What feature of the data causes the bad fit to normality? Is it that the tails of the sample distribution are too large, or is it that the sample distribution is not symmetric enough around some midpoint (equivalently, is it too skewed to the right or left)? If the χ^2 rejects, then you can check what is causing rejection by looking at those cells for which $(\hat{n}_j - n_j)/\sqrt{n_j}$ is large. If Kolmogorov–Smirnov rejects, examine those values of x around the value that maximizes $|\hat{F}(x) - F(x)|$.

However, both the χ^2 and the Kolmogorov–Smirnov test are broad spectrum tests. They test the goodness of fit of any assumed underlying distribution against all challengers. Because of this, they are very handy tools. But unless the sample size is large, when they accept, they are actually accepting all distributions that do not differ too drastically from the hypothesized distribution. Therefore, if they singly or together accept the fit of a hypothesized distribution, you have gained only some fairly crude information. If they accept, and you are interested in further analysis of the underlying distribution of the data, then the next step is to search for rejection; that is, to carefully examine the data to spot any suspicious deviations from the hypothesized distribution. The search here is for specific deviations. For example, do the data look like they are skewed around the midpoint? Are a few of the readings suspiciously large? The point is that one can devise tests for specific types of deviations that are considerably more sensitive than the χ^2 or Kolmogorov–Smirnov tests.

For example, look at the failure time data from Example **a.** of the first example. The exponential fit has been accepted for this data by both χ^2 and Kolmogorov–Smirnov. Looking at these data more carefully, notice the two high readings 262.8 and 182.4. Is it possible that these readings indicate a tail to the distribution that does not decrease as rapidly as that of an exponential distribution? One reasonable approach to answering this question is to compare the size of the largest data point with some measure of the size of most of the data points. In particular, a simple value to look at in terms of the order statistics is $x_{(n)}/\hat{\nu}$, where $\hat{\nu}$ is the median of the sample. We could also look at $x_{(n)}/\hat{\beta}$, where $\hat{\beta}$ is the sample mean. But for reasons we leave until the next chapter, $\hat{\nu}$ is a better choice. If the underlying distribution is exponential, then the distribution of $\mathsf{X}_{(n)}/\hat{\nu}$, where $\mathsf{X}_{(n)} = \max(\mathsf{X}_1, \ldots, \mathsf{X}_n)$, does not depend on β. Put $\beta = 1$, fix a level α, and determine u such that

$$P\left(\frac{\mathsf{X}_{(n)}}{\hat{\nu}} \leq u\right) = 1 - \alpha.$$

Of course, u depends on n and α. Our test is to reject exponentiality if $x_{(n)}/\hat{\nu} > u$. For $n = 50$ and $\alpha = .1$, u is 9.3. The sample median $\hat{\nu}$ is between 24.6 and 26.7, whose average is 25.7. Thus $x_{(n)}/\hat{\nu} = 262.8/25.7$ or 10.4, which leads to rejection. Now there is an interesting new development to investigate, namely, that the possibility that those items that last longer than 100 time units have a larger probability of long survival times than can be explained by an exponential model.

To carry this one step further, suppose the components being tested have an actual underlying distribution given by

$$P(\mathsf{T} \geq t) = .95e^{-t/\beta} + .05e^{-t/4\beta}.$$

This is interpreted to mean that about 1 out of every 20 components is drawn from a failure time distribution that has a mean value 4 times larger than the rest of the population. How large a sample size is needed to detect this departure from exponentiality at the .1 level, say? A computation shows that the Kolmogorov–Smirnov test will need a minimum of about 750 readings. Possibly the sample size at which detection occurs may be 50–100% larger. But the test based on $x_{(n)}/\hat{\nu}$ will detect at $n = 400$, roughly, only about half the sample size required by the Kolmogorov–Smirnov test.

There are two conclusions to draw from the above example. First, *it is the departures from the hypothesized distributions that frequently give the most informative and interesting insights into the nature of the physical system producing the data.* For example, detecting some skew in a sample distribution, some lack of symmetry about the midpoint, may lead to more understanding of the underlying phenomenon than accepting normality on the basis of a relatively insensitive but broad spectrum test. The tails of distributions should be looked at closely. (For a positive distribution, examine the values near zero, and so on.) In general, rejection gives more sharply defined information than acceptance.

The second point is that a test designed to pick up a specific type of departure, such as too long a tail or skewness, will be more sensitive to this type of departure from the hypothesized distribution than the broad spectrum tests and will therefore be more likely to detect its presence.

Summary

There are two approaches to looking at the underlying distribution and checking goodness of fit. The first and easiest, computationally, consists of arranging the data into cells. Then the exploratory steps are

a draw the histogram;
b mark in the touch confidence bars.

If, on the basis of the histogram, you think that a normal or exponential or whatever is a possible fit, then run a χ^2 test.

The second step requires ordering the data. From the ordered data a more precise histogram and touch bars can be constructed. Also the sample distribution function can be graphed and confidence bands drawn around it. The Kolmogorov–Smirnov test can be run to check goodness of fit.

Here is a summary of the properties of the two tests:

χ^2 test

a It is designed for discrete distributions.
b Continuous distributions are tested by grouping the data into cells.
c A degree of freedom is lost for every parameter estimated from the grouped data.
d It is easily used on multidimensional distributions.

Its drawbacks are:

a It requires an arbitrary grouping into cells and loses information because of the grouping.
b Requirements as to how large cells must be in order that the χ^2 approximation hold are not precisely known.

c The number of degrees of freedom is not lowered by one if a parameter is estimated from the ungrouped data.

Kolmogorov–Smirnov test

a It applies to continuous distributions only.
b It does not require grouping the data.
c It is more difficult computationally than χ^2 since it requires ordering of the data.

Its drawbacks are:

a The effect of estimated parameters on the level of the test is not well known.
b It does not generalize to testing goodness of fit of multidimensional distributions.

With each of these tests, it is important to know how much you are accepting when the test results in acceptance. Compute the minimum detection distances. For moderate sample sizes, these tests have difficulty discriminating between classes of distributions that are roughly similar. The Kolmogorov–Smirnov test, for example, needs a sample size of at least 500 to detect the difference between a normal and the closest double exponential distribution at the .1 level.

The fact that neither of these tests rejects an hypothesized distribution, does not put you on safe ground. To understand the data, serious attempts should be made to find any deviations from the hypothesized distribution.

7 Safety in Estimation

Introduction

If we want to use the maximum likelihood method to estimate the probability p that an A-type transistor lasts less than 2000 hours, and we assume that the failure time T has an exponential distribution with parameter $1/\theta$, the procedure is as follows: The parameter to be estimated is

$$p = 1 - e^{-2000/\beta}.$$

The maximum likelihood estimator for any function of β is that function of the maximum likelihood estimator of β. We know that the maximum likelihood estimator for β is

$$\hat{\beta} = \frac{T_1 + \cdots + T_n}{n},$$

where $T_1, \ldots, T_n$ are the times until failure in our sample. Hence the maximum likelihood estimator for p is

$$\hat{p} = 1 - e^{-2000/\hat{\beta}}.$$

For n large we can compute the squared error loss (see the remarks on p. 222 preceding Problem 2)

7.1 $$E(\hat{p} - p)^2 \simeq e^{-4000/\beta}\left(\frac{2000}{\beta}\right)^2 \cdot \frac{1}{n}$$

or

$$E(\hat{p} - p)^2 = [(1 - p)\log(1 - p)]^2 \cdot \frac{1}{n}.$$

Now suppose that the true distribution of the failure time T is uniform on 0 to 10,000 hours, but we still use the above estimate for $\hat{\beta}$. Then

$$\hat{\beta} = \frac{T_1 + \cdots + T_n}{n}$$

for large n will be about 5000. Therefore we get

$$\hat{p} = 1 - e^{-2/5} \simeq .33.$$

But we know that for a uniform distribution on [0, 10,000],

$$P(T < 2000) = \frac{2000}{10{,}000} = .2.$$

Thus the above method of estimation fails miserably. It does not even converge to the true value when the sample size n goes to infinity. There is no reason to be surprised—$\hat{p}$ is claimed to be a good estimator only if the underlying distribution is exponential. Now let us reanalyze the problem. To estimate the probability that a transistor fails in less than 2000 hours, we tested n of them to failure. A very appealing estimate for the desired probability is given by

$$\hat{\hat{p}} = \frac{m}{n},$$

where m is the number in the sample that failed before 2000 hours. Now let $F(t) = P(T < t)$ be the true underlying cumulative distribution. Then we are trying to estimate the parameter $p = F(2000)$.

7.2 ***Proposition*** *Under the above assumptions,*

$$E_F(\hat{\hat{p}} - p)^2 = \frac{p(1 - p)}{n}.$$

The proof here is simple. We classify the outcomes $T_1, \ldots, T_n$ either as successes, $T_k < 2000$, or failures, $T_k \geq 2000$. The probability of a success is p and we are estimating p by the proportion of successes, so this is the binomial case, for which we know from Chapter 3, p. 49, that 7.2 holds. Now since $p(1 - p) \leq \frac{1}{4}$,

$$E_F(\hat{\hat{p}} - p)^2 \leq \frac{1}{4n}.$$

Therefore, *no matter what the underlying distribution of the failure time, the estimate $\hat{\hat{p}}$ has small squared error loss for large sample size n.*

Suppose the underlying distribution is exponential; how does $\hat{\hat{p}}$ compare with $\hat{p}$? For example, suppose $\beta = 5000$; then from 7.1

$$E(\hat{p} - p)^2 \simeq e^{-4/5}\left(\frac{2}{5}\right)^2 \cdot \frac{1}{n} \simeq \frac{.07}{n}.$$

On the other hand, since

$$p = 1 - e^{-2000/\beta} = .33,$$

$$E(\hat{\hat{p}} - p)^2 \simeq \frac{.22}{n}.$$

Certainly, then, $\hat{p}$ is better than $\hat{\hat{p}}$ at $\beta = 5000$ if the underlying distribution is exponential. In fact we can easily show that $\hat{p}$ is better than $\hat{\hat{p}}$ for all values of β.

This example illustrates a fundamental principle in statistics: Suppose we assume for our underlying statistical model a set of distributions

$\{P_\theta\}$, $\theta \in \Theta$, that are fairly well specified, and then in terms of this model, find a good estimator for some characteristics of the true underlying distribution. If the true distribution of the population is not closely approximated by one of the set $\{P_\theta\}$, $\theta \in \Theta$, then the estimator can have an extremely large error, no matter how large the sample size. On the other hand, suppose we take a much larger set of underlying distributions $\{P_\theta\}$, $\theta \in \Theta'$, where Θ' includes many more distributions, and now form an estimator of the characteristic. This estimator will be good for the much larger class of states of nature Θ', and using it will avert the possibility of a disastrously bad estimate if the underlying population distribution is in Θ' but not in Θ. However, it is always true that the "nearly best" estimator relative to a smaller set of states of nature Θ will always have a smaller loss when the true state of nature is in Θ than the "nearly best" estimator for a larger set of possible states of nature Θ'. These two factors must be balanced against each other in order to construct reasonable estimators for the problem.

Going back to our example, at $\beta = 5000$, $\hat{\hat{p}}$ has an efficiency of $.07/.22 \simeq .3$, which means we are losing about 70% of our sample size if we use $\hat{\hat{p}}$ instead of $\hat{p}$. This is a substantial decrease in efficiency. Suppose we know, or feel safe in assuming, that the true distribution does not deviate very much from an exponential. Can't we come up with an estimate that has higher efficiency than $\hat{\hat{p}}$ but does not behave badly if the distribution is not exactly exponential?

Similar reasoning applies to the problem discussed earlier of measuring samples of rods to find the underlying proportion p of rods that are smaller than l. One obvious method is to keep track of the proportion $\hat{\hat{p}}$ of the sampled rods that are less than l and use $\hat{\hat{p}}$ as an estimate of p.

We asked if we discarded useful information by not recording the exact measurements of all rods sampled, instead of whether or not they are shorter than l. Suppose we record these measurements, and find them consistent with the belief that the underlying distribution for the length of the rods was $N(\mu, \sigma^2)$ with unknown μ and σ^2. Suppose further that on physical grounds we have very good reasons to believe that the lengths of the rods follow a normal distribution. Then by using the maximum likelihood estimates for μ and σ^2 we can come up with the maximum likelihood estimate $\hat{p}$ for the probability of the length being less than l. If the true underlying distribution of rod lengths is closely approximated by some normal distribution, then the estimator found this way will be more efficient than $\hat{\hat{p}}$. If not, we may be in trouble.

In this chapter we face the problem that *we never know the exact shape of the underlying distribution.* The fact is that in nature there are no normal distributions, no exponential distributions, and so on. There are only distributions that are approximately normal or approximately exponential. Furthermore, we know that if the underlying distribution looks at all like a normal distribution, it usually takes a large sample size to detect even substantial departures from normality, and similarly for detecting departures from an exponential shape or any other small family of distributions. Thus, if you assume that the underlying distribution is normal and use estimates of various parameters that are based on this assumption, you may be putting your statistical head on the block. To safeguard against this danger, there are two courses that can be taken:

a If you do not want to assume that the underlying distribution is nearly normal or nearly exponential, that is, if you want to assume as little as possible about the shape of the underlying distribution, then look for estimates that are "good" over a very large class of possible distributions. This kind of estimator we call (loosely) *nonparametric*.

b Assume some small family of possible distributions $\{P_\theta\}$, $\theta \in \Theta$, and try to find relatively high efficiency estimates whose performance or efficiency is not sensitive to relatively small departures from the assumed family $\{P_\theta\}$. This brings us to the subject of *robust estimation*.

Problem 1 Graph the efficiency of $\hat{\hat{p}}$ as a function of p for $0 \leq p \leq 1$. Where is $\hat{\hat{p}}$ least efficient? Over what range does it achieve its highest efficiency? (Assume that the underlying distribution is exponential.)

(In the next two problems, use the following technique: Suppose you are interested in estimating a parameter ϕ which is some function $g(\theta)$ of the parameter θ. Let $\hat{\theta}$ be the maximum likelihood estimate for θ; then the maximum likelihood estimate for ϕ is $\hat{\phi} = g(\hat{\theta})$. For large n we can get the squared error loss for $\hat{\phi}$ by reasoning that $\hat{\theta}$ must usually be close to θ. Hence the partial Taylor expansion

$$\hat{\phi} = g(\hat{\theta}) = g(\theta + \hat{\theta} - \theta) = g(\theta) + (\hat{\theta} - \theta)\frac{dg(\theta)}{d\theta}$$

gives a good first approximation. Since $\phi = g(\theta)$, we get

$$E_\theta(\hat{\phi} - \phi)^2 = E_\theta(\hat{\theta} - \theta)^2\left(\frac{dg(\theta)}{d\theta}\right)^2. \tag{7.3}$$

The loss in estimating θ by $\hat{\theta}$ is usually known or computable, so applying 7.3 gives the loss for $\hat{\phi}$.)

Problem 2 Suppose you are investigating the distribution of orders that come into a job-shop (a shared-time computer, for example) per unit time. In particular, suppose you want to estimate the probability p_0 that no orders arrive during unit time. To model this, assume that the numbers $x_1, \ldots, x_n$ of incoming orders in n successive unit time periods are the outcomes of n independent and identically distributed random variables. One estimate for p_0 is the proportion $\hat{\hat{p}}_0$ of x_k equal to zero. Compute the efficiency of $\hat{\hat{p}}_0$ as a function of p_0 if

a the underlying distribution is Poisson with parameter λ

b the underlying distribution is geometric, i.e.,

$$P(X = k) = (1 - r)r^k.$$

Problem 3 (Difficult!) Assume that the distribution of rod lengths is $N(\mu, \sigma^2)$ with σ^2 known, μ unknown.

a Find the maximum likelihood estimator $\hat{p}$ for $P(X < l)$, where X is the length of a rod drawn at random.

b In a sample of size n, estimate $P(X < l)$ by $\hat{\hat{p}}$, the proportion of the n rods sampled with length less than l. Find the efficiency of $\hat{\hat{p}}$ as a function of l, μ, and σ^2 and evaluate it for $l = \mu - \sigma$ and $l = \mu - 2\sigma$. (Ans. .44, .13.)

The Sample Distribution

Suppose we want estimates that are good over a wide range of underlying distributions. Then, we have to first go back to the aggravating but vital question of what it is we want to estimate. The reason this question comes up again is this: If we are dealing with a 1- or 2-dimensional parameter family of distributions $\{P_\theta\}$, $\theta \in \Theta$, then it is clear that we want to estimate the parameter θ. Thus, the reason that the sample mean is important if the underlying distribution is $N(\mu, \sigma^2)$, Poisson, or exponential, is not that the sample mean is always inherently desirable to compute, but that in these cases it is the maximum likelihood estimate for a parameter. But if we assume that our observations are outcomes of independent random variables with a common distribution $P(dx)$ and that $P(dx)$ can be any 1-dimensional distribution, then what shall we estimate? We can form the sample mean

$$\bar{x} = \frac{x_1 + \cdots + x_n}{n},$$

and this is a consistent estimation method for the first moment $\int xP(dx)$ of the underlying distribution, but why estimate the first moment?

Fundamentally, the purpose of estimation is to get significant information about the characteristics of the underlying distribution. In a parametric problem, it makes sense to estimate the parameters, because these completely determine the underlying distribution. In a nonparametric situation, the decision as to what to estimate is less clear. Often, the characteristics of the underlying distribution that you want to estimate are determined by the use that will subsequently be made of the information. The first moment of the underlying distribution may be relatively useless information, whereas knowing the median may be important. In general, though, we usually want to estimate things that determine what the underlying distribution looks like—its shape, scale, location, and so forth.

The most important example of such estimates is the estimator that was basic in the last chapter. We used the approach that no matter what the underlying distribution, the proportion $\hat{P}(I)$ of the sample falling into the interval I was a good estimate for $P(I)$, and all of our tests and estimates were based on this belief. But in what sense is $\hat{P}(I)$ a good estimate of $P(I)$? We formulate the model by assuming that $x_1, \ldots, x_n$ are the outcomes of independent identically distributed random variables $\mathsf{X}_1, \ldots, \mathsf{X}_n$ with an unknown distribution $P(dx)$, which may be anything—continuous, mixed, or discrete. Thus $\{\mathcal{P}\}$, the set of possible underlying distributions for $\mathbf{X}$, is equivalent to the set of all possible 1-dimensional distributions. As a random variable,

$$\hat{\mathsf{P}}(I) = \frac{1}{n}(\# \text{ of } \{\mathsf{X}_1, \ldots, \mathsf{X}_n\} \in I).$$

By an argument similar to the proof of 7.2, the squared error loss in $\hat{\mathsf{P}}(I)$ is

$$E[\hat{\mathsf{P}}(I) - P(I)]^2 = \frac{1}{n} P(I)[1 - P(I)].$$

7.4 ***Proposition*** *For large n, 100γ% confidence intervals for P(I) are given by*

$$\hat{P}(I) \pm \frac{z}{\sqrt{n}} \sqrt{\hat{P}(I)[1 - \hat{P}(I)]},$$

where z is determined from an $N(0, 1)$ distribution. That is, let Z *be* $N(0, 1)$ *and select z such that*

$$P(-z \leq \mathsf{Z} \leq z) = \gamma.$$

The proof uses again the simple observation that if we let $p = P(I)$, then the estimate $\hat{p} = \hat{P}(I)$ for p is simply the usual estimate for the probability p of success in biased coin-tossing. From Chapter 4, large sample size $100\gamma\%$ confidence intervals for p are

7.5
$$\hat{p} \pm \frac{z}{\sqrt{n}} \sqrt{\hat{p}(1 - \hat{p})}.$$

Substituting $\hat{P}(I)$ for $\hat{p}$ in 7.5 gives the intervals stated in the proposition.

There are two observations to be made about the above: First, notice that no matter what the underlying distribution, the estimates $\hat{\mathsf{P}}(I)$ converge to $P(I)$ as the sample size $n \to \infty$; that is, the estimation method is consistent. Second, we know that the efficiency of this method of estimating $P(I)$ may be low when applied to small families of distributions $\{P_\theta\}$, $\theta \in \Theta$. Nevertheless, we will soon see that $\hat{P}(I)$ is the maximum likelihood estimate for $P(I)$ under the model formulated above. This lends some support to our belief that $\hat{P}(I)$ is a very good estimator of $P(I)$ when we want to assume very little about the underlying distribution other than that the $\mathsf{X}_1, \ldots, \mathsf{X}_n$ are independent and identically distributed.

In the above discussion we have looked at $\hat{P}(I)$ as an estimate of $P(I)$ for I a fixed interval. For instance, we think of $\hat{P}([0, 3])$ as an estimate of $P([0, 3])$, $\hat{P}([7, 10))$ as an estimate of $P([7, 10))$, and so on. We can use the same approach to estimate the *entire* underlying distribution.

7.6 ***Definition*** *Given a sample $x_1, \ldots, x_n$ from n independent identically distributed random variables, the sample distribution $\hat{\mathsf{P}}(dx)$ is that discrete distribution which assigns probability $1/n$ to each of the points $x_1, \ldots, x_n$.*

This is consistent with our definition of $\hat{P}(I)$ above. If $\hat{P}$ assigns probability $1/n$ to each of $x_1, \ldots, x_n$, then the probability it assigns to an interval I is $1/n$ times the number of $x_1, \ldots, x_n$ in I. Furthermore, the sample cumulative distribution function $\hat{F}(x)$ defined in the last chapter is the cumulative distribution function of the discrete distribution $\hat{P}$.

We know that $\hat{P}$ is a consistent estimation method for the underlying 1-dimensional distribution P in the sense that for any interval I, the loss in $\hat{P}(I)$ goes to zero as $n \to \infty$. But there are many consistent estimators for P. To try to get the "best" estimate, we ask: Given the outcomes $x_1, \ldots, x_n$, what is the maximum likelihood estimate for the true underlying 1-dimensional distribution $P(dx)$? Suppose that all the outcomes

$x_1, \ldots, x_n$ are different, so that $x_i \neq x_j$, $i \neq j$. What distribution makes these outcomes most probable? We want to maximize

$$P(\mathsf{X}_1 = x_1)P(\mathsf{X}_2 = x_2) \cdots P(\mathsf{X}_n = x_n)$$

over all probabilities P. By the assumption of identical distribution, this means maximizing

$$P(\mathsf{X}_1 = x_1)P(\mathsf{X}_1 = x_2) \cdots P(\mathsf{X}_1 = x_n).$$

The distribution that maximizes this expression must have all of its probability concentrated on the points $x_1, \ldots, x_n$; that is,

$$P(\mathsf{X}_1 = x_1) + \cdots + P(\mathsf{X}_1 = x_n) = 1.$$

Otherwise, if there is some interval I containing none of $x_1, \ldots, x_n$ such that $P(\mathsf{X}_1 \in I) > 0$, then we can make $P(\mathsf{X}_1 = x_1) \cdots P(\mathsf{X}_1 = x_n)$ larger by taking the probability that $\mathsf{X}_1 \in I$ and putting it on the point x_1, say. Thus, all we need to consider are probabilities P under which X_1 can take only the values $x_1, \ldots, x_n$.

The rest of the argument is like the one for the multinomial distribution. We have an experiment with a finite possible number of outcomes; in n trials we see each outcome once. What is the maximum likelihood estimate for $p_k = P(\mathsf{X}_1 = x_k)$? We know this from Chapter 2; $p_k = 1/n$. Hence the maximum likelihood estimator for the distribution of X_1 is given by

$$P(\mathsf{X}_1 = x_k) = \frac{1}{n}.$$

If some of the values $x_1, \ldots, x_n$ are equal, if for instance x_1 is repeated k times, then the maximum likelihood estimate for $P(\mathsf{X}_1 = x_1)$ is k/n. But this is equivalent to having probability $1/n$ for each of the observations. We have proved the following:

7.7 ***Proposition*** *The sample distribution $\hat{P}$ as defined in 7.6 is the maximum likelihood estimator for P if the family of underlying distributions corresponds to all independent identically distributed observations.*

In a similar way, we can show that $\hat{P}(I)$ is the maximum likelihood estimate of $P(I)$.

For any parameter which is uniquely defined and finite for all possible 1-dimensional distributions, the maximum likelihood estimate is its value when P is replaced by $\hat{P}$. For example, define a truncated mean by fixing $a > 0$; then

$$\mu_a = \int_{-a}^{a} xP(dx).$$

The reason for the truncation is that the ordinary mean is infinite for some distributions. Now, the maximum likelihood estimate for μ_a is

$$\hat{\mu}_a = \int_{-a}^{a} x\hat{P}(dx)$$

$$= \frac{1}{n} \sum_{|x_k| \le a} x_k.$$

Estimates of this type will be called *sample distribution estimates*. Their most important property is that they are consistent estimators whatever the underlying distribution.

Problem 4 The following numbers were drawn from a uniform distribution on $[0, \theta]$:

0.38,	1.21,	0.54,	1.68,	0.60,
0.22,	1.32,	0.38,	0.94,	1.40,
1.54,	1.20,	0.72,	1.12,	1.38,
1.72,	1.73,	1.62,	0.52,	1.30,
0.60,	0.02,	0.54,	1.18,	1.78.

a Using the estimate $\hat{P}(I)$ defined in this section, estimate $P([0, 1])$.

b Estimate $P([\frac{1}{2}, 1])$, as in **a**.

c Compute the maximum likelihood estimates for **a.**, **b.**, given that the underlying distribution is uniform on $[0, \theta]$.

d Compare the loss of the estimate in **a.** with that of the maximum likelihood estimate for $P([0, 1])$, assuming that $\theta > 1$. (See Chapter 3, p. 69, and use 7.3.)

(Ans. **a.** 11/25, **b.** 7/25, **c.** .56, .28.)

Problem 5 Observations are being made from some distribution with density $f(x)$. It is known that the distribution has mean zero and variance one. There are two candidates for $f(x)$: the $N(0, 1)$ density, and the two-tailed exponential density

$$f(x) = \frac{1}{\sqrt{2}} e^{-\sqrt{2}|x|}.$$

To discriminate, it is proposed to use the estimate $\hat{P}([-1, +1])$. Discuss how you would use this estimate to discriminate between the two distributions, and discuss how large a sample size you will need to guarantee a given level of discrimination.

Problem 6 For the data of Problem 4, graph $\hat{F}(x)$ and draw the 90% confidence band about $\hat{F}(x)$. Do you think that the underlying distribution could have been

a uniform on [0, 3]

b uniform on [0, 2]

c $N(1, 1/4)$?

Problem 7 What do you think of the assertion that the data in Problem 4 comes from an exponential distribution with mean one? See if you can find a simple way to reject at the .1 level.

Problem 8 Find the squared error loss in $\hat{\mu}_a$ as an estimator of μ_a. Is $\hat{\mu}_a$ a consistent estimator for all possible underlying distributions?

The Median and Quantiles

Another set of numbers useful in finding the size and shape of a distribution are the median and the quantiles. The *median* ν is the "half-probability" point of the distribution. A *quantile* x_p is the point such that the probability of getting an outcome less than x_p is p. In the long run, about half the readings drawn independently from a distribution $P(dx)$ should be below the median ν (or $x_{.5}$), and half above; about 40% of the outcomes should be less than $x_{.4}$ and about 60% larger than $x_{.4}$. These numbers x_p are also called *percentiles.* For example, $x_{.4}$ is also called the 40th percentile point of the distribution; $x_{.95}$ the 95th percentile point.

In failure time characteristics, ν gives the "half-lifetime" point: Half the components die before time ν and half afterwards. If you are studying the failure time performance of complex systems containing a component with failure time distribution $P(dt)$, the mean and variance of this distribution are not as useful as some of the quantiles. Certainly information such as "5% of the components fail before 378 hours," "10% before 618," and so forth, is very relevant to the problem.

A knowledge of a few of the quantiles of a distribution gives a good working idea of how the distribution assigns its probabilities. Knowing ν tells us where the midpoint of the distribution is located. Knowing $x_{.25}$ and $x_{.75}$ gives some idea of how concentrated the distribution is to the left and right of ν, respectively. For instance, if for a continuous distribution $\nu = 8$, $x_{.25} = 5$, and $x_{.75} = 8.5$, then we know that

$$P(8 \leq X \leq 8.5) = .25, \quad P(X \geq 8.5) = .25,$$
$$P(5 \leq X \leq 8) = .25, \quad P(X \leq 5) = .25.$$

The distribution gives fairly large probability (.25) to the narrow range between 8 and 8.5 and gives .25 probability to the wide range from 5 to 8. The real usefulness of the quantiles is that:

a We can estimate them nonparametrically. That is, there are estimators for x_p which are "good" over a large class of underlying distributions.

b The estimates for x_p can be easily computed from the sample. The only necessary computation is counting.

c The $100\gamma\%$ confidence intervals for x_p do not depend on the underlying distribution (as long as it is continuous).

Before developing these points, we have to go back and examine our definitions more carefully. Look at a random variable X with distribution

$$P(X = -1) = \tfrac{1}{2}$$
$$P(X = 1) = \tfrac{1}{2}.$$

What is the median of this distribution? The temptation is to say zero, because

$$P(X \geq 0) = \tfrac{1}{2}, \qquad P(X \leq 0) = \tfrac{1}{2}.$$

But notice that

$$P(X \geq .2) = \tfrac{1}{2}, \qquad P(X \leq .2) = \tfrac{1}{2},$$

So .2 is a half-probability point also, and so is any other number in the interval between -1 and $+1$. Do we get this multiplicity of medians because the distribution is discrete? Not at all. Look at a distribution with a density like that shown in Figure 7.1. This distribution spreads probability $\frac{1}{2}$ uniformly on each of the intervals [0, 1] and [2, 3]. Take x to be any number between 1 and 2. Then

$$P(X \leq x) = \tfrac{1}{2}, \qquad P(X \geq x) = \tfrac{1}{2}.$$

Not only may there not be a unique median, but if the distribution is discrete, there may not even be a half-probability point. For example,

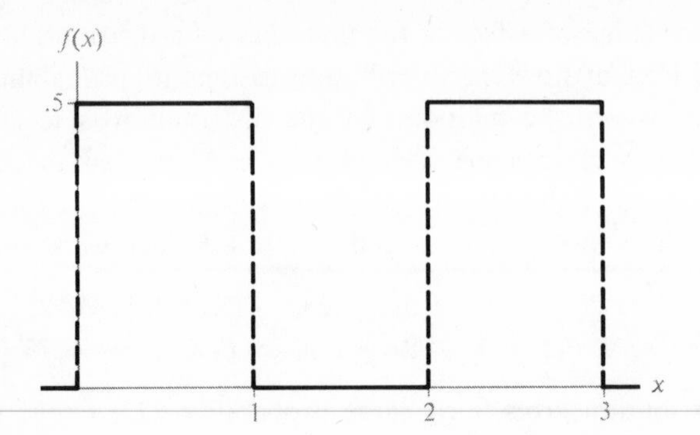

Figure 7.1

take $P(\mathrm{X} = 0) = \frac{1}{4}$, $P(\mathrm{X} = 1) = \frac{1}{2}$, $P(\mathrm{X} = 2) = \frac{1}{4}$. One might expect the point $x = 1$ to be the median, but

$$P(\mathrm{X} \leq 1) = \tfrac{3}{4}, \qquad P(\mathrm{X} \geq 1) = \tfrac{3}{4}.$$

Usually, the median is the solution of the equation $F(x) = \frac{1}{2}$. To locate v, graph $F(x)$, and draw the line at height $\frac{1}{2}$. Then the x-coordinate of the intersection point is v (see Figure 7.2). The reason this does not work in the example graphed in Figure 7.1 is illustrated in Figure 7.3.

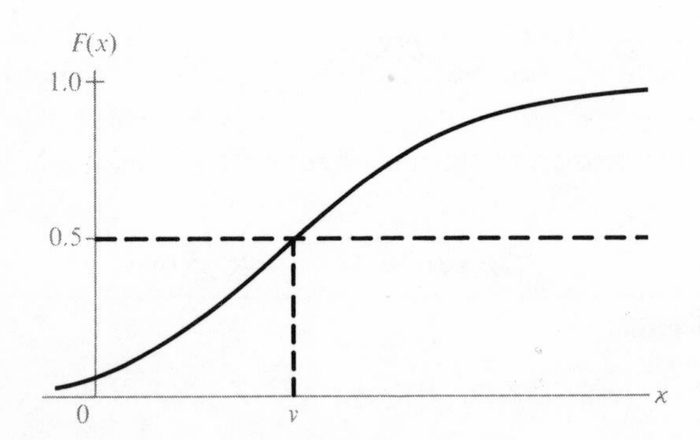

Figure 7.2 *Graphic determination of the median*

If we wanted to be more picky, when there are many possible values for the median we could choose one by some well-defined rule, for example, by taking the middle value. Instead, we will work in this section with distributions for which all quantiles x_p are uniquely defined. First of all, assume that the *distribution is continuous.* Then our basic definitions are:

7.8 ***Definition*** *The median v of a distribution (or $\mathrm{x}_{.5}$) is any number satisfying*

$$P(\mathrm{X} \leq v) = \tfrac{1}{2}.$$

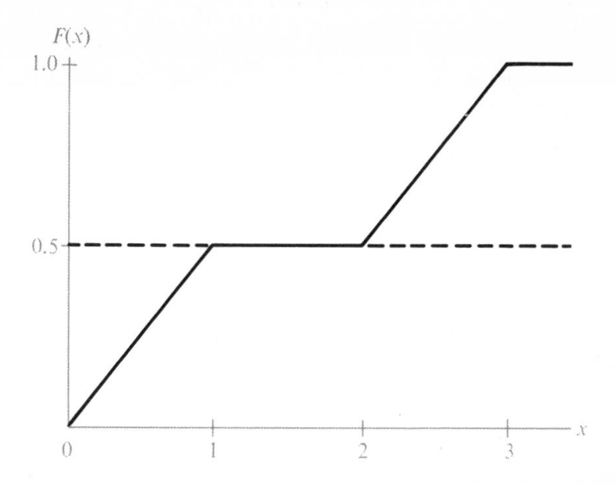

Figure 7.3 *A cumulative distribution function for the density of Figure 7.1*

7.9 ***Definition*** *The quantile* x_p *of a distribution is any number satisfying*

$$P(\mathsf{X} \leq \mathsf{x}_p) = p.$$

This says that x_p is any solution of the equation $F(x) = p$. Let the range of the variable X be defined as the smallest interval $[x_0, x_1]$ such that

$$P(x_0 \leq \mathsf{X} \leq x_1) = 1.$$

Looking at Figure 7.4, you can see that there is a unique quantile value x_p for any p, $0 < p < 1$, if $F(x)$ has no "flat spots" between x_0 and x_1. That is, $F(x)$ steadily increase from the value zero at x_0 (which may be $-\infty$) to the value one at x_1 (which may be $+\infty$), and have no interval in this range on which it is constant. If the density $f(x)$ is positive for all

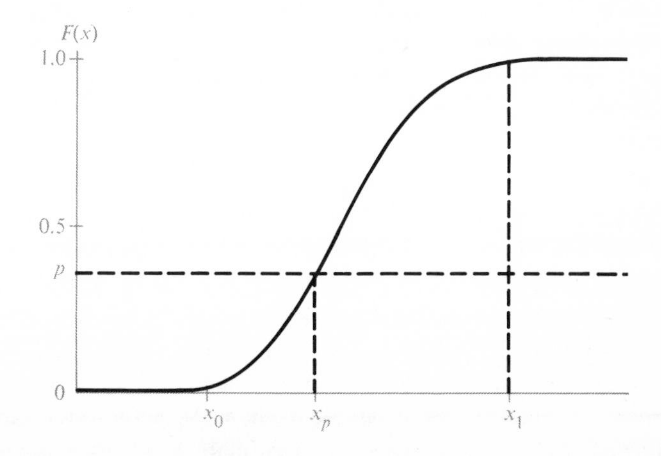

Figure 7.4 *Graphic determination of a quantile*

x between x_0 and x_1, then $F(x)$ has no "flat spots" in its range. For the remainder of this section, assume that all distributions we work with increase everywhere on their range, so that all quantiles are uniquely and simply defined.

Even when there is a unique median, or half-probability point, with

$$P(\mathsf{X} \leq \nu) = P(\mathsf{X} \geq \nu) = \tfrac{1}{2},$$

the median does not have to be equal to the mean. For example, let F be exponential: $P(\mathsf{X} \geq x) = e^{-x}$. We know the mean is unity (the parameter $\beta = 1$). But the median is the value of x that satisfies

$$e^{-x} = \tfrac{1}{2},$$

which equals log 2. There is an important special case where the median equals the expectation. If the distribution has a density function that is symmetric around some value μ; that is, if

$$f(\mu - x) = f(\mu + x),$$

then both median and expectation equal μ. Therefore, for the $N(\mu, \sigma^2)$ distribution, the median equals μ.

Suppose we have sampled from a distribution n times, getting the values $x_1, \ldots, x_n$. Then how do we estimate the median and quantiles from this sample data? The answer is truly simple—define the sample median $\hat{\nu}$ as the "midvalue" of the sample: half the sample lies above this value and half below. Similarly, define the sample quantile as that value $\hat{\mathsf{x}}_p$ such that a fraction p of $x_1, \ldots, x_n$ fall below $\hat{\mathsf{x}}_p$ and a fraction $1 - p$ above. Of course, $\hat{\nu}$ will be our estimate for ν and $\hat{\mathsf{x}}_p$ for x_p. Another way of looking at these estimates is that they are estimates based on the sample distribution. The median ν of P is estimated by $\hat{\nu}$, the median of the sample distribution $\hat{P}$. Similarly, the $\hat{\mathsf{x}}_p$ are the quantiles of the sample distribution.

We need to make these definitions more precise. Arrange $x_1, \ldots, x_n$ in increasing order, getting the order statistics

$$x_{(1)} < x_{(2)} < \cdots < x_{(n)}.$$

Incidentally, the reason that there is strict inequality is that for a continuous distribution the probability is zero that a sample drawn from it will have any repeated values.

Recall that if n is odd ($n = 2m + 1$), then $\hat{\nu}$ is uniquely defined as $x_{(m+1)}$. But if n is even ($n = 2m$), any number between $x_{(m)}$ and $x_{(m+1)}$ is as good a candidate as any other. By convention, the midpoint between $x_{(m)}$ and $x_{(m+1)}$ is taken as the sample median. We get into the same difficulty in defining $\hat{\mathsf{x}}_p$. Suppose we want to estimate $\mathsf{x}_{1/3}$ from a sample of size 4, ordered as $x_{(1)}, x_{(2)}, x_{(3)}, x_{(4)}$. Now $\hat{\mathrm{x}}_{1/3}$ should be that value such that $\frac{2}{3}$ of the sample is above it and $\frac{1}{3}$ below it. Do we choose $x_{(1)}$ or $x_{(2)}$ or an intermediate value? For a large sample size, the range of possible values for $\hat{\mathrm{x}}_p$ will be close together and it will not matter much which one we select.

For definiteness, we settle on the following:

7.10 ***Definition*** *Let* $[np]$ *denote the smallest integer greater than or equal to np. Then define the sample pth quantile as*

$$\hat{\mathrm{x}}_p = x_{([np])}.$$

For example, if $n = 4$ and $p = \frac{1}{3}$, then $np = 1 + \frac{1}{3}$, $[np] = 2$, and $\hat{\mathrm{x}}_{1/3} = x_{(2)}$. If $n = 20$, $p = \frac{2}{3}$, then $np = 13 + \frac{1}{3}$, $[np] = 14$, and $\hat{\mathrm{x}}_p = x_{(14)}$.

The estimates $\hat{\mathrm{x}}_p$ *are consistent for all underlying distributions that are continuous and have no "flat spots."* In this sense they are nonparametric estimates relative to the given class of distributions. In fact, it is possible to compute their loss by some straightforward but complicated computations.

7.11 ***Theorem*** *For a distribution with density* $f(x)$,

$$E(\hat{\mathrm{x}}_p - \mathrm{x}_p)^2 \simeq \frac{1}{n} \cdot \frac{p(1-p)}{[f(\mathrm{x}_p)]^2}.$$

This result will turn out to be handy. Using 7.11 we can illustrate the effects of using the estimates $\hat{\mathrm{x}}_p$ in a small family of underlying distributions. Suppose the distributions $\{P_\theta\}$ are exponential and we want to estimate the median. For an exponential with parameter β, the median is the solution of

$$P_\beta(\mathrm{X} \leq x) = \tfrac{1}{2},$$

or

$$1 - e^{-x/\beta} = \tfrac{1}{2},$$

which gives

$$v = \beta \log 2.$$

The maximum likelihood estimate for v is $\hat{\beta} \log 2$, where $\hat{\beta}$ is the sample mean. Its loss is

$$\begin{aligned} E_\beta(\hat{\beta} \log 2 - v)^2 &= (\log 2)^2 E_\beta(\hat{\beta} - \beta)^2 \\ &= (\log 2)^2 \cdot \beta^2/n \end{aligned}$$

By 7.11, with $\hat{v}$ the sample median,

$$E_\beta(\hat{v} - v)^2 = \frac{1}{4n} \frac{1}{[f_\beta(v)]^2}.$$

Now $f_\beta(x) = \beta^{-1}e^{-x/\beta}$, so

$$f_\beta(\nu) = \frac{1}{\beta}\, e^{-\nu/\beta} = \frac{1}{2\beta}\,.$$

Substituting, we get

$$E_\beta(\hat{\nu} - \nu)^2 = \beta^2$$

This gives $(\log 2)^2 \simeq .49$ as the efficiency of $\hat{\nu}$ if $\{P_\beta\}$ is the family of exponential distributions. We lose half the sample size if we want to use $\hat{\nu}$ instead of $\hat{\beta} \log 2$. What we gain is the assurance of knowing we are using an estimator which is good for any underlying distribution that is continuous and steadily increasing.

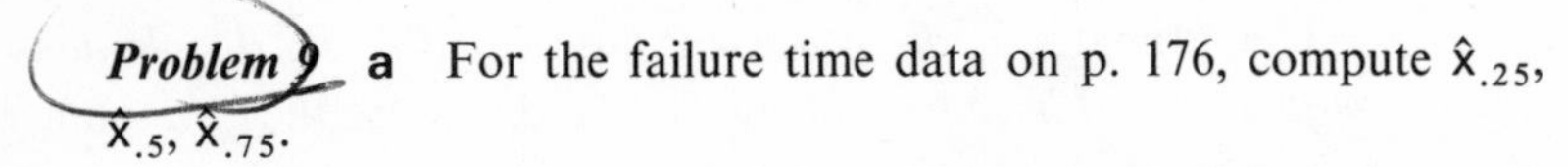

Problem 9 **a** For the failure time data on p. 176, compute $\hat{x}_{.25}$, $\hat{x}_{.5}$, $\hat{x}_{.75}$.

b If the underlying distribution is exponential, what is the ratio $x_{.5}/x_{.25}$? $x_{.75}/x_{.5}$? Does this help you to form any opinion about the underlying distribution of this sample?

c To get some idea of orders of magnitude, assume the underlying distribution is exponential with parameter 41.1. Compute $x_{.25}$, $x_{.5}$, $x_{.75}$ and the asymptotic loss $\hat{x}_{.25}$, $\hat{x}_{.5}$, $\hat{x}_{.75}$ using Theorem 7.11. Now what do you think about the compatibility of the values computed in **a.** with the hypothesis that the underlying distribution is exponential?

Problem 10 Compute the efficiency of the median $\hat{\nu}$ as an estimator of the mean of an $N(\mu, \sigma^2)$ distribution. (Ans. $2/\pi$.)

Problem 11 **a** If the underlying family of distributions is assumed to be uniform on $[0, l]$, will the sample median $\hat{\nu}$ be a good estimate for $l/2$?

b The numbers in Table 10 are drawn at random from a uniform distribution on $[0, 1]$. For the second 50 of these numbers, compute $\hat{\nu}$ and the maximum likelihood estimate of the mean.

Problem 12 **a** Compare the squared error loss for $\hat{x}_{.95}$ and $\hat{x}_{.5}$ if the underlying distribution is exponential with parameter 1.

b Repeat assuming the underlying distribution is $N(0, 1)$.

c In nonparametric situations, the estimation of the location of the "tails" of a distribution is generally much more difficult than the estimation of the location of the "central portion." How is this disparity reflected in 7.11?

Confidence Intervals for Quantiles

For finding confidence intervals, 7.11 is practically useless. The quantiles are useful when we know very little about the underlying distributions. Certainly, then, we would not expect to have knowledge as detailed as the density of the distribution evaluated at x_p. So even though we have used the squared error loss to compute confidence intervals in parametric situations; here this is a futile approach. Fortunately, one of the most pleasant things about the quantiles is that there is a simple and universal system of confidence intervals for them. Let $\mathsf{X}_{(1)} < \mathsf{X}_{(2)} < \cdots < \mathsf{X}_{(n)}$ be the random variables whose outcomes are the order statistics for the sample. That is, if the outcomes of $\mathsf{X}_1, \ldots, \mathsf{X}_n$ are $x_1, \ldots, x_n$, then

$$\mathsf{X}_{(1)} = x_{(1)}, \ldots, \mathsf{X}_{(n)} = x_{(n)}.$$

Reason this way: We have a quantile x_p. We estimate it by $x_{([np])}$. Letting $[np] = m$, can we find confidence intervals of the form

$$x_{(m-k)} \leq \mathsf{x}_p \leq x_{(m+k)}?$$

In other words, can we get $100\gamma\%$ confidence intervals by using the order statistics? For instance, if $n = 100$, can we choose k such that the probability that

$$\mathsf{X}_{(50-k)} \leq v \leq \mathsf{X}_{(50+k)}$$

is always at least γ? If we could do this, then we could compute confidence intervals simply by ordering and counting. For moderate sample sizes our need for a desk calculator or computer would be eliminated.

Formally, can we find m_1 and m_2 such that

7.12 $$P(\mathsf{X}_{(m_1)} \leq \mathsf{x}_p \leq \mathsf{X}_{(m_2)}) \geq \gamma$$

for all underlying distributions? At a first look, it seems as though m_1 and m_2 must depend on the unknown underlying distribution, so we would have to calculate m_1 and m_2 for a given underlying distribution and then try to show that we could get an m_1 and m_2 that make 7.12 work for all distributions. Fortunately, our first impression is wrong:

7.13 ***Theorem*** *For fixed m_1, m_2, n, p, the value of*

7.14 $$P(\mathsf{X}_{(m_1)} \leq \mathsf{x}_p \leq \mathsf{X}_{(m_2)})$$

is the same for all continuous underlying distributions.

With 7.13, we could proceed by taking $X_1, \ldots, X_n$ to have any continuous distribution, say, uniform on [0, 1]. Then compute m_1 and m_2 so that 7.12 holds. By Theorem 7.13 it then holds for all underlying continuous distributions. Hence $[x_{(m_1)}, x_{(m_2)}]$ form $100\gamma\%$ confidence intervals for x_p.

The proof of 7.13 rests on the same transformation used to get confidence bands for $F(x)$, the cumulative distribution function of the underlying distribution. Denote again,

$$y_k = F(x_k), \qquad Y_k = F(X_k).$$

Assume $F(x)$ is steadily increasing; then $y_k < y_j$ if and only if $x_k < x_j$. Therefore, for the ordered sample and order statistics, we have

$$y_{(k)} = F(x_{(k)}), \qquad Y_{(k)} = F(X_{(k)}).$$

The inequality

$$X_{(m_1)} \leq x_p \leq X_{(m_2)}$$

is equivalent to

$$F(X_{(m_1)}) \leq F(x_p) \leq F(X_{(m_2)}),$$

or

$$Y_{(m_1)} \leq p \leq Y_{(m_2)}.$$

In Chapter 6, p. 211 we showed that $Y_1, \ldots, Y_n$ have a uniform distribution on [0, 1]. Thus $Y_{(m_1)}$ and $Y_{(m_2)}$ are order statistics from a uniform distribution, and of course, for a uniform distribution on [0, 1], $x_p = p$. Thus if

$$\gamma = P(Y_{(m_1)} \leq p \leq Y_{(m_2)})$$

for $Y_1, \ldots, Y_n$ having a uniform distribution on [0, 1], then for any continuous underlying distribution

$$\gamma = P(X_{(m_1)} \leq x_p \leq X_{(m_2)}).$$

Actually, we have proved 7.13 only for steadily increasing $F(x)$, but slight modifications will remove this condition.

Now fix n, p, γ and try to find m_1 and m_2 such that

$$P(X_{(m_1)} \leq x_p \leq X_{(m_2)}) \geq \gamma$$

with m_1 and m_2 as close together as possible. The above theorem states that m_1 and m_2 depend only on n, p, γ.

For estimating the median ($p = .5$), exact values of m_1 and m_2 have been computed for small sample sizes (see *The Chemical Rubber Co.*

Handbook of Tables for Probability and Statistics, 1966, p. 122). To get some idea, look at this table for $\gamma = .95$:

n	(m_1, m_2)
10	(2, 9)
20	(6, 15)
30	(10, 21)
40	(14, 27)
50	(18, 33)
60	(22, 39)

To compute the values of m_1 and m_2 for large n, we need an elementary identity:

7.15 ***Proposition*** *Let S_n be the number of heads in n tosses of a biased coin with P(Heads) = p. Then*

$$P(X_{(m_1)} \le x_p \le X_{(m_2)}) = P(m_1 \le S_n < m_2).$$

The proof of this is enjoyable. For a continuous distribution

$$P(X_{(m_2)} = x_p) = 0,$$

so it is enough to prove that 7.15 is true for $P(X_{(m_1)} \le x_p < X_{(m_2)})$. By 7.13 we may as well assume that $X_1, \ldots, X_n$ are uniform on $[0, 1]$, so $x_p = p$. Now the inequalities

a $X_{(m_1)} \le p$

b $X_{(m_2)} > p$

will occur *if and only if*

i m_1 or more of $\{X_1, \ldots, X_n\}$ are $\le p$;
ii less than m_2 of $\{X_1, \ldots, X_n\}$ are $\le p$.

Say we get a head on the kth toss if $X_k \le p$, otherwise a tail, and let S_n be the total number of heads. Then the above conditions become

i $S_n \ge m_1$
ii $S_n < m_2$,

and this proves 7.15.

For n large, take $m_1 = np - k$, $m_2 = np + k$. We want to determine k to get $100\gamma\%$ intervals. Using 7.15 the condition on k is translated into

$$P(np - k \le S_n \le np + k) = \gamma.$$

By the usual central limit approximation (see p. 40) this equals

$$P\left(-\frac{k}{\sqrt{npq}} \leq \mathsf{Z} \leq \frac{k}{\sqrt{npq}}\right),$$

where $q = 1 - p$ and Z is an $N(0, 1)$ variable. Define z by

$$P(-z \leq \mathsf{Z} \leq z) = \gamma,$$

and then take k such that

$$\frac{k}{\sqrt{npq}} = z,$$

or

7.16
$$k = z\sqrt{npq}.$$

7.17 ***Proposition*** *For n large, $100\gamma\%$ confidence intervals for x_p are given (approximately) by*

$$[x_{([np-k])}, x_{([np+k])}],$$

where k is defined by 7.16.

For example, with $\gamma = .95$, $z \simeq 2$, so for the median ($p = .5$), $k = 2\sqrt{n/4} = \sqrt{n}$. With $n = 100$, the 95% confidence interval for ν is given by

$$[x_{(40)}, x_{(60)}].$$

How large are these confidence intervals for $n \gg 1$? The length l_γ of the $100\gamma\%$ interval is given by

$$\mathsf{l}_\gamma = \mathsf{X}_{([np+k])} - \mathsf{X}_{([np-k])}.$$

To get some idea of the size of the intervals, we take the expected value of their lengths, that is, we compute

$$E\mathsf{l}_\gamma = E\mathsf{X}_{([np+k])} - E\mathsf{X}_{([np-k])}.$$

If the distribution has a density $f(x)$, then

7.18
$$E\mathsf{l}_\gamma \simeq 2\frac{z}{\sqrt{n}}\frac{\sqrt{p(1-p)}}{f(\mathsf{x}_p)}.$$

Notice how this is related to the squared error loss given in 7.11.

Problem 13 For the failure data on p. 211, using the tabled values of m_1, m_2 given on p. 237, find the 95% confidence interval for $\hat{\nu}$.

Problem 14 Suppose you want to estimate $x_{.95}$ to within $\pm .1$ in sampling from an $N(\mu, 1)$ distribution by using the sample quantile $\hat{x}_{.95}$. Interpret this to mean that the expected length of the 95% confidence interval for $x_{.95}$ is .2. How large a sample size do you need? (Ans. 1790.)

Problem 15 A river dam is customarily built to a taller height than that of the highest river water level in the last 100 years (the 100 years' flood level). Assuming that the yearly maximum river heights for a certain river are independently drawn from the same distribution, what is the probability of exceeding the 100 years flood level in the 10 year period following? (Ans. 1/11.)

Two and More Dimensions

Suppose the data consist of pairs of points $(x_1, y_1), \ldots, (x_n, y_n)$ assumed to be independent outcomes of the random vector (X, Y). How does one go about getting some hold on the joint distribution of (X, Y)? If we suspect that X and Y are independent, tests for independence should be run (see the next chapter). If they are independent, then the joint distribution reduces to a product of the individual distributions, and we know how to deal with these. If the tests reject independence, or if it is obvious that a strong dependence exists, then we have a genuinely 2-dimensional problem on our hands.

The visual display which gives the most insight into the distribution is the *scatter diagram*. This is a plot of every one of the points (x_k, y_k), $k = 1, \ldots, n$, on a piece of graph paper. For any 2-dimensional set R, the proportion $\hat{P}(R)$ of the n points $(x_1, y_1), \ldots, (x_n, y_n)$ that fall into R is a good estimate of $P(R)$ with squared error loss and confidence intervals given by (7.2) and (7.4) with R in place of I. This implies that the density of points in the scatter diagram, as judged visually, gives a rough indication of the distribuuon of probability over the various regions of the plane.

However, in two or more dimensions there is no analogue of the order statistics of the sample. Consider—is (3, 5) < (4, 4), or is (4, 4) < (3, 5)? In other words, there is no natural ordering of the points on the plane. There is no simple and natural generalization of the idea of the quantiles x_p. In general, for a continuous distribution with cumulative distribution

$$F(x, y) = P(X < x, Y < y),$$

there are an infinite number of points satisfying $F(x, y) = p$.

What can be done is this: From an examination of the scatter diagram select a one-parameter family of regions. The idea behind the selection is to have the regions roughly reflect the distribution of probability. For instance, if the points seem symmetrically distributed around some center at about (x_0, y_0), take the family of regions to be, say, the interiors of the circles with center at (x_0, y_0) or the interiors of the squares with center at (x_0, y_0). If the points are nonsymmetrically arranged, then consider a one-parameter family of ellipses, or perhaps of rectangles with axes at a fixed angle and a fixed ratio between the sides.

Now suppose, for instance, that the regions C_r are the interiors of circles of radius r around the point (x_0, y_0). Count the number of points inside any circle. The proportion $\hat{P}(C_r)$ of points inside the circle C_r gives an estimate of its probability. By looking at $\hat{P}(C_r)$ as a function of the one variable r, we get an idea of how the probability is spread over the family of circles around (x_0, y_0). Actually, we are ordering the sample by means of an *order function* of the sample. That is, we define

$$r_k = \sqrt{(x_k - x_0)^2 + (y_k - y_0)^2},$$

so r_k is the distance of (x_k, y_k) from (x_0, y_0). The $r_1, \ldots, r_n$ are independent outcomes selected from the random variable

$$\mathsf{R} = \sqrt{(\mathsf{X} - x_0)^2 + (\mathsf{Y} - y_0)^2}.$$

We can order the values $r_1, \ldots, r_n$ as

$$r_{(1)} \leq r_{(2)} \leq \cdots \leq r_{(n)}$$

and use these order statistics to get estimates of the quantiles of the distribution of R as well as confidence intervals. Then these estimates and intervals can be translated back into statements about the joint distribution of (X, Y). For instance, the 95% confidence interval for the median with $n = 51$ is $[r_{(19)}, r_{(33)}]$. This translates into: Let C_r be the circle such that

$$P[(\mathsf{X}, \mathsf{Y}) \in C_r] = \tfrac{1}{2}.$$

Then 95% of the time, C_r lies between the circles of radius $r_{(19)}$ and $r_{(33)}$.

Similarly, if our regions are squares S_r of sides r parallel to the coordinate axes and with center at (x_0, y_0), then each point (x_k, y_k) falls on one square. Denote the corresponding value of r by r_k. The values $r_1, \ldots, r_n$ are drawn from some random variable R, which can be expressed in terms of X and Y. That is really not necessary. Simply order the values $r_1, \ldots, r_n$ as $r_{(1)}, \ldots, r_{(n)}$ and use the estimates and confidence intervals for the quantiles. Then translate back by saying, for instance, if $n = 51$, let S_r be the square such that $P((\mathsf{X}, \mathsf{Y}) \in S_r) = \frac{1}{2}$. Then, 95% of the time, the square S_r lies between the two squares $S_{r_{(19)}}$ and $S_{r_{(33)}}$.

For three or more dimensions, a simultaneous visual display of the data is difficult, and there is no scatter diagram to guide us. But the idea of an ordering function follows just as before: Let

$$R = \phi(X, Y, Z)$$
$$r_k = \phi(x_k, y_k, z_k),$$

where $\phi(x, y, z)$ is some conveniently chosen function. For example, $\phi(x, y, z)$ could be the distance of the point (x, y, z) from some fixed point (x_0, y_0, z_0). Now repeat the procedures outlined above on the data $r_1, \ldots, r_n$. Incidentally, the regions S_r are given by

$$S_r = \{(x, y, z);\quad \phi(x, y, z) < r\}.$$

Problem 16 Your data consist of the 50 pairs of points given below. Construct the scatter diagram, select ordering regions, and do whatever estimation and confidence interval construction you feel is necessary to get a rough picture of the joint distribution.

1st	1.51	1.91	1.97	1.03	1.09	1.59	1.54	1.57	1.03	1.40
2nd	1.40	1.40	1.36	0.09	0.92	0.68	1.03	1.65	0.76	1.15
1st	1.95	1.24	1.88	1.01	1.60	1.86	1.39	1.42	1.65	1.49
2nd	1.51	1.00	1.78	0.98	0.00	1.77	1.05	0.54	1.31	1.46
1st	1.93	1.79	1.21	1.97	1.91	1.73	1.39	1.86	1.04	1.23
2nd	1.18	1.05	0.81	0.13	1.57	0.88	1.67	1.54	1.41	0.56
1st	1.15	1.37	1.96	1.38	1.19	1.43	1.10	1.64	1.84	1.47
2nd	0.33	0.76	0.99	0.97	0.60	0.46	0.68	0.91	0.94	0.70

The Midpoints of Slightly Deviant Distributions

Every so often, the results of some independently repeated measurement may show most of the data nicely clustered around some central section but with a few readings (the wild shots) scattered far away. The first impulse is to regard these few extremely large or extremely small values as products of some malfunction and throw them out. But in repeated sampling from any distribution, occasionally we are bound to get readings near the top and bottom of the range. For example, suppose in testing exponentially distributed failure times of some component, one sees that the component with the smallest failure time lasted less than one-fourth as long as the next component to fail. But the probability of this occurrence can be computed and is close to .25 for large n. The point of this example is that the few largest and smallest values in a sequence of observations, even though they may seem outrageously large or small, may be quite consistent with the underlying distribution. Thus, the voice of statistical caution argues that the outlying values may be genuine data points, and throwing them out would be destroying information. On the other hand, some wrong button may have been pushed, and one or more of the observations may be from a completely different underlying distribution. There is some established theory as to when these outliers can be rejected. But what we wish to point out here is that some estimates are much better in the presence of possible contamination than others.

Here is a simple model for the occurrence of wild shots. Suppose a proportion p of the time you are sampling from an $N(\mu, \sigma^2)$ distribution with μ, σ^2 unknown, and you want to estimate μ. Infrequently, something slips, and a proportion $q = 1 - p$ of the time you get a wild shot drawn from the distribution $N(p, k\sigma^2)$ where $k > 1$. In other words, the wild shots are drawn from a normal population with the same mean but larger variance. Let $f_0(x)$ be the density of the $N(\mu, \sigma^2)$ distribution and $f_1(x)$ that of the $N(\mu, k\sigma^2)$ distribution. Suppose that whether you get a wild shot or not on the next reading is decided by tossing a coin with probability q of heads where heads means "wild shot occurs." Then the probability that on the next reading X falls in the range dx is given by

$$\begin{aligned} P(\mathsf{X} \in dx) &= P(\mathsf{X} \in dx \mid \mathsf{X} \text{ is not a wild shot})\, P(\mathsf{X} \text{ is not a wild shot}) \\ &\quad + P(\mathsf{X} \in dx \mid \mathsf{X} \text{ is a wild shot})\, P(\mathsf{X} \text{ is a wild shot}), \end{aligned}$$

or

$$P(\mathsf{X} \in dx) = pf_0(x)\, dx + qf_1(x)\, dx.$$

Hence X has the density

$$f(x) = pf_0(x) + qf_1(x).$$

Thus the model where wild shots occur randomly according to independent tosses of a biased coin is equivalent to a sequence of independent identically distributed observations $\mathsf{X}_1, \ldots, \mathsf{X}_n$ selected from the distribution with density $pf_0(x) + qf_1(x)$. Notice that for p nearly one, this distribution is approximately $N(\mu, \sigma^2)$. The chief difference is in the tails of the distribution, since $N(\mu, k\sigma^2)$ gives more probability to readings further away from the mean. The graph in Figure 7.5 illustrates the difference between $f_0(x)$ and $pf_0(x) + qf_1(x)$ for $\mu = 0$, $\sigma = 1$, $k = 9$, $p = .9$, $q = .1$. Suppose now we decide to use

$$\hat{\mu} = \frac{\mathsf{X}_1 + \cdots + \mathsf{X}_n}{n}$$

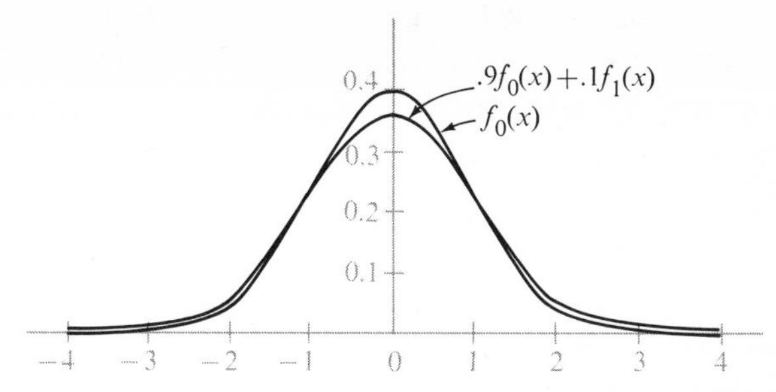

Figure 7.5 *$f_0(x)$ and .9 $f_0(x)$ + .1 $f_1(x)$ for $\mu = 0$, $\sigma = 1$, $k = 9$*

as our estimate for μ. Why not? After all, it is the maximum likelihood estimator if the underlying distribution is normal and the distribution in our model is very close to normal. Notice that the expectation of $\hat{\mu}$ is μ. If X has the density $f(x)$, then its variance is σ^2 with probability p, and its variance is $k\sigma^2$ with probability q. Hence

7.19
$$E(\hat{\mu} - \mu)^2 = V\left(\frac{X_1 + \cdots + X_n}{n}\right)$$
$$= \frac{1}{n} V(X_1) = \frac{1}{n}(p\sigma^2 + qk\sigma^2)$$
$$= \frac{\sigma^2}{n}(p + qk).$$

If qk is large, then the squared error loss is large. Even a very infrequent wild shot, if it is wild enough, gets $\hat{\mu}$ into sizable error. We want an estimate that is not as sensitive to wild shots.

A reasonable question at this point is to ask: Why not derive the form of the maximum likelihood estimate using the model density $f(x) = pf_0(x) + qf_1(x)$? This will give an estimate which is as efficient as possible in this model. This is not desirable nor satisfactory because our assumption that the wild shots follow the distribution $N(\mu, k\sigma^2)$ is largely manufactured out of thin air. We rarely know much about the distribution of outliers except perhaps some order of magnitude ideas about how frequently they occur and about how big they are. What we want is an estimate that is insensitive to the presence of wild shots, no matter what their distribution.

Since the median ν of the distribution is also its mean, why not use $\hat{\nu}$, the sample median, to estimate μ? The reason $\hat{\nu}$ might be better than $\hat{\mu}$ is shown by the following: Take the sample $x_1, \ldots, x_n$ and add one enormously large reading x_{n+1}. This last value will have a large effect on the sample mean, but the sample median will generally change only slightly. If we use the median to estimate μ in an $N(\mu, \sigma^2)$ distribution, its efficiency is $2/\pi = .64$. However, we expect $\hat{v}$ to be less perturbed by contamination. To compare it with $\hat{\mu}$ in our contamination model we compute its loss:

$$f(v) = f(\mu) = pf_0(\mu) + qf_1(\mu)$$
$$= \frac{1}{\sqrt{2\pi}\sigma}\left(p + \frac{q}{\sqrt{k}}\right).$$

By 7.11 the squared error loss is about

$$\frac{\pi\sigma^2}{2n} \frac{1}{(p + q/\sqrt{k})^2}.$$

Notice here that the loss is not sensitive to the size of k as long as p is near one. For $k = 1$ it equals $\pi\sigma^2/2n$, and as $k \to \infty$ it converges to $\pi\sigma^2/2np^2$. The ratio of these two values is $1/p^2 \approx 1$. In other words, the sample median is not much affected by how wild the wild shots are. Contrast this with the squared error loss of $\hat{\mu}$ (see 7.19) which goes to infinity as $k \to \infty$. To compare $\hat{\mu}$ and $\hat{v}$: At the values $k = 9$, $q \simeq .085$; their losses are equal,

$$R(\hat{\mu}) = R(\hat{v}) \simeq \frac{1.8\sigma^2}{n}.$$

Their efficiencies at this point are around 70%. As k increases past 9 or q becomes larger than .085, then $\hat{v}$ becomes a better estimator of μ than $\hat{\mu}$.

Actually, we can do much better. For a symmetric distribution, the mean is the average of any two quantiles equidistant from the median. That is,

$$\mu = \tfrac{1}{2}(x_{.5-d} + x_{.5+d}).$$

So another possible estimate for μ is

$$\tfrac{1}{2}(\hat{x}_{.25} + \hat{x}_{.75}).$$

This has an efficiency of 81% in the case of normality ($q = 0$). But its efficiency increases to around 90% for $k = 9$, $q = .085$. We can even do better. Call the *r per cent trimmed mean* $\hat{\mu}_r$ the estimate we get by removing the upper r% and the lower r% and then taking the mean of the remaining data values. These can be highly efficient estimates of the mean. For instance, the 6% trimmed mean is 97% efficient in the normal, and for $k = 9$ its efficiency never drops below 95% as the contamination increases up to $q = .1$. For larger values of q, its efficiency begins to drop, and eventually becomes much lower.

It is important to notice that unless the sample size is extremely large it is impossible to decide whether the underlying distribution is normal or is a distribution whose tail differs significantly from that of the normal. For example, let $F(x)$ be the normal cumulative distribution and let $G(x)$ be the cumulative for the contaminated normal distribution with $k = 9$, $q = .1$. Then

$$\max_x |F(x) - G(x)| \simeq .024.$$

These distribution functions, even for $q = .1$, are fairly close to each other. To detect the contamination at the .05 level, a sample size of at least 1200 is needed if the Kolmogorov–Smirnov test is used. If a continuous distribution is symmetric about its mean μ and if its density decreases monotonically from a single maximum value at μ, then usually we can find a value of σ^2 so that, as far as we can tell from moderate sample sizes, the underlying distribution is $N(\mu, \sigma^2)$.

In particular, since relatively few of the observations are drawn from the tails of the underlying distribution, it is usually impossible to decide on the basis of a sample of moderate size whether the underlying distribution is normal (or any other hypothesized distribution) or simply a distribution that appears normal in its central section, but has tails which behave in a much different way than the tails of a normal distribution. But the "best" estimators for the mean and variance of a normal distribution are very sensitive to the tail behavior of the underlying distribution. As we saw in our example, if the tail of the underlying distribution is more spread out than the tail of a normal distribution, then using the sample mean as an estimator may be bad estimation policy.

What we would really like is this: For the class of all continuous distributions symmetric around their medians, find an estimator for the median (which also equals the mean in this class) which is good for the entire class. We know that the sample median works, but it has pretty low

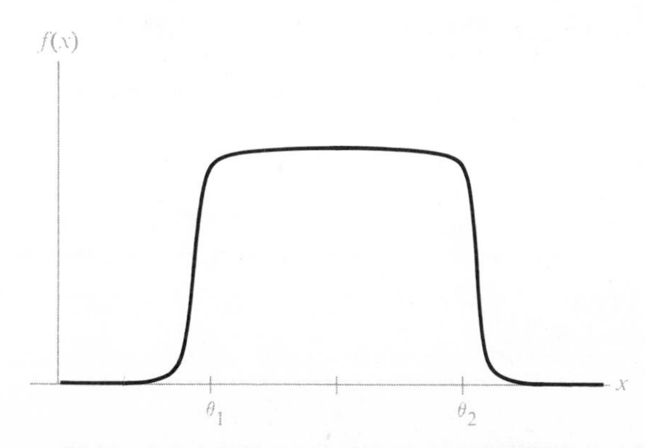

Figure 7.6 *Density of a symmetric distribution approximately uniform on* $[\theta_1, \theta_2]$

efficiency (64%) when used on normal distributions. We also know that $\frac{1}{2}(\hat{x}_{.25} + \hat{x}_{.75})$ is probably better, and the trimmed means are probably even better still. Recent work has shown that there is an estimate that seems to have universally very high efficiency. But the natural context in which to present this estimate is in the next chapter, so we defer discussion until then.

Problem 17 **a** Consider a problem where the underlying distributions are double exponential with parameter $\beta = 1$ centered at μ; that is, with densities

$$f_\mu(x) = \tfrac{1}{2}e^{-|x-\mu|}, \qquad -\infty < x < \infty.$$

Recall that the maximum likelihood estimate of μ is the sample median $\hat{v}$. Since $\hat{v}$ is maximum likelihood, use

$$E(\hat{v} - v)^2 \simeq \frac{1}{nI(\mu)},$$

where

$$I(\mu) = \int_{-\infty}^{+\infty} \left[\frac{\partial}{\partial\mu} \log f_\mu(x)\right]^2 f_\mu(x)\, dx,$$

to find the squared error loss for $\hat{v}$. Compare this with the result using (7.11).

b Suppose that the actual distribution from which a sample is drawn is double exponential with scale parameter $\beta = 1$. This is not known to us. We look at the sample, mutter to ourselves that everything in nature is normally distributed, and conclude that the data come from a normal distribution. If we use the sample mean to estimate the location parameter, what is the efficiency of the estimation? Discuss the effects of our mistaken conclusion on the confidence intervals for the location parameter.

Problem 18 Suppose that the underlying distribution is nearly uniform over $[\theta_1, \theta_2]$ and symmetric about the midpoint $(\theta_1 + \theta_2)/2$ (Figure 7.6). The parameters θ_1, θ_2 are unknown. Why is the *mid-range* $\hat{m}$

$$\hat{m} = \frac{x_{(1)} + x_{(n)}}{2},$$

likely to be a far superior estimate of the midpoint of the distribution than either the sample mean or median?

Problem 19 A colony of one-celled organisms that reproduce by splitting was observed, and the time from the birth of each organism (the splitting of its parent) until it split was recorded. These times seemed to follow an exponential distribution except that it was suspected that extremely large values appeared more frequently than an exponential would account for. You are called upon to estimate the median lifetime v of an organism. How might you proceed in view of the following facts?

a If the distribution is really exponential, then the sample median $\hat{v}$ has an efficiency of only 48% in estimating v compared with $\hat{\beta} \log 2$, where $\hat{\beta}$ is the sample mean.

b If the distribution has a larger tail than an exponential, then $\hat{\beta} \log 2$ may be an inefficient estimator for μ.

Robust Estimates for the Scale of a Normal Distribution

The mean μ of a normal distribution is a location parameter and specifies the location of its peak. The standard deviation σ is a scale parameter and governs how peaked or broad the distribution is, whether it spreads its probability over a wide range or concentrates it closely around the mean.

If $\hat{\sigma}^2$ is the sample variance,

$$\hat{\sigma}^2 = \frac{1}{n} \sum_1^n (x - \bar{x})^2,$$

where $\bar{x}$ is the sample mean, then the maximum likelihood estimator for σ is

$$\hat{\sigma} = \sqrt{\hat{\sigma}^2}.$$

For large samples, we can use (7.3) to compute the expected loss in $\hat{\sigma}$. Use $\theta = \sigma^2$, $g(\theta) = \sqrt{\theta}$ to get

$$E(\hat{\sigma} - \sigma)^2 \cong E(\hat{\sigma}^2 - \sigma^2) \cdot \left(\frac{1}{2\sigma}\right)^2$$

$$\cong \frac{1}{4\sigma^2} E(\hat{\sigma}^2 - \sigma^2)^2.$$

A straightforward computation gives $E(\hat{\sigma}^2 - \sigma^2)^2 \simeq 2\sigma^4/n$; hence

$$E(\hat{\sigma} - \sigma)^2 \simeq \frac{\sigma^2}{2n}. \tag{7.20}$$

So far, so good. But what happens if the distribution is not normal, but we still use the above estimator for the standard deviation σ of the distribution? Define the fourth central moment μ_4 of the distribution as

$$\mu_4 = \int (x - \mu)^4 P(dx),$$

where μ is the mean. Then some computation gives the more general result

$$E(\hat{\sigma} - \sigma)^2 \simeq \frac{1}{4\sigma^2 n}(\mu_4 - \sigma^4). \tag{7.21}$$

We will use this to show that the effects on $\hat{\sigma}$ of even slight nonnormality can be disastrous. Use the model

$$f(x) = pf_0(x) + qf_1(x)$$

introduced in the last section, where $f_0(x)$ is the $N(\mu, \sigma_0^2)$ density and $f_1(x)$ the $N(\mu, k\sigma_0^2)$ density. For a normal distribution $N(\mu, \sigma^2)$, a quick computation gives $\mu_4 = 3\sigma^4$. Hence for the density above,

$$\mu_4 = p(3\sigma_0^4) + q(3k^2\sigma_0^4) = 3\sigma_0^4(p + k^2 q).$$

Let σ^2 be the variance of the given distribution and notice that because of the contamination $\sigma^2 \neq \sigma_0^2$; actually,

$$\sigma^2 = p\sigma_0^2 + qk\sigma_0^2.$$

We are interested in the effects of very small contamination; that is, $p \simeq 1$, $q \ll 1$. Hence, in our computations we assume qk small, so that $\sigma_0 \simeq \sigma$. Using 7.21 we get

$$\begin{aligned} E(\hat{\sigma} - \sigma)^2 &= \frac{1}{4\sigma^2 n}(3\sigma_0^4(p + k^2 q) - \sigma^4) \\ &\simeq \frac{\sigma^2}{2n}\left(1 + \frac{3}{2}k^2 q\right). \end{aligned} \tag{7.22}$$

Take $k = 9$, and $q = .02$. This means that on the average only *one reading in fifty* is contaminated. The right-hand side above becomes

$$\frac{\sigma^2}{2n}(3.4).$$

The squared error loss has *increased by a factor greater than three* with only slight contamination.

Our conceptual difficulties are also increased. We have $\sigma = 1.08\sigma_0$. Therefore, if it is σ_0 we are trying to estimate, then $\hat{\sigma}$ is a biased estimator; moreover, the bias does not go to zero as the sample size increases. For the squared error loss we then get

$$\begin{aligned} E(\hat{\sigma} - \sigma_0)^2 &= E(\hat{\sigma} - \sigma + \sigma - \sigma_0)^2 \\ &= E(\hat{\sigma} - \sigma)^2 + (\sigma - \sigma_0)^2 \\ &\simeq 1.7\frac{\sigma^2}{n} + .064\sigma^2. \end{aligned}$$

The bias becomes the dominant term for $n \geq 270$.

This raises the question of why we should be trying to estimate σ_0 instead of the standard deviation of the contaminated distribution. The reason is that our basic purpose is to find the normal distribution which best fits the data. The unusually large values produced by the contamination are generally quite infrequent. The data look for all practical purposes as though they came from a normal distribution and we are interested in estimating the parameters of the best fitting normal. Using virtually any criterion, the normal which best fits $pf_0(x) + qf_1(x)$ has a standard deviation very close to σ_0. In fact, let

$$F_c(x) = pF_0(x) + qF_1(x)$$

be the cumulative of the contaminated normal distribution. Then the normal distribution whose cumulative $F(x)$ minimizes the Kolmogorov–Smirnov distance

$$\max_x |F_c(x) - F(x)|$$

has standard deviation

$$\sigma \simeq \sigma_0(1 + 1.4q).$$

Of course, the reason for this is that in the central portion, the contaminated distribution for q small looks very much like $N(\mu, \sigma_0^2)$. In fact, as we mentioned, for $q = .02$, 49 out of 50 readings come from the $N(\mu, \sigma_0^2)$ distribution. In a sample of size 500, only 10 readings come from the $N(\mu, k\sigma_0^2)$ distribution (on the average). But of this 10, about 50% will fall in the range $-2\sigma_0$ to $2\sigma_0$. Therefore, an $N(\mu, \sigma_0^2)$ distribution will be a good fit to 495 of the readings.

Now we go to a more extreme case: Let $q = .1$. The the standard deviation σ of the contaminated model is $1.35\sigma_0$. If we use the estimate $\hat{\sigma}$ on a large sample, we get about $1.35\sigma_0$. But the normal with standard deviation $1.35\sigma_0$ gives a bad fit to the underlying distribution. In fact, we know that this normal distribution has .68 of its probability between $\mu - 1.35\sigma_0$ and $\mu + 1.35\sigma_0$. But the contaminated distribution with $q = .1$ has .77 of its probability between these two limits. Furthermore, the distance $d(F_c, F)$ between the cumulative F_c of the contaminated normal distribution and the cumulative F of the normal with $\sigma = 1.35\sigma_0$ is .05. If we take $\sigma = 1.14\sigma_0$, we get a much better fit, with the distance $d(F_c, F)$ dropping to .01—about *one-fifth* as large!

So what we want is an estimate of σ that is fairly efficient if the underlying distribution is normal, but still gives good estimates of σ_0 under conditions of slight contamination. The first candidate is the sample deviation $\hat{d}$ suitably normalized:

$$\hat{d} = \frac{1}{n}\sum_{1}^{n} |x_i - \bar{x}|.$$

If the underlying distribution is normal, then assuming the sample size is large enough so that $\bar{x} \simeq \mu$, we get

$$\begin{aligned} E\hat{d} &= \frac{1}{\sqrt{2\pi}\sigma}\int_{-\infty}^{+\infty} |x - \mu| e^{-(x-\mu)^2/2\sigma^2}\,dx \\ &= \sqrt{\frac{2}{\pi}}\,\sigma. \end{aligned}$$

This suggests that we try an estimate for σ of the form

$$\hat{\sigma}^* = \sqrt{\frac{\pi}{2}}\,\hat{d}.$$

The estimate $\hat{\sigma}^*$ has an efficiency of 88% compared with $\hat{\sigma}$, if there is no contamination. In the presence of contamination $\hat{\sigma}^*$ very rapidly becomes a much better estimator of σ_0 than $\hat{\sigma}$. In fact, it becomes better for values of q as low as .01. Furthermore, its bias is significantly smaller; for $k = 9$,

$$E\hat{\sigma}^* = \sigma_0(1 + 2q).$$

The reason for the improvement is clear: $\hat{\sigma}^*$ does not weight the large values of the sample as highly as $\hat{\sigma}$.

Our next candidate is based on the quantiles. For any variable X, recall that x_p is defined by

$$P(\mathsf{X} \le \mathsf{x}_p) = p.$$

If X has an $N(\mu, \sigma^2)$ distribution, then

$$P(\mathsf{X} \leq x) = P\left(\mathsf{Z} \leq \frac{x - \mu}{\sigma}\right),$$

where Z is $N(0, 1)$. Hence letting z_p be the quantile for an $N(0, 1)$ distribution, we have

$$\frac{\mathsf{x}_p - \mu}{\sigma} = \mathsf{z}_p,$$

or

$$\mathsf{x}_p = \sigma \mathsf{z}_p + \mu.$$

So for any $p_2 > p_1$,

$$\sigma = \frac{1}{\mathsf{z}_{p_2} - \mathsf{z}_{p_1}} (\mathsf{x}_{p_2} - \mathsf{x}_{p_1}).$$

This suggests the estimators for σ given by

$$\hat{\sigma}^{**} = \frac{1}{\mathsf{z}_{p_2} - \mathsf{z}_{p_1}} (\hat{\mathsf{x}}_{p_2} - \hat{\mathsf{x}}_{p_1}).$$

A good choice is $p_2 = .93$, $p_1 = .07$. From Table 1, verify that

$$\frac{1}{\mathsf{z}_{p_2} - \mathsf{z}_{p_1}} = .339,$$

and consider the estimate

$$\hat{\sigma}^{**} = .339(\hat{\mathsf{x}}_{.93} - \hat{\mathsf{x}}_{.07}).$$

The efficiency of this estimate is only 65% at $q = 0$. As the contamination increases, it becomes better. At $q \simeq .04$, its variance becomes less than that of the deviation estimate $\hat{\sigma}^*$. Its bias for, say, $k = 9$, is small for q small:

$$E\hat{\sigma}^{**} \simeq \sigma_0(1 + 1.2q).$$

However, if the fraction of contamination q becomes larger than .07, this estimate becomes inefficient and we should use an estimate

$$C(\hat{\mathsf{x}}_{1-r} - \hat{\mathsf{x}}_r),$$

where $r > q$.

Another set of good estimates are the $100r\%$ trimmed standard deviations $\hat{\sigma}_r$. These are gotten by trimming the top and bottom $100r\%$ from the sample and then forming the sample variance $\hat{\sigma}^2$ from the remaining data. If $r > 2q$ these estimates are superior to either $\hat{\sigma}^*$ or $\hat{\sigma}^{**}$.

The moral is clear: Think hard before using $\hat{\sigma}$ as an estimate of σ.

Problem 20 Take any 50 consecutive numbers from Table 11. For these numbers, estimate σ by

a $\hat{\sigma} = \left(\sqrt{\frac{1}{n}\sum_{1}^{n}(x_k - \hat{\mu})^2}\right)$

b $\hat{\sigma}^* = \left(\sqrt{\frac{\pi}{2}}\,\hat{d}\right)$

c $\hat{\sigma}^{**} = [.339(\hat{x}_{.93} - \hat{x}_{.07})]$

d $\hat{\sigma}_{.1}$.

Take any two of the 50 numbers and replace them by the values $+4$, -3. What is the effect on the four estimates?

Problem 21 (Difficult) Consider trying to estimate the parameter of an exponential distribution by $\hat{\hat{\beta}} = C_p\hat{x}_p$.

a Show that taking C_p as

$$C_p = \frac{1}{\log(1 - p)},$$

makes $\hat{\hat{\beta}}$ a consistent estimator.

b Show that the loss in $\hat{\hat{\beta}}$ as an estimate of β for large n is

$$E_\beta(\hat{\hat{\beta}} - \beta)^2 \simeq \frac{\beta^2}{n}\left(\frac{p}{1 - p}\cdot\frac{1}{[\log(1 - p)]^2}\right).$$

c Show that the value of p which minimizes the above loss satisfies

$$\log(1 - p) = -2p$$

and use a log table to solve this graphically. (Ans. The minimizing p is .80.)

d For $p = .80$ show that the efficiency of the estimate $\hat{\hat{\beta}}$ is 64%.

e Why would you expect $\hat{\hat{\beta}}$ to be a more robust estimate of β than the sample mean? Construct a model for a contaminated exponential distribution and compute losses to verify your beliefs.

Summary

The information that can be extracted from moderate sample sizes regarding goodness of fit of various families of distributions is not very precise. There are two courses of action open:

a Use estimates that are consistent for a large family of underlying distributions. Such estimates are called nonparametric.

Examples of nonparametric estimates are $\hat{P}(I)$, and the sample quantiles $\hat{x}_p$; associated with these are the nonparametric confidence intervals.

The usual difficulty with nonparametric estimates is that if the underlying distribution really is in some small, parametric family, then the best estimates relative to the small family are considerably more efficient than the wide-spectrum estimates. An alternative which avoids this difficulty is:

b If the distribution is "nearly" normal, "nearly" exponential, etc., use estimators that are fairly efficient for the parametric family in question, and are not overly sensitive (in terms of loss of efficiency) to small departures from the assumed distributions. Such estimates are called robust.

We focused on robust estimates for two problems, estimating the midpoint and the scale of normal distributions that were subject to varying amounts of contamination. Unless you are confident that the tails of the underlying distribution are well behaved (that is, follow the normal distribution), the standard estimates $\hat{\mu}$ and $\hat{\sigma}$ should be treated with doubt and suspicion. There are other estimates, just as easy or easier to compute, which are more robust in the presence of contamination.

8 Nonparametric One-Sample Tests

Introduction

The material in this chapter arises out of the study of two problems: paired comparisons and independence. To give a feeling for these problems, here are two examples:

Example a In order to test the effect of a certain drug on weight, pairs of rats were selected in some way, various doses of the drug administered to one rat in each pair, and the other rat used as a control. The way that the pairs were selected is important, and we will come back to this later. For n pairs of rats, our data at the end of the test period consist of n pairs $(x_1, y_1), \ldots, (x_n, y_n)$, where x_i, y_i are the weight gains for the test rat and control rat, respectively, in the ith pair. If due precaution has been taken, we can assume that the $2n$ weight gains $x_1, \ldots, x_n, y_1, \ldots, y_n$ are independent outcomes. The question is, of course, how to decide whether the drug has had any effect. This translates into: Construct a suitable model for testing the hypothesis that *the drug has no effect*.

If the pairs of rats were selected purely at random from a population of rats, then a natural assumption is that the pairs (x_i, y_i) are the outcomes of identically distributed random vectors $(\mathsf{X}_i, \mathsf{Y}_i)$, where X_i has distribution $P_1(dx)$ and Y_i has distribution $P_2(dy)$. But notice that now the pairing is irrelevant. We have a set of n random variables $(\mathsf{X}_1, \ldots, \mathsf{X}_n)$ with one underlying distribution $P_1(dx)$, and a set $(\mathsf{Y}_1, \ldots, \mathsf{Y}_n)$ with another, possibly different, distribution $P_2(dy)$. How the x_i's and y_j's are paired has no bearing on the problem. The question is to decide whether $\mathbf{x}$ and $\mathbf{y}$ came from the same underlying distribution; the hypothesis becomes: *P_1 and P_2 are the same.* But suppose that the pairs have been matched by size, age, etc., in an attempt to match all factors relevant to the rate of weight increase. Then one pair of rats may have characteristics quite different from another pair, and we cannot reasonably assume that $(x_1, y_1), \ldots, (x_n, y_n)$ all come from the same underlying distribution. What can be done?

Example b Recall the one-minute count data on two adjacent lanes of a freeway given on p. 181, Chapter 6. The normality of these counts is in doubt. But whether the count in one lane is independent of the corresponding count in the other lane is significant information. See Figure 8.1 for the cluster diagram of the data. We are looking for an indication that low counts in one lane accompany low counts in the other

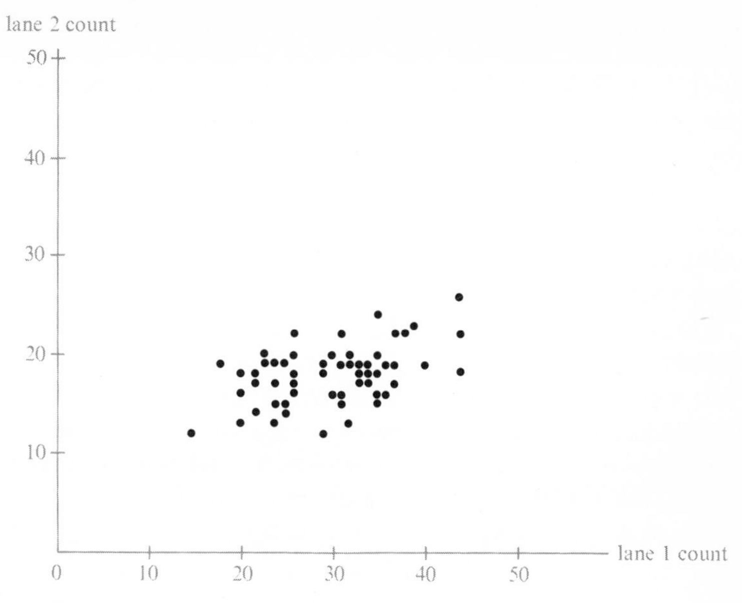

Figure 8.1 *Cluster diagram for pairs of lane counts*

lane, and similarly for high counts. If the assumption that the counts have a bivariate normal distribution is suspect, then we need to construct a more general model. Assume that the pairs of counts are independent outcomes of the vector variable $(\mathsf{N}_1, \mathsf{N}_2)$. Then the hypothesis is: N_1 *and* N_2 *are independent*, or for all I_1, I_2,

$$P(\mathsf{N}_1 \in I_1, \mathsf{N}_2 \in I_2) = P(\mathsf{N}_1 \in I_1)P(\mathsf{N}_2 \in I_2).$$

We want to test this hypothesis, assuming very little about the underlying joint distribution of $(\mathsf{N}_1, \mathsf{N}_2)$. Ideally, we want to take $\mathscr{P}$ to be the set of all bivariate distributions and the hypothesis set Θ_0 to be the set of all bivariate distributions corresponding to independent variables. Now the problem is to find a test that will discriminate between the distributions in Θ_0 and those not in Θ_0.

As we attempt to set up procedures for dealing with the above two problems, we will discover useful by-products, among which are a highly efficient estimator for the median of symmetric distributions, and tests for "trend" in a sequence of observations.

Paired Comparison

Pairing is done to reduce the variability due to irrelevant factors; for instance, in testing the skill of bridge players a duplicate tournament gives all tables the same hands. Conceivably, a tournament could be run by random unduplicated deals at each table, and eventually the best players would pull ahead. But over shorter periods, the point spread between players at different tables would be due to some combination of the differences in their skills and the variation in hands between the tables.

Getting back to our rats, here is a simple model whose purpose is to give a crude formulation of the paired comparison situation. Think of a very large rat population divided into categories 1, 2, . . . , J, where the division is based on all the measurable characteristics that might affect weight change. For instance, the 8th category might consist of those rats with present weight between 8–10 oz., age between 30–33 days, and so on. For all rats in the jth category, suppose that w_j is the average or median weight gain over the test period under the controlled (undrugged) conditions. Let the distribution of weight gain in the jth category be given by

$$w_j + \mathsf{Z}_j,$$

where Z_j has mean zero and variance σ_j^2. By selecting a rat at random from the large population we mean putting all the rats into a large hat, mixing them up thoroughly, and pulling one out. We can break this random selection down into an equivalent two-stage procedure. First, select a category with probability p_j equal to the proportion of rats in the jth category. Then select a rat at random from within the selected category. The weight gain of a randomly selected rat is the sum of W, the mean or median weight gain for the category selected, and Z, the weight gain fluctuation within the selected category. More precisely,

$$\mathsf{X} = \mathsf{W} + \mathsf{Z},$$

where $\mathsf{W} = w_j$ with probability p_j, and if the jth category is selected, then Z has the distribution of Z_j. To compute variances, we need

$$\begin{aligned} E\mathsf{WZ} &= \sum_j E(\mathsf{WZ} \mid j\text{th category selected})\, p_j \\ &= \sum_j w_j p_j E\mathsf{Z}_j \\ &= 0. \end{aligned}$$

Since $E\mathsf{Z} = 0$,

$$\begin{aligned} V(\mathsf{X}) &= E(\mathsf{W} - E\mathsf{W} + \mathsf{Z})^2 \\ &= V(\mathsf{W}) + V(\mathsf{Z}) + 2E(\mathsf{Z})(\mathsf{W} - E\mathsf{W}) \\ &= V(\mathsf{W}) + V(\mathsf{Z}). \end{aligned}$$

Now, what happens when two rats are both selected at random, as compared with matching them by category? In the first situation, their weight changes are outcomes of independent variables X, Y with the same distribution as X above. Then the variance of the difference in X and Y is

$$\begin{aligned} V(X - Y) &= E[(X - \mu) - (Y - \mu)]^2 \\ &= V(X) + V(Y) - 2E(X - \mu)(Y - \mu) \\ &= 2V(X) \\ &= 2V(W) + 2V(Z). \end{aligned}$$

On the other hand, consider the scheme where the first rat is selected at random, and then matched by selecting the second rat at random *from the same category*. The difference in weight changes X, Y of the two rats has the distribution of $Z - Z'$, where Z and Z' are independent with the same distribution as Z above. Another computation then gives

$$V(X - Y) = 2V(Z).$$

Thus, matching eliminates the category variance term $2V(W)$ from the variance of the difference of the pairs of observations. The point is that the smaller the variance between pairs without any drug treatment, the easier it will be to detect any systematic changes due to treatment. For instance, suppose that the variance between pairs is zero. Then if the drug invariably caused an increase in weight change, the treated rat in the pair would *always* have a larger weight gain than the control rat. But if the variance between pairs is large even when the treatment has no effect, then a small but systematic increase in weight due to the drug diet will be difficult to detect.

At any rate, suppose you have paired the rats by some method: either by random selection or by careful matching, or by some intermediate procedure. Assume that the weight gains $(x_1, y_1), \ldots, (x_n, y_n)$ of the rats are outcomes of random vectors $(X_1, Y_1), \ldots, (X_n, Y_n)$, not necessarily identically distributed. If the pairs were chosen at random, then it is sensible to assume that all pairs of variables have the same underlying distribution. But if the rats are classified and matched, there is no reason to believe, for instance, that pairs of larger rats will have the same distribution of weight gain as smaller rats. At any rate, to reap the rewards of any matching that may have been done, we cancel the "category effect" by working with the differences

$$v_1 = x_1 - y_1, \ldots, v_n = x_n - y_n.$$

These are the outcomes of

$$V_1 = X_1 - Y_1, \ldots, V_n = X_n - Y_n,$$

and it is these differences that are analyzed in matched comparisons. The hypothesis that the drug has no effect translates into: *For each* $i = 1, \ldots, n$, X_i *and* Y_i *are independent with the same distribution.* Of course, the common distribution of X_i, Y_i may be different from that of X_j, Y_j, $j \neq i$. But the distributions of the differences V_i, $i = 1, \ldots, n$, have a common property.

8.1 ***Proposition*** *Let* X *and* Y *be independent variables. If they have the same distribution, then* $V = X - Y$ *has a distribution symmetric about zero.*

In other words, $X - Y$ has as much probability of being between x and $x + \Delta x$ as of being between $-x - \Delta x$ and $-x$. The proof is simple:

$$P(x < X - Y < x + \Delta x) = P(-x - \Delta x < Y - X < -x).$$

But $Y - X$ has the same distribution as $X - Y$. Hence the last expression equals

$$P(-x - \Delta x < X - Y < -x).$$

Now we can state the hypothesis of no drug effect as: $V_1, \ldots, V_n$ are independent variables, each one having a distribution symmetric about zero. On the other hand, if the drug usually has a salutary effect on weight gain, then $V_1, \ldots, V_n$ will tend to have positive outcomes. If the drug has an inhibiting effect on weight gain, they will tend to have negative outcomes. This leads to the very simple *sign test*: Assume that the distributions of X_i, Y_i are continuous. Then $V_1, \ldots, V_n$ have continuous distributions. Under the hypothesis of no effect, their medians are zero, so that

8.2 $$P(V_1 > 0) = \tfrac{1}{2}, \ldots, P(V_n > 0) = \tfrac{1}{2}.$$

Now look at the signs of $v_1, \ldots, v_n$. The sign test is based on the fact that when 8.2 is true, the plus and minus signs are distributed like heads and tails in fair coin tossing. Therefore, let S_n be the number of positive values in the sequence $v_1, \ldots, v_n$. If the hypothesis is true, S_n is distributed like the number of heads in fair coin tossing. If the drug has a positive effect, then the x_i will tend to be larger than the y_i and the v_i will usually be positive. Conversely, if the drug has an inhibiting effect, the v_i will tend to be negative. The sign test for no effect against the alternatives of any systematic effect

accepts the hypothesis if S_n *is neither too large nor too small.*

The one-sided sign test for no effect as against a positive effect

accepts the hypothesis if S_n *is not too large.*

The boundaries of the acceptance region are determined by the requirement that the test be at level α. For example, if we are testing the no-effect hypothesis against the two-sided alternative at level α, the acceptance region is

$$\left|\mathsf{S}_n - \frac{n}{2}\right| \leq k,$$

where k is determined by the condition that under the binomial distribution with $p = \frac{1}{2}$,

$$P\left(\left|\mathsf{S}_n - \frac{n}{2}\right| \leq k\right) \simeq 1 - \alpha.$$

If we want to test the no-effect hypothesis against the one-sided alternative at level α, then we accept if $\mathsf{S}_n \leq (n/2) + k$, and fix k by the requirement that under the binomial distribution, $P(\mathsf{S}_n - n/2 \leq k) \simeq 1 - \alpha$. As usual, if n is large enough, we can determine k by using the central limit theorem.

Problem 1 Add θ for $\theta = .1, .5, 1.0$ in turn to the fifth column of random $N(0, 1)$ numbers in Table 11, and pair the resulting numbers with the corresponding numbers in the sixth column. At the 5% level, does the one-sided sign test detect the additive increment of size θ when $\theta = .1$? when $\theta = .5$? when $\theta = 1.0$?

Problem 2 Another application of matched comparisons is to two batches of items, say types I and II, where you want to check whether there is any significant difference in some numerical characteristics of the two batches under a variety of conditions. For example, you may be testing fatigue lifetimes of two different types of aluminum at various temperatures, and the data consist of pairs (x_i, y_i) of durations until failure, where the ith pair was tested at a different temperature from any other pair. Why might you not want to compare the durations $x_1, \ldots, x_n$ of type I aluminum to the durations $y_1, \ldots, y_n$ of the type II aluminum, say, by taking some overall measure of the two populations? For example, why not base your decision on the difference between the overall average lifetimes $\bar{x}, \bar{y}$ of the two types divided by the sample standard deviation of the combined set of x and y readings? Consider using this latter procedure on the following data.

temperature	t_1	t_2	t_3	t_4	t_5	t_6	t_7	t_8
x_i	2681	2935	3150	3764	3010	2520	1887	969
y_i	2423	2798	3337	3587	3121	2899	2038	895

What's wrong? What cogent reasons does this example suggest for using the matched comparison approach and basing your decision only on the differences $x_i - y_i$? Describe other situations in which matched comparisons would be the appropriate approach.

The Wilcoxon Test

The sign test is easy to apply and is valid under a minimum of assumptions. The trouble with it is that it has low detection capability—there must be a very marked effect before the sign test will reject no effect. For example, for $n = 50$ the 5% level test is (approximately)

$$\textit{Accept if } 18 \leq S_{50} \leq 32$$

(see Table 7). The one-sided test at the 5% level for no effect is

$$\textit{Accept if } S_{50} \leq 31.$$

Therefore to detect a positive effect, at least 32 of the differences have to be positive. Generally, the sign test requires a substantially larger sample size to detect a given positive shift than more specific tests tailored for smaller parametric families.

The difficulty with the sign test is that it discards a lot of information. Suppose that in the sample of 50 pairs, only 30 differences are positive. But suppose that the positive differences are generally very large, and that the negative differences are usually close to zero. The sign test ignores the relative sizes of the differences, counts 30 positive differences, and accepts the no-effect hypothesis at the 5% level.

The problem is how to incorporate the information about the size of the differences in a simple nonparametric test. That is, we want some way of comparing the relative sizes of the positive v_i and the negative v_i. The most effective methods known for accomplishing this are based on the simple notion of ranking. Given n numbers $x_1, \ldots, x_n$, form the order statistics

$$x_{(1)} \leq x_{(2)} \leq \cdots \leq x_{(n)}.$$

8.3 ***Definition*** *The place of x_k in the ordering $x_{(1)} \leq \cdots \leq x_{(n)}$ is called its rank.*

For instance, if x_k is the smallest of the numbers, its rank is one. If it is the largest, its rank is n.

The Wilcoxon test for matched comparisons goes this way: Order the absolute values $|v_1|, \ldots, |v_n|$. Let u^+ be the sum of the ranks of the positive v_i in this ordering. Let u^- be the sum of the ranks of the negative v_i. For example, look at this sample of size 10:

$$v_i\colon -3.1, 2.5, 4.2, -.8, 2.7, -5.3, -.2, 2.3, 1.8, 6.5.$$

The ordering by absolute values (with the signs indicated above) is

	−	−	+	+	+	+	−	+	−	+
	.2,	.8,	1.8,	2.3,	2.5,	2.7,	3.1,	4.2,	5.3,	6.5
rank	1,	2,	3,	4,	5,	6,	7,	8,	9,	10.

The ranks of the positive v_i are 3, 4, 5, 6, 8, 10; of the negative v_i, 1, 2, 7, 9. Hence

$$u^+ = 36, \qquad u^- = 19.$$

Notice that

$$u^+ + u^- = 1 + 2 + \cdots + 10 = 55,$$

and in general,

$$u^+ + u^- = 1 + 2 + \cdots + n = \frac{n(n+1)}{2}.$$

The Wilcoxon test accepts

a *no effect against a positive effect if the sum u^+ of the ranks of the positive differences is not too large,*

b *no effect against any systematic effect if u^+ is neither too large nor too small.*

Let $\mathsf{U}^+, \mathsf{U}^-$ be the random variables whose outcomes are u^+, u^-. The maximum value of u^+ or u^- is $N = n(n+1)/2$, and the minimum value is zero. To determine the size of the acceptance regions for a level α test we want to determine the maximum values of k_1, k_2 such that

a $$P(\mathsf{U}^+ < N - k_1) \geq 1 - \alpha$$

8.4

b $$P(k_2 < \mathsf{U}^+ < N - k_2) \geq 1 - \alpha,$$

with these inequalities holding for all probabilities in the set corresponding to "no effect." In other words, we want to determine k_1, k_2 so that whenever $\mathsf{V}_1, \ldots, \mathsf{V}_n$ are independent variables with distributions $P_1(dv), \ldots, P_n(dv)$ symmetric around zero, 8.4 holds. The surprising result is this:

8.5 ***Theorem*** *Take $\mathsf{V}_1, \ldots, \mathsf{V}_n$ to be independent, each one distributed uniformly on $[-\frac{1}{2}, \frac{1}{2}]$. Determine k_1, k_2 such that for this distribution*

$$P(\mathsf{U}^+ < N - k_1) = 1 - \alpha$$
$$P(k_2 < \mathsf{U}^+ < N - k_2) = 1 - \alpha.$$

Then these same equalities hold for all distributions such that $V_1, \ldots, V_n$ *are independent with continuous distributions (not necessarily the same) symmetric around zero.*

One part of this result is more surprising than the other. If the $V_1, \ldots, V_n$ have a common distribution, then it is not surprising that any test based on the ranks has the same probability of acceptance whenever the underlying distribution is symmetric around zero (see Problem 5). The novel element in this theorem is that the same thing holds true even if the variables are not identically distributed. We will not give a proof.

The table of values of k_1 and k_2 are constructed as follows: Since $U^+ + U^- = N$, we can write

$$P(U^+ < N - k_1) = P(U^- > k_1) = 1 - P(U^- \leq k_1),$$

and

$$P(k_2 < U^+ < N - k_2) = P(U^+ > k_2, U^- > k_2).$$

Define the variable

$$T = \min(U^+, U^-).$$

Then for $k_1 < N/2$,

$$\begin{aligned} P(U^- \leq k_1) &= P(T \leq k_1, T = U^-) \\ &= \tfrac{1}{2}P(T \leq k_1). \end{aligned}$$

Also

$$\begin{aligned} P(U^+ > k_2, U^- > k_2) &= P(T > k_2) \\ &= 1 - P(T \leq k_2). \end{aligned}$$

Therefore, values of both k_1 and k_2 can be determined from a table of the distribution of T. For the level α one-sided test, k_1 is the largest integer such that

$$P(T \leq k_1) = 2\alpha.$$

For the level α two-sided test k_2 is determined by

$$P(T \leq k_2) = \alpha.$$

Table 8 tabulates the values of k for common values of the level up to $n = 25$. For larger values of n, T is approximately normal with

$$ET = \frac{n(n+1)}{4}$$

$$\sigma^2(T) = \frac{n(n+1)(2n+1)}{24}.$$

This last fact can be used in the customary way to find values of k such that $P(T \leq k) = \alpha$.

Because the underlying distributions are assumed continuous, equal values of $|v_i|$ should not occur and the ranks are unambiguously defined. But limited accuracy of measurement may sometimes lead to a few ties. Suppose that two or more of the v_i have a common absolute value. Then assign to each one a rank number equal to the average of the ranks they would have if we altered each one slightly to break the tie. For instance, suppose the four readings with the smallest absolute value are .7, -1.0, -1.0, 1.0. With a slight alteration the last three outcomes would have ranks 2, 3, 4. Hence they are all assigned rank 3, since

$$\frac{2 + 3 + 4}{3} = 3.$$

Problem 3 Apply the Wilcoxon test for positive effect in Problem 1. What value of the positive shift θ does it detect, $\theta = .1, .5$, or 1.0? What value of θ would be detected if the two-sided Wilcoxon test were used?

Problem 4 Take the first 10 numbers from the first column of Table 10, add θ, and pair them with the first 10 numbers of the second column. What is the minimum positive value of θ at which the sign test rejects the no-effect hypothesis when the level is 0.1? What is the minimum positive value of θ at which the Wilcoxon test rejects (at the same level)?

Problem 5 For random variables $X_1, \ldots, X_n$, let T_j be the rank of X_j in the ordered sample. If $X_1, \ldots, X_n$ are independent with the same continuous distribution, what is the distribution of $\mathbf{T} = (T_1, \ldots, T_n)$? (Note that $\mathbf{T}$ must equal one of the permutations of $(1, 2, \ldots, n)$.) What does this say about the distribution of any test based on the value of $\mathbf{T}$?

How Well Do Nonparametric Tests Perform?

Nonparametric models for hypothesis testing and nonparametric tests are quite different from parametric models and tests. Their distinctive features are

a While the hypothesis set of probabilities is always clearly defined, the alternative set of probabilities may be incompletely defined.

b The operating characteristic function is virtually impossible to compute for most of the probabilities in the alternative set.

To illustrate the first point, let $V_1, \ldots, V_n$ be the differences in a matched comparison. For the sign and Wilcoxon tests the "no effect" set of probabilities $\mathscr{P}_0$ consists of all those distributions under which $V_1, \ldots, V_n$ are independent, with each variable having a continuous distribution symmetric about zero. The set of *all* possible underlying probabilities for the problem is partially specified by the assumption that $V_1, \ldots, V_n$ are independent. But how can we select the set of distributions $\mathscr{P}$ that models the statement: There is either a positive effect or no effect. If all the distributions are assumed normal with the same unknown μ and known σ^2, say, then the model is clear. Take $0 \le \mu < \infty$ as the parameter space, and use a one-sided test to check for positive μ. For the nonparametric analogue, there are various things we might try. For instance, take $\mathscr{P}$ to consist of all probabilities under which $V_1, \ldots, V_n$ are independent with distributions symmetric around ν, and restrict $0 \le \nu < \infty$. However, this makes $\mathscr{P}$ unduly restrictive in the service of trying to get a precise definition. There is really no satisfactory single definition that specifies the distribution of a random variable as more positive than negative. Think about the large number of different possible definitions. For a nonparametric test *it is not necessary or usual to completely define* $\mathscr{P}$. The hypothesis set $\mathscr{P}_0 \subset \mathscr{P}$ must be completely defined in order to make sure that the acceptance region S for the test gives a level α test. That is, we always require that

8.6
$$P(\mathbf{X} \in S) \ge 1 - \alpha$$

holds for all P in $\mathscr{P}_0$.

As for the second point, the operating characteristic function is given by

$$\phi(P) = P(\mathbf{X} \in S)$$

for all $P \in \mathscr{P}$. Since $\mathscr{P}$ is, in a nonparametric situation, an infinite dimensional space, it is usually impossible to compute $\phi(P)$ for all P in $\mathscr{P}$, even if $\mathscr{P}$ is completely defined. Therefore, it is usually difficult to tell how well the test discriminates between probabilities in $\mathscr{P}_0$ and those outside $\mathscr{P}_0$. In fact, there may be probabilities P outside of $\mathscr{P}_0$ such that $P(\mathbf{X} \in S)$ is nearly 1. In other words, there may be distributions outside of $\mathscr{P}_0$ that, with very high probability, fool the test into believing that they are in $\mathscr{P}_0$.

The consequence of this is that, even more than in the parametric case, you stand on shaky ground when the data accept the hypothesis. This is particularly clear in the sign test. Here, any set of continuous distributions with medians at zero is just as acceptable as a set of distributions symmetric around zero. Hence when you accept, the precise statement is not that you accept "no effect," but that you accept "zero median." The Wilcoxon test is more complicated, but what is true is that there are distributions not at all symmetric around 0 such that for variables drawn from these distributions, the distributions of the rank sums U^+, U^- look as though they were drawn from distributions symmetric around zero. If you reject when using a nonparametric test you are on solid ground. For instance, if either the sign or Wilcoxon test rejects at the 1% or 5% level, then you can be reasonably sure that the underlying distributions are not symmetric around zero. If they were, we would have accepted with probability at least $1 - \alpha$.

However, most nonparametric tests are designed with a rough idea of what alternatives we want it to be effective against. That is, we design the test to detect certain types of departures from the null hypothesis. Both the sign and Wilcoxon tests are aimed at detecting *shift alternatives.* In this context, this means those distributions which are not too skewed and have most of their probability located to one side of the origin. We will not be too disturbed by the fact that there are some skewed distributions which deceive the sign or Wilcoxon tests into accepting the null hypothesis. But we would be concerned if these tests could not detect distributions whose sampled values are usually positive (or negative).

The next question is: How good are nonparametric tests? We saw that nonparametric estimates could be considerably less efficient when applied to small parametric families than the estimate best tailored to that family. Suppose that we set up a classical normally distributed model for a matched comparison. How well do the sign and Wilcoxon tests perform compared with the best tailored-to-fit test?

Here is the model: Assume that the ith pair (x_i, y_i) are drawn from the distributions

$$N(\mu_i + \Delta, \sigma_0^2), \quad N(\mu_i, \sigma_0^2)$$

where μ_i is the mean for the ith category, Δ is the shift in mean due to treatment effect, and the variance σ_0^2 is unknown but is the same for every category. Then $V_i = X_i - Y_i$ has an $N(\Delta, \sigma^2)$ distribution, $\sigma^2 = 2\sigma_0^2$. The hypothesis of no effect translates into $\Delta = 0$. For a two-sided model Δ ranges over $(-\infty, \infty)$; it ranges over $[0, \infty)$ for the one-sided, zero or positive effect model. The best tailored-to-fit test for either model is based on the Student's t test. Form the ratio

$$\hat{t} = \sqrt{n}\bar{v}/\hat{s},$$

where $\bar{v}$ is the sample mean of $v_1, \ldots, v_n$ and $\hat{s}$ the unbiased sample variance. The two-sided test for no effect rejects when $|\hat{t}|$ is too large; the one-sided, when $\hat{t}$ is too large.

Here is a way to see how good the sign and Wilcoxon tests are in this situation as compared to the t test. Take $|\Delta|$ small but not zero, and compute for each test the sample size needed to detect the treatment effect Δ at level α. Denote the sizes by $n_t(\Delta)$, $n_s(\Delta)$, $n_w(\Delta)$ for the t, sign, and Wilcoxon tests, respectively. One can show that as $\Delta \to 0$, the limits of the ratios

$$\frac{n_t(\Delta)}{n_s(\Delta)}, \qquad \frac{n_t(\Delta)}{n_w(\Delta)}$$

exist and do not depend on α. Call these limit ratios *the asymptotic efficiencies of the sign and Wilcoxon tests relative to normal shift alternatives.* The sample size $n_t(\Delta)$ needed by Student's t test will always be the smallest. The efficiency gives the measure of how much larger a sample size will be needed by the sign and Wilcoxon tests to do the same detection job as the t test. The results are that the efficiency of the sign test is

$$\frac{2}{\pi} = .64,$$

and surprisingly, that of the Wilcoxon test is

$$\frac{3}{\pi} = .95.$$

Efficiencies have been computed for a wide variety of models where the underlying distributions are taken to be a location or location-scale family of distributions symmetric about the median v. The hypothesis of no effect is taken to be $v = 0$ versus the alternative of a shift in the median of amount Δ. In almost all cases, the efficiency of the Wilcoxon as compared with the best tailored-to-fit tests has been close to one. This justifies the statement that use of the t test in matched comparisons is virtually obsolete. The Wilcoxon test works almost as well with normally distributed alternatives and usually behaves better when the underlying distributions are not normal. However, because the sign test is so easy to compute, it is usually easier to try it first. If the sign test doesn't reject, try the Wilcoxon test.

Problem 6 In a matched comparison, suppose that there is actually a positive effect and that the true underlying distribution of $V_1, \ldots, V_n$ is uniform on $[-.9l, 1.1l]$, where l is unknown. What sample size is required by the one-sided sign test to detect this effect at the 5% level? (Use the normal approximation.) (Ans. $n \simeq 1{,}100$.)

Problem 7 Neither the sign test nor the Wilcoxon test really test fully whether the underlying distributions are symmetric around zero. Assuming that $V_1, \ldots, V_n$ are identically distributed, a χ^2 test can be set up. Divide the axis into $2J$ cells $I_1, I_{-1}, I_2, I_{-2}, \ldots, I_J, I_{-J}$ symmetric around zero. That is, if $I_k = [a_k, b_k)$, then $I_{-k} = (-b_k, -a_k]$. Under the hypothesis of a continuous distribution $P(dx)$ symmetric around zero,

$$p_k = P(I_k) = P(I_{-k}) = p_{-k}.$$

Let $\hat{n}_k$ be the observed frequencies in cell I_k. Show that the maximum likelihood estimate for p_k is

$$\hat{p}_k = \frac{\hat{n}_k + \hat{n}_{-k}}{2n}.$$

Substituting $\hat{p}_k$ for p_k in the χ^2 statistic gives

$$\sum_{-J}^{J} \frac{(\hat{n}_k - np_k)^2}{n\hat{p}_k} = \sum_{1}^{J} \frac{(\hat{n}_k - \hat{n}_{-k})^2}{\hat{n}_k + \hat{n}_{-k}}.$$

How many degrees of freedom should be assigned to the distribution of this expression?

Problem 8 In an experiment on perception, a subject is presented with a small square white card with two lines on it, one darker than the other. The subject is then asked whether the two lines have the same length. The answer *yes* is scored as a zero; *no* as a one. Some time later, the subject is handed a much larger white card, rectangular in shape, with the same two lines on it. The same question is asked and the response scored the same way. For each subject, one of the four possible couples (0, 0), (1, 0), (0, 1), (1, 1) of scores are recorded. Of the 100 subjects tested, the results are

Score	*Number*
(0, 0)	16
(0, 1)	32
(1, 1)	29
(1, 0)	23

Construct a matched comparison test for the hypothesis of no effect due to card size.

Problem 9 Discuss the general shape of a skewed distribution with a positive median, say $v = 1$, for which the Wilcoxon test will usually lead to the acceptance of a negative effect.

Rank Estimates of the Median

Suppose $x_1, \ldots, x_n$ are independently drawn from a continuous distribution assumed symmetric about an unknown median v^*. If a variable X has median v^*, then $X - v^*$ has median zero. (Why?) Hence, suppose we subtract any number v from each data reading to get the revised set of values

$$x_1 - v, \qquad x_2 - v, \ldots, \qquad x_n - v.$$

Then if v is close to v^*, these revised values behave as though they came from a continuous symmetric distribution with median zero. The Wilcoxon test is based on the idea that if a continuous distribution is symmetric around zero, then $u^+ \simeq u^-$. This suggests estimating v^* by that number v which makes the sum of the ranks of the positive $x_i - v$ in the ordered absolute values $|x_i - v|$ as nearly equal as possible to the sum of the ranks of the negative $x_i - v$. Denote these two sums by $u^+(v)$, $u^-(v)$. For instance, look at $x_i - v$ for the ten sample data of p. 261.

	v					
	0	1.0	2.0	0.5	1.2	1.1
$x_1 - v$	−3.1	−4.1	−5.1	−4.6	−4.3	−4.2
$x_2 - v$	2.5	1.5	0.5	1.0	1.3	1.4
$x_3 - v$	4.2	3.2	2.2	2.7	3.0	3.1
$x_4 - v$	−0.8	−1.8	−2.8	−2.3	−2.0	−1.9
$x_5 - v$	2.7	1.7	−0.3	2.2	1.5	2.6
$x_6 - v$	−5.3	−6.3	−7.3	−6.8	−6.5	−6.4
$x_7 - v$	−0.2	−1.2	−2.2	−1.7	−1.4	−1.3
$x_8 - v$	2.3	1.3	0.3	0.8	1.1	1.2
$x_9 - v$	1.8	0.8	−0.2	0.3	0.6	0.7
$x_{10} - v$	6.5	5.5	4.5	5.0	5.3	5.4
$u^+(v)$	(36)	29	$22\frac{1}{2}$	26	27	28
$u^-(v)$	(19)	26	$32\frac{1}{2}$	29	28	27

If you work with these some more, you will see that for $1.05 < v < 1.25$, $|u^+ - u^-| = 1$ and that for v outside of this range, $|u^+ - u^-| > 1$. Estimate v as 1.15, the midpoint of this range. Denote this estimate of the median by $\hat{\hat{v}}$ to distinguish it from $\hat{v}$, the sample median.

While we can get the value of $\hat{\hat{v}}$ fairly easily when $n = 10$, the computational difficulty involved in a direct evaluation of this estimate for large n is obvious. Here is an easier way: Take the original sample $x_1, \ldots, x_n$ and form all the values

$$\frac{x_i + x_j}{2}, \qquad 1 \leq i \leq j \leq n.$$

This will give $N = n(n + 1)/2$ distinct numbers. Denote these by $z_1, \ldots, z_N$ (in any order).

8.7 ***Proposition*** *u^+ equals the number of positive z_l, $l = 1, \ldots, N$.*

For example, for the ten sample data we used above, we have

i	$\# j \geq i$ such that $(x_i + x_j)/2 > 0$
1	2
2	8
3	7
4	4
5	5
6	1
7	3
8	3
9	2
10	1
55	36

Notice that since there are N values of z_l altogether, and u^+ is the number of positive values, then u^- must be the number of negative values.

Now, the plan is to apply 8.7 to the revised x-values $x_1 - v, \ldots, x_n - v$. First, form the corresponding z-values

$$\frac{(x_i - v) + (x_j - v)}{2} = \frac{x_i + x_j - 2v}{2}.$$

Then, the revised z values are $z_1 - v, \ldots, z_N - v$, where $z_1, \ldots, z_N$ are the original z values. We estimate v^* by selecting that v which makes $u^+(v) \simeq u^-(v)$. By 8.7, this means that we want, as closely as possible, as many of the $z_1 - v, \ldots, z_N - v$ to be positive as negative. This implies that $\hat{\hat{v}}$ is the median of the values $z_1, \ldots, z_N$.

8.8 ***Proposition*** *The Wilcoxon test leads to the estimator*

$$\hat{\hat{\nu}} = \text{median of} \left\{ \frac{x_i + x_j}{2}, 1 \leq i \leq j \leq n \right\}.$$

Equivalently,

$$\hat{\hat{\nu}} = \tfrac{1}{2} \text{ median of } \{x_i + x_j, 1 \leq i \leq j \leq n\}.$$

We use the midpoint rule on medians.

We can also get nonparametric confidence intervals using the inversion method of the last section in Chapter 5. Determine k so that

$$P(k < \mathsf{U}^+ < N - k) \simeq \gamma$$

for all "no-effect" distributions. Assume that $x_1, \ldots, x_n$ are drawn from a continuous distribution symmetric about an unknown median ν^*. To use the Wilcoxon test for $\nu^* = \nu$, form the shifted values $x_1 - \nu, \ldots, x_n - \nu$ and test these for median zero. The level $1 - \gamma$ acceptance region is

$$\{k < u^+(\nu) < N - k\}.$$

Recall that $100\gamma\%$ confidence intervals are given by

8.9 $$J(\mathbf{x}) = \{\nu; k < u^+(\nu) < N - k\}.$$

To simplify, form the N z values $(x_i + x_j)/2$, $1 \leq i \leq j \leq n$. Then from 8.7,

$$u^+(\nu) = \{\# \, z_l > \nu\}.$$

Putting this expression into 8.9 gives

$$J(\mathbf{x}) = \{\nu; k < \{\# \, z_l > \nu\} < N - k\},$$

which means that for every value of ν in the interval $J(\mathbf{x})$, there must be

a at least k values of z greater than ν,

b no more than $N - k$ values of z greater than ν.

Together **a.** and **b.** imply that

8.10 $$J(\mathbf{x}) = (z_{(k)}, z_{(N-k)}),$$

where $z_{(1)} < z_{(2)} < \cdots < z_{(N)}$ is the ordering of the $z_1, \ldots, z_N$ values.

The most systematic way to get $\hat{\hat{\nu}}$ (called the Hodges-Lehmann estimator) and confidence intervals is to form the order statistics $z_{(1)}, \ldots, z_{(N)}$ of the N values $(x_i + x_j)/2$, $1 \leq i \leq j \leq N$. Then $\hat{\hat{\nu}}$ equals the median, and the $100\gamma\%$ intervals can be quickly gotten by using (8.10.) Actually, you can save some effort by ordering the values $x_i + x_j$, $1 \leq i \leq j \leq n$, and dividing the resulting median and confidence interval boundaries in two. But if $n = 100$, then getting the values $x_i + x_j$, $1 \le i \le j \le 100$, still involves $N = (100 \cdot 101)/2 = 5050$ additions and a z sample of size 5050.

In the light of this massive computation, why use this estimate? It is the most efficient wide spectrum estimator of the median of a symmetric distribution known. For example, it has 95% efficiency as an estimate for the mean of a normal distribution, and its efficiency remains high under contamination and for distributions as disparate as the rectangular and the Cauchy. Assuming that the computations are justified, we can recommend its use. But if computational time is a serious consideration, some of the robust estimates considered in the last chapter are more desirable.

Notice that these computational difficulties do not apply to the Wilcoxon test.

Problem 10 For the 10 sample data used in the above section, write out the values $(x_i + x_j)$, $1 \le i \le j \le n$, find $\hat{\nu}$, and 90% confidence intervals.

Independence

The second major problem of this chapter is independence. Given n samples $(x_1, y_1), \ldots, (x_n, y_n)$ drawn from the same underlying bivariate distribution, are the corresponding x and y values independent? More precisely, let $(x_1, y_1), \ldots, (x_n, y_n)$ be the outcomes of independent identically distributed 2-dimensional vectors $(\mathsf{X}_1, \mathsf{Y}_1), \ldots, (\mathsf{X}_n, \mathsf{Y}_n)$. Are X_j, Y_j independent? Dependence is complex and varied. The relationship between two variables X and Y can vary from independence to slight dependence up to complete dependence. Recall our measure of dependence: Let $g(\mathsf{X})$ be the best predictor of Y based on X, and see how much subtracting $g(\mathsf{X})$ from Y decreases the variance of Y. When X and Y are independent, there is no decrease. We define X and Y to be completely dependent when $\mathsf{Y} \equiv g(\mathsf{X})$, or equivalently, when the revised variance is zero.

Furthermore, the functional form of the dependence can be very intricate. This refers partly to the form of the function $g(x)$. For instance, we might have a situation in which values of X in the range around $x = 1$ tends to make the values of Y cluster around $y = 3$ and $y = 5$. If $g(x)$ oscillates and behaves in an erratic fashion, then it will generally take a very large sample size to disentangle the threads of the dependency.

We will generally confine ourselves to looking for tests that will detect a *monotone dependency* between X and Y. By monotone dependency we mean that either

a small values of X and Y are associated and large values are associated,

or

b small values of X are associated with large values of Y and conversely.

Roughly this is equivalent to a monotone predictor function. That is, $g(x)$ is either an everywhere increasing function or an everywhere decreasing function. Suppose that X and Y have a joint normal distribution. Then the best predictor of Y based on X is

$$g(X) = c + dX,$$

where $d = \rho\sigma_1/\sigma_2$ and c depends on the means and variances of X and Y. Thus $g(x)$ is always linear for bivariate normal variables. If the correlation coefficient ρ is positive, then X large and negative (positive) tends to make Y large and negative (positive). For $\rho < 0$, X large in one direction (positive or negative) tends to make Y large in the other direction.

The classic test for independence is based on the assumption that the underlying distribution is bivariate normal. Then the ρ test is the best tailored-to-fit test (see Chapter 5, p. 150 ff.). Estimate the correlation coefficient by

$$\hat{\rho} = \frac{\sum_{j=1}^{n} (x_j - \bar{x})(y_j - \bar{y})}{n\hat{\sigma}_1\hat{\sigma}_2}.$$

Accept the hypothesis of independence if $|\hat{\rho}|$ is small enough; otherwise, reject.

The estimate $\hat{\rho}$ for ρ may behave badly if the underlying distribution deviates slightly from normality. We need some robust or nonparametric tests.

Problem 11 Let X, Z be independent variables each having an $N(0, 1)$ distribution. Form the variable

$$Y = Z + cX^2,$$

where c is any real number. Then show that

a X and Y are not independent

b the best predictor function for Y based on X is $g(X) = cX^2$

c the correlation coefficient between X and Y is zero.

What is the nature of the dependence between X and Y? What is the ratio of the variances of $\mathrm{Y} - c\mathrm{X}^2$ and Y?

Contingency Tables and χ^2 Again

A classical nonparametric test for independence is provided by the ubiquitous χ^2. Suppose X ranges over the values $1, \ldots, J$, and Y over $1, \ldots, K$. The joint distribution is specified by the JK probabilities

$$P(\mathrm{X} = j, \mathrm{Y} = k) = p_{jk}.$$

Let $(x_1, y_1), \ldots, (x_n, y_n)$ be n samples drawn from the same distribution as (X, Y), denote by $\hat{n}_{jk}$ the number of couples such that $x_i = j$, $y_i = k$, and let $n_{jk} = np_{jk}$ be the expected frequencies. By the χ^2 theorem, if n_{jk} is large enough for all j, k, then

$$\mathrm{D} = \sum_{j=1}^{J} \sum_{k=1}^{K} \frac{(\hat{n}_{jk} - n_{jk})^2}{n_{jk}}$$

has a χ^2_{JK-1} distribution. Suppose we assume only that X and Y are independent, but that otherwise the p_{jk} are unknown. How many parameters do we have to estimate? Let

$$P(\mathrm{X} = j) = p_j^{(1)}, \qquad j = 1, \ldots, J,$$

$$P(\mathrm{Y} = k) = p_k^{(2)}, \qquad k = 1, \ldots, K.$$

Independence implies $p_{jk} = p_j^{(1)} p_k^{(2)}$, and thus reduces the unknown parameters to the $J + K$ probabilities $\{p_j^{(1)}\}$ and $\{p_k^{(2)}\}$. Since

$$\sum_1^J p_j^{(1)} = 1 \qquad \text{and} \qquad \sum_1^K p_k^{(2)} = 1,$$

there are $(J - 1) + (K - 1)$ parameters left to estimate. Let

$$\hat{n}_j^{(1)} = \sum_k \hat{n}_{jk} = \{\# \; x_i = j\}$$

$$\hat{n}_k^{(2)} = \sum_j \hat{n}_{jk} = \{\# \; y_i = k\}.$$

The maximum likelihood estimates are

$$\hat{p}_j^{(1)} = \frac{\hat{n}_j^{(1)}}{n}, \qquad \hat{p}_k^{(2)} = \frac{\hat{n}_k^{(2)}}{n}.$$

Under the hypothesis of independence, therefore, the expression

8.11 $$D = n \sum_{j,k} \frac{[\hat{n}_{jk} - (\hat{n}_j^{(1)}\hat{n}_k^{(2)}/n)]^2}{\hat{n}_j^{(1)}\hat{n}_k^{(2)}}$$

has a χ^2 distribution with

$$JK - 1 - (J - 1) - (K - 1) = (J - 1)(K - 1)$$

degrees of freedom.

This χ^2 expression can be simplified by multiplying out the square and using the identities

$$\sum_{j,k} \hat{n}_{jk} = n$$

$$\sum_{j,k} \hat{n}_j^{(1)}\hat{n}_k^{(2)} = n^2.$$

Then

8.12 $$D = n\left(\sum_{j,k} \frac{\hat{n}_{jk}^2}{\hat{n}_j^{(1)}\hat{n}_k^{(2)}} - 1\right).$$

Usually, the $\hat{n}_{jk}$ are tabled in a two-way layout. This table is called a *contingency table*, and D as given in 8.12 is the common statistic used to test for association between the two variables in the table. Contingency tables are useful when the variables involved are difficult to measure by a continuous magnitude, but can be classified into a finite number of categories. Problems 13 and 14 illustrate this.

The χ^2 approach applies to continuous distributions. Break the x-axis into cells $I_j^{(1)}$ and the y-axis into cells $I_k^{(2)}$, then subdivide the plane into the rectangles R_{jk} with horizontal sides $I_j^{(1)}$ and vertical sides $I_k^{(2)}$. Under independence,

$$P[(X, Y) \in R_{jk}] = P(X \in I_j^{(1)}, Y \in I_k^{(2)})$$

factors into the product $P(X \in I_j^{(1)}) \cdot P(Y \in I_k^{(2)})$. Hence the situation becomes exactly like the discrete case and 8.11 applies again.

If you have made a cluster diagram of the points (x_i, y_i), then the χ^2 test will be easy to set up and compute: Draw the grid generated by the cells $I_j^{(1)}$, $I_k^{(2)}$ on the diagram. The number of points falling into R_{jk} is $\hat{n}_{jk}$. If there are some rectangles such that $\hat{n}_{jk}$ is too small, say less than 5, then group together *cells*, not rectangles, to get increased frequencies. Grouping rectangles together will not generally preserve the independence relationship $p_{jk} = p_j^{(1)} p_k^{(2)}$.

Problem 12 In the perception experiment given in Problem 8, a question can be raised as to whether the first and second responses of the subjects were independent. If the couples of responses are assumed to be drawn from the *same* underlying distribution, then 8.11 and 8.12 are applicable. Assuming the same underlying distribution, run a χ^2 test for independence. If the underlying distributions may be different for the different couples, then there is no sensible way to test for independence. Can you see why?

Problem 13 The table below gives some results obtained on the association between left-handedness and "left-eyedness." The X categories 1, 2, 3 are left-handed, ambidextrous, and right-handed. The Y categories 1, 2, 3 are left-eyed, "ambiocular," and right-eyed.

		Y		
		1	2	3
	1	34	62	28
X	2	27	28	20
	3	57	105	52

Test for dependence between the two variables.

Problem 14 A study is being made of the possible association between a certain type A_1 of mental illness and the presence in the body of traces of a certain chemical. The examination and diagnosis of 1015 mentally ill patients gave the following data:

Occur together	187
Chemical but no illness A_1	193
Illness A_1 but no chemical	490

Test this data for dependence.

Problem 15 Test the traffic count data given on p. 181 for independence. Group the data where necessary.

From Independence to Trend

To detect a monotone dependence between the variables X and Y given the data

$$(x_1, y_1), (x_2, y_2), \ldots, (x_n, y_n),$$

begin by ordering $x_1, \ldots, x_n$ into the order statistics $x_{(1)}, \ldots, x_{(n)}$. Let y'_k be the y value coupled with $x_{(k)}$. For instance, if x_1 is the third lowest of the x's, then $y'_3 = y_1$. Look at the sequence $y'_1, \ldots, y'_n$. If there is no dependence between X and Y, then this reordering of the $y_1, \ldots, y_n$ values according to the sizes of the coupled x_i should produce only a random shuffling of the y values. But if lower values of X are associated with lower values of Y and similarly for the higher values, then the sequence $y'_1, \ldots, y'_n$ should show a tendency to increase. Similarly, if the relationship between X and Y is an attraction of opposites, then the sequence $y'_1, \ldots, y'_n$ should tend to decrease.

Thus, the problem of testing for independence is connected with the problem of detecting *trend.* That is, given an ordered sequence of observations $z_1, \ldots, z_n$, we want to test the hypothesis that these are outcomes of independent, identically distributed random variables versus the alternative that they have a general tendency to increase or decrease. There are various tests for trend; the simplest is the Mood-Brown median test. Assume n is even, and denote by $\hat{\nu}$ the median of $z_1, \ldots, z_n$. Let $\hat{N}$ be the number of readings in the first half of the observations that are greater than the median. That is,

$$\hat{N} = \left\{\# \; z_i > \hat{\nu}, \quad i = 1, \ldots, \frac{n}{2}\right\}.$$

The distribution of $\hat{\mathsf{N}}$ under the hypothesis is easily computable when the underlying distribution is continuous.

8.13 ***Proposition*** *If* $\mathsf{Z}_1, \ldots, \mathsf{Z}_n$, *n even, are independent with the same continuous distribution, then* $\hat{\mathsf{N}}$ *has the hypergeometric distribution*

$$P(\hat{\mathsf{N}} = j) = \frac{\binom{m}{j}\binom{m}{m-j}}{\binom{2m}{m}}, \qquad n = 2m. \tag{8.14}$$

Here is a quick sketch of the proof. Look at the order statistics $\mathsf{Z}_{(1)}, \mathsf{Z}_{(2)}, \ldots, \mathsf{Z}_{(n)}$. One way to get the variables $\mathsf{Z}_1, \ldots, \mathsf{Z}_n$ is to drop all the $\mathsf{Z}_{(k)}$, $k = 1, \ldots, n$ into an urn and pull them out at random. Now "paint" $\mathsf{Z}_{(1)}, \ldots, \mathsf{Z}_{(m)}$ black and $\mathsf{Z}_{(m+1)}, \ldots, \mathsf{Z}_{(n)}$ red. Then, the number of values we get in the first m trials that are greater than the median is exactly equal to the number of reds we pull out of the urn in m trials. For an urn containing m reds and m blacks, the probability of j reds in m draws is given by the hypergeometric distribution 8.14.

But 8.13 does not help much when n is large. What is more useful is the fact that for $n \geq 25$, $\hat{\mathsf{N}}$ is approximately normal with mean $n/4$ and standard deviation $\sqrt{n}/4$.

Use $\hat{N}$ to test against any trend increasing or decreasing, by rejecting the hypothesis if $\hat{N}$ is too large or too small. For testing no trend versus an upward trend, reject the hypothesis if $\hat{N}$ is too small, and analogously for testing against a downward trend. If n is odd, or if there are a few values of z_i at the median, split points in half. This means that if n is odd and the middle reading in the sequence $z_1, \ldots, z_n$ is larger than $\hat{\nu}$, increase $\hat{N}$ by $\frac{1}{2}$; otherwise, decrease it by $\frac{1}{2}$. If $z_i = \hat{\nu}$, $i < n/2$, then count it as $\frac{1}{2}$ in computing $\hat{N}$. But if there are too many $z_i = \hat{\nu}$, the level of the test will be disturbed.

This test, called the median test, only takes into account the sign of $z_i - \hat{\nu}$, not its size. In this sense it is similar to the sign test. Since it does not use much of the available information, its efficiency is suspect. Furthermore, it rather arbitrarily splits the sample in two and compares the first half with the second half.

There are two common nonparametric tests for trend, both based on the ranking of $z_1, \ldots, z_n$ by size. The first one is called Kendall's Tau test. Suppose that the sequence $z_1, \ldots, z_n$ is generally increasing. Then z_i is usually larger than most of the preceding members of the sequence. That is, usually $z_i > z_j$, for $j < i$. To formalize this idea, we introduce the "reversal" quantities

$$u_{ij} = \begin{cases} 1 & z_i > z_j \\ -1 & z_i < z_j, \\ 0 & z_i = z_j \end{cases} \qquad i > j.$$

That is, u_{ij} is 1 if the rank of z_i is greater than the rank of z_j, and -1 if the reverse is true. Now fix i and look at the sum s_i of all u_{ij} for $j < i$:

$$s_i = u_{i1} + u_{i2} + \cdots + u_{i,i-1}.$$

This sum counts the number of $z_j, j < i$, that are less than z_i, and subtracts the number that are greater than z_i. If the z_i are tending to increase, then the sums s_i tend to be large and positive. Hence

8.15 $$\hat{S} = \sum_{i=1}^{n} s_i = \sum_{j<i} u_{ij}$$

will be large and positive. In fact $\hat{S}$ has its maximum value when all the $u_{ij}, j < i$, are $+1$; equivalently, when

$$z_1 < z_2 < \cdots < z_n.$$

In this case

$$\hat{S}_{\max} = \sum_{i=1}^{n} (i - 1) = \frac{n(n-1)}{2}.$$

Similarly, when the z_i are tending downward, $\hat{S}$ is large and negative. If $z_1 > z_2 > \cdots > z_n$, $\hat{S}$ achieves its minimum value of

$$\hat{S}_{\min} = -\frac{n(n-1)}{2}.$$

This suggests the definition

8.16 $$\hat{\tau} = \frac{\hat{S}}{(n/2)(n-1)}.$$

This quantity is called *Kendall's rank correlation coefficient* or *Kendall's Tau*. By its definition, $-1 \leq \hat{\tau} \leq 1$. Positive values of $\hat{\tau}$ near $+1$ occur when the z_i are tending upward and negative values near -1 when the z_i are tending downward. The τ test for trend is

a Accept no trend versus the alternative of any trend if $|\hat{\tau}|$ is not too large.

b Accept no trend versus an upward trend if $\hat{\tau}$ is not too large.

c Accept no trend versus a downward trend if $\hat{\tau}$ is not too small.

For $n > 10$, the distribution of $\hat{\tau}$ under the hypothesis of no trend is adequately approximated by a normal distribution with mean zero and variance

8.17 $$V(\hat{\tau}) = \frac{2}{9}\frac{2n+5}{n(n-1)}.$$

The other test for trend is based on *Spearman's rank correlation coefficient*. Recall that for a sample $(x_1, y_1), \ldots, (x_n, y_n)$ from some bivariate distribution, the sample correlation coefficient is given by the expression

$$\sum_{i=1}^{n} (x_i - \bar{x})(y_i - \bar{y})$$

divided by $n\hat{\sigma}_1\hat{\sigma}_2$. Now think of computing the correlation between the place that a z value occupies in the sequence $z_1, \ldots, z_n$ and its rank. By definition z_i is in the ith place in the sequence $z_1, \ldots, z_n$. The rank of z_i is its place in the ordered sample $z_{(1)} < \cdots < z_{(n)}$. Denote the rank of z_i by t_i. If there is no trend, then the numbers i and t_i should be uncorrelated—knowing what place a value occupies in the unordered sequence should give no information about its rank.

To check this, take the couples

$$(1, t_1), (2, t_2), \ldots, (n, t_n)$$

and compute the correlation sum

8.18 $$\hat{R} = \sum_{1}^{n} (i - \bar{\imath})(t_i - \bar{t}).$$

Here,

$$\bar{\imath} = \frac{1}{n}\sum_{1}^{n} i = \frac{1}{n}\frac{n(n+1)}{2} = \frac{n+1}{2},$$

and

$$\bar{t} = \frac{1}{n}\sum_{1}^{n} t_i.$$

Since $t_1, \ldots, t_n$ is some permutation of $1, \ldots, n$,

$$\sum_{1}^{n} t_i = 1 + 2 + \cdots + n = \frac{n(n+1)}{2},$$

and therefore $\bar{t} = (n+1)/2$. Notice that

$$\sum_{1}^{n} \bar{\imath}\,(t_i - \bar{t}) = \bar{\imath}\sum_{1}^{n} (t_i - \bar{t}) = 0,$$

so that 8.18 simplifies to

8.19
$$\hat{R} = \sum_{1}^{n} i(t_i - \bar{t}).$$

If there is an upward trend in the sequence $z_1, \ldots, z_n$, then the larger ranking z_i will tend to occur near the end of the sequence. This implies that large values of i and t_i tend to occur together, and also small values of i and t_i. We get the maximum of $\hat{R}$ when there is perfect correlation between i and t_i; that is, when $t_i = i$ for all i. This happens if and only if the sequence is increasing: $z_1 < z_2 < \cdots < z_n$. (Can you show this?) The maximum value of $\hat{R}$ equals

$$\sum_{1}^{n} i\left(i - \frac{n+1}{2}\right) = \sum_{i=1}^{n} i^2 - \left(\frac{n+1}{2}\right)\sum_{i=1}^{n} i$$

$$= \frac{n}{6}(n+1)(2n+1) - \frac{n}{4}(n+1)^2$$

$$= \frac{n}{12}(n^2 - 1).$$

Similarly, the minimum value of $\hat{R}$ occurs when $z_1 > z_2 > \cdots > z_n$. In this case,

$$t_1 = n,\ t_2 = n - 1,\ \ldots,\ t_n = 1$$

and

$$\hat{R} = -\frac{n(n^2-1)}{12}.$$

These facts lead us to define

8.20 $$\hat{\rho}_s = \frac{12\hat{R}}{n(n^2 - 1)}.$$

This number is known as Spearman's rank correlation coefficient, or Spearman's Rho. By definition $-1 \leq \hat{\rho}_s \leq 1$, and values of $\hat{\rho}_s$ near $+1$ or -1 imply upward or downward trend respectively. Naturally, Spearman's test for no trend against either upward or downward trend accepts if $|\hat{\rho}_s|$ is not too large. It is clear what the one-sided tests are. For $n \leq 30$, use Table 9 to get the extent of the acceptance region for various levels. For $n \geq 30$, the distribution of $\hat{\rho}_s$ is adequately approximated, under the hypothesis of no trend, by a normal distribution with mean zero and variance

8.21 $$V(\hat{\rho}_s) = \frac{1}{n - 1}.$$

Consider $z_1, \ldots, z_n$ to be outcomes of $\mathsf{Z}_1, \ldots, \mathsf{Z}_n$, where

$$\mathsf{Z}_i = \beta i + \mathsf{X}_i, \qquad i = 1, \ldots, n,$$

β is an unknown parameter, and $\mathsf{X}_1, \ldots, \mathsf{X}_n$ are independent variables with the same $N(\mu, \sigma^2)$ distribution, μ and σ^2 unknown. Here, the hypothesis of no trend is the hypothesis $\beta = 0$. There is a classical test for $\beta = 0$ which is the best tailored-to-fit test for this parametric model. (See Problem 17.) Again, we compute the efficiency of another test by dividing the sample size it requires to detect some small nonzero β at level α by the sample size required by the tailored-to-fit test. The median test has efficiency .83. Both the Kendall and Spearman tests have efficiency .98. These are surprisingly high. In particular, the efficiency .83 for the median test seems remarkably high. But notice that the linear increase term βi in Z_i is almost tailored to be detected by a first-half, second-half comparison. Regardless, the fact that the median test is so simple computationally makes it almost always worthwhile as a first try. If it rejects, then the situation is clear cut. If not, it may be useful to apply a more efficient test.

For any sequence $z_1, \ldots, z_n$ for which you suspect a trend, first make a sequence graph; that is, plot the values of z_i versus i. A look at this graph should give some hint of the presence, direction, and magnitude of trend, although, at times, the random fluctuations may be very deceptive to the eye.

The three tests in this section for trend can be largely applied by using a horizontal straightedge on the sequence graph. For example, to get $\hat{N}$ for the median test, move the straightedge upward until half the points fall above it. Then count the number of points above the straightedge in the left half of the graph. To compute $\hat{S}$, put the straightedge through each point z_i, and let s_i^+, s_i^- be the number of points to the left of z_i above and below the straightedge, respectively. Then

$$s_i^+ + s_i^- = i - 1$$

and

$$\begin{aligned} s_i &= s_i^- - s_i^+ \\ &= 2s_i^- + (i - 1). \end{aligned}$$

Count only the points s_i^- below the straightedge. Then letting

$$S = \sum_1^n s_i^-,$$

$$\begin{aligned} \hat{S} &= \sum_1^n s_i = 2 \sum_1^n s_i^- - \sum_1^n (i - 1) \\ &= 2S - \frac{n(n-1)}{2} \end{aligned}$$

and

8.22 $$\hat{\tau} = \frac{4S}{n(n-1)} - 1.$$

Finally, placing a straightedge through z_i, counting the number of points below the straightedge, and adding one gives the rank t_i of z_i.

Problem 16 The data below reads sequentially from left to right. Make a sequence graph for the data and compute $\hat{N}$, $\hat{\tau}$, $\hat{\rho}_s$. What are your conclusions about the presence of trend in the data?

0.890	0.494	0.391	0.061	0.833
0.664	0.846	0.897	1.077	1.005
1.102	0.738	0.569	0.506	0.411
1.031	0.887	0.579	0.433	0.944
0.581	0.791	0.427	0.993	0.806
0.096	0.496	0.0826	0.744	0.755
1.015	0.807	0.296	0.883	1.134
0.965	0.945	0.574	0.502	1.034
1.171	1.161	1.259	1.311	0.395
0.480	1.032	0.720	0.557	0.679

Problem 17 (Difficult!) For the model

$$Z_i = \beta i + X_i, \qquad i = 1, \ldots, n,$$

where $X_1, \ldots, X_n$ are independent with the distribution $N(\mu, \sigma^2)$, μ and σ^2 unknown, find the form of the likelihood ratio test for the hypothesis $\beta = 0$, and simplify. Assume $n = 2m - 1$ is odd. (Ans.

$$\textit{Accept if} \quad \left| \frac{\sum_j (j - m) z_j}{\hat{s}} \right| \leq d,$$

where

$$\hat{s}^2 = \frac{1}{n-1} \sum_j (z_j - \bar{z})^2$$

and $\bar{z}$ denotes the sample mean of the $z_1, \ldots, z_n$.)

Problem 18 **a** For the test of Problem 17, assume n large and $\hat{s} \simeq \sigma$ if $\beta = 0$. Evaluate d for a .05 level test.

b For β very small, assume again that $\hat{s} \simeq \sigma$ and find the minimum value of β that can be detected at the .05 level.
(Ans. Let

$$c_n^2 = \sum_1^n (j - m)^2,$$

then

a $d = 2.0c_n$

b $\Delta\beta = 3.6\sigma/c_n$.)

Independence Again

Now that three tests for trend have been set up, the next problem is to see how these tests work in the context of testing for independence. Once again, assume our data consist of couples $(x_1, y_1), \ldots, (x_n, y_n)$ independently drawn from the same underlying bivariate distribution. To apply the trend tests, order the x's as $x_{(1)} < x_{(2)} < \cdots < x_{(n)}$ and let the corresponding y values be $y'_1, \ldots, y'_n$. Denote the median of the $y'_1, \ldots, y'_n$ (or equivalently of the $y_1, \ldots, y_n$) by $\hat{v}_2$ and assume n is even, $n = 2m$. The median test uses the number

$$\hat{N} = \{\# \ y'_j > \hat{v}_2, \ j = 1, \ldots, m\}.$$

In other words, $\hat{N}$ is the number of couples (x_i, y_i) such that

$$x_j < \hat{\nu}_1, \quad y_j > \hat{\nu}_2,$$

where $\hat{\nu}_1$ is the median of the $x_1, \ldots, x_n$ values. This number can be quickly read from the cluster diagram. Draw a horizontal line through $\hat{\nu}_2$, a vertical line through $\hat{\nu}_1$, then $\hat{N}$ is the number of points in the second quadrant.

The $\hat{S}$ value for the couples (x_i, y_i), $i = 1, \ldots, n$, is gotten by scoring every possible pair of couples (x_i, y_i), (x_j, y_j), $i \neq j$, as $+1$ if the couple that has the largest x value also has the largest y value and assigning the score -1 to the pair otherwise. Then $\hat{S}$ is the sum of the scores for all distinct pairs of couples. This recipe is no help computationally. The simplest way to compute $\hat{\tau}$ is to look at the cluster diagram and treat it as though it were the sequence graph. Using a horizontal straightedge, for each point (x_i, y_i) count the number of points s_i^- to the left of (x_i, y_i) and below the straightedge. Then proceed to compute $\hat{\tau}$, using $S^- = \sum_i s_i^-$, exactly as in the last section.

Spearman's rank correlation statistic can be neatly expressed in terms of the original data couples. Suppose that the rank of some x_j is i; then in the sequence $y'_1, \ldots, y'_n$ the y'_i value is y_j and the t_i appearing in 8.18 is the rank of y_j. Hence the couple (x_j, y_j) contributes the term

$$\left(\text{rank } x_j - \frac{n+1}{2}\right)\left(\text{rank } y_j - \frac{n+1}{2}\right)$$

to the sum $\hat{R}$, and 8.18 is equivalent to

$$\hat{R} = \sum_1^n \left(\text{rank } x_j - \frac{n+1}{2}\right)\left(\text{rank } y_j - \frac{n+1}{2}\right). \tag{8.23}$$

Written in this form you can see why $\hat{\rho}_s$ is called a rank correlation coefficient. It is similar to the ordinary correlation coefficient, but computed between the ranks of the readings instead of between their actual values.

For detecting dependence between two variables when they are assumed to have a bivariate normal distribution, the best tailored-to-fit test rejects if the sample correlation coefficient $\hat{\rho}$ is too large. The efficiencies of the median test and the two rank correlation tests are .41, .91, .91, respectively.

The efficiency of the median test is disappointingly low. But the efficiencies of the two rank tests hold up quite well. The performance of these tests in nonnormal situations is similar, except that the median test can sometimes have fairly high as well as fairly low efficiencies. The two rank tests have consistently high efficiencies.

Problem 19 Use each of the three given tests for independence on the 20 pairs of numbers consisting of the first 20 entries of columns 1 and 2, Table 11. Use $\alpha = .1$.

Summary

This chapter revolved around two problems, matched comparisons and independence.

Matched comparisons lead to the problem of testing the hypothesis that independent variables $V_1, \ldots, V_n$, not necessarily identically distributed, each have a distribution symmetric around zero. The alternatives to the hypothesis are that $V_1, \ldots, V_n$ tend to be larger than zero or to be smaller than zero. Two tests were suggested:

a *The sign test*: Count the number of positive values in $v_1, \ldots, v_n$. Under the null hypothesis, this has the same distribution as the number of heads in fair coin tossing.

b *The one-sample Wilcoxon test*: Order $v_1, \ldots, v_n$ according to their absolute values and take the sum of the ranks of the positive v_i. Accept if this sum is neither too large nor too small (two-sided test).

The Wilcoxon test has high efficiency compared with the best tailored-to-fit tests for detecting nonzero medians in various parametric families of distributions symmetric about their medians. This observation leads to the Hodges-Lehmann estimate for the median of such a distribution:

$$\hat{v} = \textit{median of} \left\{ \frac{x_i + x_j}{2}, \quad 1 \leq j \leq i \leq n \right\}.$$

This estimate, while difficult to compute, has uniformly high efficiency.

The problem of testing for independence has one universal, fairly quick, and nonparametric answer: Break the plane up into rectangular cells and use a χ^2 test. This requires large sample size, and for continuous data faces the usual objection against the arbitrary selection of cells. If the data are discrete by nature, then the χ^2 approach is a natural and convenient test for independence.

We restrict the problem of testing for independence to those situations where the alternatives are monotone dependence. Then the problem can be translated into the problem of detecting trend in a sequence $z_1, \ldots, z_n$. Three tests are discussed:

a *The median test*: Compute $\hat{N}$, the number of z_i in the first half of the sequence that are larger than the median of $z_1, \ldots, z_n$. Accept no trend if $\hat{N}$ is not too large nor too small (two-sided test).

b *Kendall's Tau test*: Let s_i be the number of z_j, $j < i$, that are smaller than z_i minus the number that are larger, compute

$$\hat{S} = \sum_{1}^{n} s_i,$$

and divide $\hat{S}$ by its maximum value to get $\hat{\tau}$. Accept no trend if $|\hat{\tau}|$ is not too large (two-sided).

c *Spearman's Rho test*: Compute the "correlation" $\hat{R}$ between the place a value z_i has in the unordered sequence and its place in the ordered sequence. That is,

$$\hat{R} = \sum_{1}^{n} i(t_i - \bar{t}),$$

$\bar{t} = (n + 1)/2$. Divide $\hat{R}$ by its maximum value to get $\hat{\rho}_s$. Accept no trend if $|\hat{\rho}_s|$ is not too large (two-sided).

For testing against normal alternatives with a linearly increasing mean, these tests have efficiencies .83, .98, .98. When they are used as tests of independence and checked in a bivariate normal model, their efficiencies are .41, .91, .91. Even though the sign test and median test have low efficiencies compared to the other tests discussed, they are so simple to compute that it is usually worthwhile to try them first.

9 Testing Whether Underlying Distributions Are the Same

Introduction

This chapter is built around a single problem: Given a number of samples from different populations, decide whether the populations have the same distribution. Precisely, suppose $x_1, \ldots, x_n$ is a sample drawn from one underlying distribution and $y_1, \ldots, y_m$ a sample drawn, independently of the first, from some other underlying distribution. How can these data be used to decide whether the two underlying distributions are the same? Or suppose the data consist of three samples $x_1, \ldots, x_n$, $y_1, \ldots, y_m$, and $z_1, \ldots, z_l$ drawn from three different populations. Are the three underlying distributions the same?

Recall the example we posed in the first chapter. Suppose that a culture-free IQ test is given to two different African tribal groups of children and a group of American middle-class children of similar ages. Do the three sets of scores indicate any systematic differences in test performance between the three populations? Can we conclude that except for random fluctuations, the three sets of scores look as though they were all drawn from the same underlying distribution? Or, does one set of scores look like it comes from a distribution with generally higher outcomes—and therefore higher mean. Or perhaps the means look about the same, but one suspects that one of the African groups has a much larger dispersion around the mean than the other two groups.

The problem of deciding whether a number of samples come from the same underlying distribution is a fundamental and frequent problem in statistics. It comes up in a variety of applications. For example, consider testing samples of the same component manufactured by different companies or testing tensile strength of samples of alloys with varying percentages of the constituent metals. Or, consider the comparison of one group of subjects which have been given a certain treatment with an untreated control group. The list of uses is large.

There are two approaches to the problem, depending on what alternative you want to test against. Specifically, suppose we are in the two-sample problem with two underlying distributions P_1 and P_2. The hypothesis is that the two samples come from the same distribution; that is, that

$$P_1 = P_2.$$

In a *shift alternative problem*, the alternatives to $P_1 = P_2$ are that one distribution has most of its probability either to the right or to the left of the other. Thus, in a shift alternative problem, we are trying to detect whether the values in one sample are running consistently higher or lower than in the other sample. For instance, at a very crude level, is the mean of one sample "substantially" higher than the mean of the other sample? If there are three or more samples, then the *shift alternative problem* is to test whether the underlying distributions are the same versus the alternative that some of the underlying distributions have outcomes generally larger than the others.

In the classical model where all distributions are assumed normal, the best test to use on a two-sample alternative problem is an old friend, Student's t test. We also want efficient nonparametric tests for this situation, and some ideas of the last chapter can be applied. The median test gives a test that is quick computationally, but not very efficient. To take a step up in efficiency, we use a Wilcoxon test on the ranks; even if there are three or more samples, a Wilcoxon test on the ranks is available and works well. We can just about say that if the data are not discrete, then unless we are firmly convinced that the distributions are normal, the most effective test in a shift alternative problem is a Wilcoxon test procedure.

In the second type of multisample problem, the hypothesis again is that the underlying distributions are the same, but the alternative is that the underlying distributions differ in some way, shape, or form. For instance, in a two-sample situation, you may suspect that the two underlying distributions P_1 and P_2 have about the same mean or median but that one of them is much more tightly concentrated about the mean than the other. Or, you might suspect that one of them is much more skewed to the left than the other. Since they have about the same mean or median, there is no sense in applying a shift test. A Wilcoxon rank test will not detect the differences in shape you are looking for. We call a problem such as this, where we are trying to design a test that will detect almost *any* significant difference, a *general alternative problem*.

There are two widely used methods for handling this problem. If the data are discrete, use χ^2. If they are continuous, then using cells will make χ^2 applicable. Of course, there are the usual objections to grouping continuous data into a χ^2 mold. The other test available is designed for continuous data. It is simply a two-sample version of the Kolmogorov-Smirnov test.

Whether the alternatives are shift or general, this kind of problem is sometimes referred to as testing for *homogeneity of populations*. The hypothesis that all underlying distributions are the same is called the *hypothesis of homogeneity*. You can see that, in the main, the methods used in testing homogeneity are familiar ones, revised to apply in this new context.

Problem 1 The standard method for measuring a surface film of deposited metal on some electronic components is expensive and destructive. A proposed method using X-ray diffraction is inexpensive and much faster. Discuss how to set up an experiment to test whether the proposed method is comparable in accuracy to the standard method. Under what conditions would you set up a matched comparison experiment? Under what conditions would you set up a two-sample experiment?

Problem 2 How would you design an experiment to test the effect of an oil additive on piston wear? How many different samples might be desirable? What type of multisample problem is this?

Problem 3 List some situations from your own field which can be tested in terms of a two or more sample situation.

The Two-Sample Student *t* Test

Suppose you have a sample $x_1, \ldots, x_n$ from one distribution P_1 and an independent sample $y_1, \ldots, y_m$ from another distribution P_2. Suppose further that P_1 is normal $N(\mu_1, \sigma_1^2)$ and P_2 is normal $N(\mu_2, \sigma_2^2)$, where the four parameters are unknown. Then the model is that the data vector $\mathbf{x}, \mathbf{y}$ is the outcome of independent random variables $\mathsf{X}_1, \ldots, \mathsf{X}_n$, $\mathsf{Y}_1, \ldots, \mathsf{Y}_m$, where the first n are normal $N(\mu_1, \sigma_1^2)$ and the last m are normal $N(\mu_2, \sigma_2^2)$. The hypothesis of homogeneity is that the 4-dimensional parameter $\theta = (\mu_1, \mu_2, \sigma_1^2, \sigma_2^2)$ lies in the 2-dimensional subspace defined by $\mu_1 = \mu_2$, $\sigma_1 = \sigma_2$. Let $\hat{\mu}_1, \hat{\mu}_2$ be the sample means, and look at the difference $\hat{\mu}_1 - \hat{\mu}_2$. Under the hypothesis of homogeneity, $\hat{\boldsymbol{\mu}}_1 - \hat{\boldsymbol{\mu}}_2$ has mean zero. Furthermore, under homogeneity, since $\sigma_1 = \sigma_2 = \sigma$ and $\hat{\boldsymbol{\mu}}_1$ and $\hat{\boldsymbol{\mu}}_2$ are independent, the variance of $\hat{\boldsymbol{\mu}}_1 - \hat{\boldsymbol{\mu}}_2$ is

9.1
$$V(\hat{\boldsymbol{\mu}}_1 - \hat{\boldsymbol{\mu}}_2) = V(\hat{\boldsymbol{\mu}}_1) + V(\hat{\boldsymbol{\mu}}_2)$$
$$= \frac{\sigma^2}{n} + \frac{\sigma^2}{m} = \sigma^2\left(\frac{1}{n} + \frac{1}{m}\right).$$

Thus, the scatter of $\hat{\mu}_1 - \hat{\mu}_2$ depends on the unknown common variance σ^2. To get rid of this parameter divide $\hat{\mu}_1 - \hat{\mu}_2$ by an estimate of its standard deviation. Let $\hat{s}_1^2, \hat{s}_2^2$ be the unbiased estimates of σ_1^2, σ_2^2. Since $\sigma_1^2 = \sigma_2^2 = \sigma^2$, we get a combined unbiased estimate $\hat{s}^2$ for σ^2 by taking

9.2 $$\hat{s}^2 = \frac{N\hat{s}_1^2 + M\hat{s}_2^2}{N + M}, \qquad N = n - 1, M = m - 1.$$

The standard deviation of $\hat{\mu}_1 - \hat{\mu}_2$ is

$$\sigma\sqrt{\frac{1}{n} + \frac{1}{m}}.$$

Therefore, look at the ratio

9.3 $$\hat{t} = \frac{\hat{\mu}_1 - \hat{\mu}_2}{\hat{s}\sqrt{1/n + 1/m}}.$$

It is easy enough to show that 9.3 has a Student's t distribution with $N + M$ degrees of freedom under the hypothesis of homogeneity. It is clear how to use 9.3: Accept the hypothesis of homogeneity if the absolute value $|\hat{t}|$ of the t ratio 9.3 is not too large. The exact bound can be found from Table 3 of the t distribution if $N + M \leq 30$. For larger samples, $\hat{t}$ is approximately $N(0, 1)$ if homogeneity holds. This gives the two-sided test. The one-sided tests are the usual variations on the two-sided one.

Using the t test in this context gives a test against shift alternatives. If $\mu_1 = \mu_2$ then the t test will not generally detect even large differences between σ_1 and σ_2—it is not designed to do so.

Suppose we are in a shift situation where the two distributions have about the same variances but $\mu_1 = \mu + \Delta\mu$, $\mu_2 = \mu$. Assuming equal sample sizes, $m = n$, what sample size is needed to detect $\Delta\mu$ at level α? For large sample sizes, use the approximation $\hat{s} \simeq \sigma$. Then a computation gives (for $\alpha = .05$)

9.4 $$n = \frac{26}{(\Delta\mu/\sigma)^2}.$$

To detect a shift in mean of a whole σ-unit means that we need samples of about size 26 from each population. To detect a shift of $(1/2)\sigma$, about 100 samples from each population are required. The detection capability is disappointingly low, but this is actually the best we can do.

The t test is a good test. In contrast to many other tests based on normally distributed models, it is surprisingly robust. It maintains its stated level accurately even if the underlying distributions are not exactly normal. The larger the sample size, the more robust it is, and the two-sided test is more robust than the one-sided.

Problem 4 It is claimed that a chemical compound retards the flammability of paper. Twenty sheets of paper, ten treated and ten untreated, are set on fire at one corner and timed until the flame goes out. These are the data:

Treated	9.8,	11.0,	6.4,	9.3,	6.8,	5.0,	3.8,	9.1,	8.1,	7.8
Untreated	5.8,	2.4,	8.7,	8.9,	3.1,	6.2,	2.3,	5.7,	3.2,	8.3.

Run a t test on these data and see what you conclude. Is a one-sided or two-sided test more appropriate? What can be said about the hypothesis of normality?

Problem 5 Show that the likelihood ratio test for $\mu_1 = \mu_2$ in a two-sample situation, with underlying distributions $N(\mu_1, \sigma^2)$, $N(\mu_2, \sigma^2)$, σ unknown, is the t test based on 9.3. (See Chapter 5, p. 153 ff., and simplify the two-sample F test.)

The Median Test

As robust as the t test may be to some departure from normality, if the actual underlying distributions are markedly nonnormal in shape and if the sample size is not very large, then the t test may not work as tabulated.

We will look at two nonparametric competitors to the t test. The first of these, the median test, generally is not too efficient, but can be computed much more easily than the t test or any other common nonparametric tests.

This test is very similar to the median test for trend used in the last chapter. Assume $n + m$ is even, and let $\hat{v}$ be the median of the combined sample $x_1, \ldots, x_n, y_1, \ldots, y_m$. Let $\hat{N}$ be the number of $x_1, \ldots, x_n$ that are less than $\hat{v}$. If the two samples come from the same distribution, then $\hat{N}$ should be around $n/2$. For a two-sided test, we accept the hypothesis of homogeneity if $\hat{N}$ is neither too large nor too small. If we are testing against the one-sided alternative that the distribution underlying the x values has generally larger outcomes, then we accept the hypothesis if $\hat{N}$ is not too small—and analogously for testing against the alternatives that the y values are generally larger. The exact distribution of $\hat{N}$ is easy to find when the hypothesis holds. Consider drawing half of the balls out of an urn containing n red balls and m black balls. Then the probability that you will draw j red balls is the probability that $\hat{N} = j$. Therefore,

$$P(\hat{\mathsf{N}} = j) = \frac{\binom{n}{j}\binom{m}{\frac{n+m}{2} - j}}{\binom{n+m}{\frac{n+m}{2}}} \tag{9.5}$$

(the hypergeometric distribution). For larger sample sizes N has an approximately normal distribution with

$$E\hat{\mathsf{N}} = \frac{n}{2}, \qquad V(\hat{\mathsf{N}}) = \frac{1}{4}\left(\frac{nm}{n + m - 1}\right). \tag{9.6}$$

Now look at the computational speed of the median test. Essentially, the only difficult part is in evaluating $\hat{\nu}$. This is followed by an easy counting procedure. Compare this with the t test, which required the computation of means and variances for both samples.

The drawback of the median test is that its efficiency is not up to par. When used in the normal model of the last section and compared to the t test, its efficiency is .64. This implies that to detect a shift of amount $\Delta\mu$ at a given level, it generally needs a sample size 57% larger than that for the t test.

Another Wilcoxon Test

We can get a highly efficient nonparametric test by using a Wilcoxon test on the ranks. Suppose we have two samples $x_1, \ldots, x_n$ and $y_1, \ldots, y_m$ drawn from continuous distributions. Combine the two samples into one sample of size $n + m$ and order this combined sample. For instance, you might get

$$x_{10} < y_3 < y_7 < x_5 < \cdots < y_{19} < x_{52}.$$

Assume $n \leq m$ and let T be the sum of the ranks of the x values in the combined ordering. The idea is to accept homogeneity if T is neither too small nor too large, with the obvious modifications for one-sided tests.

The distribution of T under the hypothesis is fairly simple. The combined sample consists of $N = n + m$ draws from the same distribution. The assigned ranks range over the integers $1, 2, 3, \ldots, N$. The ranks of the

x values have the same distribution as n numbers drawn at random from the numbers $1, 2, 3, \ldots, N$. Then T is the sum of the numbers drawn. From this, it is not difficult to get the mean of T (see Problems 7 and 9) or its exact distribution for n, m small.

There is another way of looking at this test which is reminiscent of Kendall's Tau. Let U be the number of times that an x value precedes a y value in the combined ordering. One could compute U by counting, for every x value, how many y values follow it in the combined ordering, and then summing these counts. If the x values are generally smaller, then they come early in the ordering; many y values follow them, and U will be large. Also T will be small. If the x values are larger, then U will be small and T will be large. This identity is slightly surprising;

9.7 $$U + T = mn + \frac{m(m+1)}{2}.$$

To get the value of U, it is usually easier to compute T and then use 9.7. Seeing why 9.7 is true is an interesting byroad that we leave to the reader as Problem 8. At any rate, because of 9.7, it makes no difference whether we base our tests on U or on T. The U function was introduced by Mann and Whitney, and the two-sample Wilcoxon test is frequently called the Mann-Whitney test.

The tables are usually based on the value of U. For large sample size, under homogeneity, the distribution of U is approximately normal with

9.8 $$E\mathrm{U} = \frac{nm}{2}, \qquad V(\mathrm{U}) = \frac{nm(n+m+1)}{12}.$$

The approximation is adequate for $n \geq 9$, $m \geq 9$.

The efficiency of this test in the normal shift problem of the second section as compared to the t test is .95. Furthermore, its efficiency as compared with the best tailored-to-fit tests has been computed in a variety of parametric families. It usually has high efficiency.

The t test really tests to see whether the means of the two underlying distributions are equal. The median test is actually a test for equality of the two medians. What the Mann-Whitney test does is this: By its definition U is the number of all couples (x_i, y_j), $i = 1, \ldots, n$, $j = 1, \ldots, m$, such that $x_i < y_j$. Let X be a random variable having the underlying distribution of the first sample, and Y an independent random variable having the underlying distribution of the second sample. Then U/mn, the proportion of couples for which $x_i < y_j$, is a consistent estimate for

$$P(\mathrm{X} < \mathrm{Y}).$$

The Mann-Whitney test accepts the hypothesis if U is not too far from $mn/2$; therefore it is testing whether

$$P(\mathrm{X} < \mathrm{Y}) = \tfrac{1}{2},$$

that is, whether an outcome chosen at random according to the first distribution has probability $\frac{1}{2}$ of being less than an outcome independently selected from the second distribution. Writing

$$P(\mathrm{X} < \mathrm{Y}) = P(\mathrm{X} - \mathrm{Y} < 0)$$

shows that equivalently we are testing the hypothesis that the distribution of $\mathrm{X} - \mathrm{Y}$ has median zero.

Problem 6 The first row of numbers below is taken from a table of random numbers drawn from a distribution uniform on [0, 1]. The second row is also taken from that table. The remaining rows are the second row entries with an amount θ added, where $\theta = .10$, .20, .30, .40.

Row									
1	0.14,	0.55,	0.49,	0.76,	0.35,	0.44,	0.59,	0.03,	0.87,
2	0.09,	0.86,	0.07,	0.42,	0.55,	0.97,	0.16,	0.41,	0.01,
3	0.19,	0.96,	0.17,	0.52,	0.65,	1.07,	0.26,	0.51,	0.11,
4	0.29,	1.06,	0.27,	0.62,	0.75,	1.17,	0.36,	0.61,	0.21,
5	0.39,	1.16,	0.37,	0.72,	0.85,	1.27,	0.46,	0.71,	0.31,
6	0.99,	1.26,	0.47,	0.82,	0.95,	1.37,	0.56,	0.81,	0.41,

Continued

Row									
1	0.52,	0.48,	0.19,	0.02,	0.94,	0.01,	0.92,	0.09,	0.04,
2	0.10,	0.69,	0.34,	0.13,	0.30,	0.94,	0.31,	0.80,	0.70,
3	0.20,	0.79,	0.44,	0.23,	0.40,	1.04,	0.41,	0.90,	0.80
4	0.30,	0.89,	0.54,	0.33,	0.50,	1.14,	0.51,	1.00,	0.90
5	0.40,	0.99,	0.64,	0.43,	0.60,	1.24,	0.61,	1.10,	1.00
6	0.50,	1.09,	0.74,	0.53,	0.70,	1.34,	0.71,	1.20,	1.10

Comparing row 1 with rows 3, 4, 5, 6, at which row does the one-sided median test at level .05 detect the shift? At which row does the one-sided Mann-Whitney test at level .05 detect the shift?

Problem 7 Suppose that an urn contains slips of paper with N numbers 1, 2, 3, ..., N written on them. Let M_k be the number drawn on the kth draw, and let $\mathsf{n}_1, \ldots, \mathsf{n}_{N-k}$ be the numbers remaining after the kth draw. At the $(k + 1)$th draw, each of $\mathsf{n}_1, \ldots, \mathsf{n}_{N-k}$ may be selected with probability $1/(N - k)$. Therefore

$$E(\mathsf{M}_{k+1} \mid \mathsf{n}_1, \ldots, \mathsf{n}_{N-k}) = \frac{\mathsf{n}_1 + \cdots + \mathsf{n}_{N-k}}{N - k}.$$

Let $\mathsf{T}_k = \mathsf{M}_1 + \cdots + \mathsf{M}_k$. Then since

$$1 + 2 + \cdots + N = \frac{N(N+1)}{2},$$

$$E(\mathsf{M}_{k+1} \mid \mathsf{n}_1, \ldots, \mathsf{n}_{N-k}) = \frac{1}{N-k}\left(\frac{N(N+1)}{2} - \mathsf{T}_k\right).$$

Take expectations of both sides to get

$$E\mathsf{M}_{k+1} = \frac{N(N+1)}{2(N-k)} - \frac{1}{N-k} E\mathsf{T}_k.$$

Since $\mathsf{T}_{k+1} = \mathsf{T}_k + \mathsf{M}_{k+1}$, this gives

$$E\mathsf{T}_{k+1} = \frac{N(N+1)}{2(N-k)} + \frac{N-k-1}{N-k} E\mathsf{T}_k.$$

Use this to prove that

$$E\mathsf{T}_n = \frac{n}{2}(N+1).$$

How does this square with 9.8? For an easier way to get ET, see Problem 9.

Problem 8 The number U can be computed this way: In the combined ordering, let s_j be the number of y values following x_j and let r_j be the number of x values following x_j. Then the rank of x_j is $N - r_j - s_j$ and

$$T = \sum_{j=1}^{n} (N - r_j - s_j).$$

Since $U = \sum_1^n s_j$,

$$U + T = nN - \sum_{j=1}^{n} r_j.$$

Use this identity to prove 9.7.

Problem 9 Here is a much easier way to get the result of Problem 7. Define a function

$$h(u, v) = \begin{cases} 1, & u < v \\ 0, & u \geq v. \end{cases}$$

Then

$$U = \sum_{k=1}^{m} \sum_{j=1}^{n} h(x_j, y_k),$$

so that

$$EU = \sum_{k,j} Eh(X_j, Y_k).$$

Use this to show that $EU = mn/2$ and use 9.7 to get ET. The variance of U can be gotten by computing EU^2 using the same sum representation as above.

Problem 10 Given a single sample $x_1, \ldots, x_N$, the Mann-Whitney two-sample test can be adapted to give a test for trend. Divide the sample into a first part $x_1, \ldots, x_n$ and a second part $x_{n+1}, \ldots, x_N$. Now run the two-sample test on these two groups. Use this approach on the data of Problem 16, Chapter 8, by splitting the sample in two.

omit

The Wilcoxon Test for More Than Two Samples

Suppose we have samples

$$x_1, \ldots, x_n$$
$$y_1, \ldots, y_m$$
$$z_1, \ldots, z_l$$

from three populations, and we want to know whether the three underlying distributions are the same. One approach is to do a two by two comparison. That is, do $x_1, \ldots, x_n$ and $y_1, \ldots, y_m$ come from the same distribution? Do $y_1, \ldots, y_m$ and $z_1, \ldots, z_l$ come from the same distribution? And so on. This leads to a test whose probability of rejecting when the hypothesis is true is actually much larger than the level α you are working with. If you are comparing a number of samples two by two, then even if they all come from the same underlying distribution, just by random fluctuations you are liable to find two samples that deviate considerably from each other (see Problem 12).

It is preferable to first run an overall test for homogeneity on all samples. Then if this test rejects, begin examining the data to see where the lack of homogeneity is manifested. There is a classical test for a normal model. This is the F test developed in Chapter 5, p. 153. But why bother to use the F test when there is a highly efficient nonparametric test available that is usually computationally simpler? This test takes the samples, combines them into one large sample, and orders by size. Let R_1 be the sum of the ranks of the first sample, R_2 the sum of the ranks of the second sample, and so on. Specifically, in the three-sample case, R_1 is the sum of the ranks of $x_1, \ldots, x_n$ in the ordering of the $n + m + l$ numbers $x_1, \ldots, x_n, y_1, \ldots, y_m, z_1, \ldots, z_l$, R_2 the sum of the ranks of $y_1, \ldots, y_m$, and R_3 the sum of the ranks of $z_1, \ldots, z_l$. Let $N = n + m + l$. Then

$$R_1 + R_2 + R_3 = \frac{N(N+1)}{2}.$$

If all three samples come from the same distribution, then their ranks are scattered at random among the numbers 1, 2, 3, ..., N. For instance, we get R_1 by mixing the numbers 1, 2, ..., N in an urn, picking out n of them at random, and summing the numbers selected. From the results of Problem 7 or 9,

$$E\mathsf{R}_1 = \frac{n}{2}(N+1), \qquad E\mathsf{R}_2 = \frac{m}{2}(N+1), \qquad E\mathsf{R}_3 = \frac{l}{2}(N+1).$$

These three expectations, denoted μ_1, μ_2, μ_3, are the expected values of the sums of the ranks *if the samples are drawn from the same distribution.* If the underlying distributions are different, then one or more of R_1, R_2, R_3 will tend to be too small and one or more will tend to be too large. We can get some measure of their deviation from homogeneity by subtracting their means, μ_1, μ_2, μ_3 and looking at the squares of the differences

$$(\mathsf{R}_1 - \mu_1)^2, \qquad (\mathsf{R}_2 - \mu_2)^2, \qquad (\mathsf{R}_3 - \mu_3)^2.$$

Then we can get an overall measure of deviation by taking some weighted sum of these squares. If $n = m = l$, then the appropriate measure would just be the sum of all three squares. But if the sample sizes are unequal, the selection of the weighting may be done in many different ways. The particular weighting that is selected by the Kruskal-Wallis test follows the ideas that led us to the χ^2 test.

By a variation of the central limit theorem it turns out that for l, m, n large enough, the variables

$$\mathsf{Z}_1 = \mathsf{R}_1 - \mu_1, \qquad \mathsf{Z}_2 = \mathsf{R}_2 - \mu_2, \qquad \mathsf{Z}_3 = \mathsf{R}_3 - \mu_3$$

have a joint normal distribution. Furthermore, they satisfy the linear relationship

$$\mathsf{Z}_1 + \mathsf{Z}_2 + \mathsf{Z}_3 = 0.$$

This suggests looking for a weighting such that

$$aZ_1^2 + bZ_2^2 + cZ_3^2$$

has the same distribution as $U_1^2 + U_2^2$, where U_1, U_2 are independent $N(0, 1)$ variables. In other words, look for a weighting such that for large sample size, *if the hypothesis is true*, then the weighted sum has a χ^2 distribution with two degrees of freedom. A weighting that does this is

$$a = \frac{6}{N\mu_1}, \qquad b = \frac{6}{N\mu_2}, \qquad c = \frac{6}{N\mu_3}.$$

Therefore, look at

9.9 $$H = \frac{6}{N}\left(\frac{Z_1^2}{\mu_1} + \frac{Z_2^2}{\mu_2} + \frac{Z_3^2}{\mu_3}\right).$$

The test based on this H value is called the *Kruskal-Wallis test*; it rejects if H is too large. There is no one-sided version—it is always a two-sided test.

The expression for H can be simplified by squaring $(R_j - \mu_j)^2$ and using the equation

$$R_1 + R_2 + R_3 = \mu_1 + \mu_2 + \mu_3 = \frac{N(N+1)}{2}$$

to get

9.10 $$H = \frac{12}{N(N+1)}\left(\frac{R_1^2}{n} + \frac{R_2^2}{m} + \frac{R_3^2}{l}\right) - 3(N+1).$$

Now, suppose we have J samples of size $n_1, \ldots, n_J$ from J populations. Let R_j, $j = 1, \ldots, J$, be the sum of the ranks of the jth sample in the ordering of the combined sample of size $N = n_1 + \cdots + n_J$. Compute the value of

9.11 $$H = \frac{12}{N(N+1)}\left(\sum_{j=1}^{J} \frac{R_j^2}{n_j}\right) - 3(N+1).$$

Then, for $n_1, \ldots, n_J$ large enough, the distribution of H is χ^2 with $J - 1$ degrees of freedom. Thus, the Kruskal-Wallis test easily extends to the case of J different samples.

Relevant values of H for the three-sample case for small sample sizes are tabled. For larger samples use the χ^2 approximation.

The Kruskal-Wallis test retains the high efficiency associated with Wilcoxon tests. For normal shift models, it has an efficiency of .95 compared with the best tailored-to-fit test.

Problem 11 Look at the following three samples:

1. 1.43, 1.65, 1.58, 3.64, 2.48, 2.82, 2.94, 2.62, 1.76, 3.32
2. 3.15, 3.12, 2.97, 4.16, 3.93, 2.93, 4.10
3. 2.91, 2.16, 4.12, 0.64, 2.53, 4.23, 1.96

Test for homogeneity at the .05 level using the Kruskal-Wallis test. Repeat the test after subtracting 0.5 from every value in the second sample.

Problem 12 Let $x_1, \ldots, x_n$ be drawn from an $N(\mu_1, 1)$ distribution and $y_1, \ldots, y_n$ from an $N(\mu_2, 1)$ distribution, μ_1 and μ_2 unknown. The hypothesis of homogeneity here reduces to $\mu_1 = \mu_2$. The appropriate test is to reject homogeneity if $|\bar{x} - \bar{y}|$ is too large.

a Show that the .05 level acceptance region is given by

$$|\bar{x} - \bar{y}| \leq \frac{2.77}{\sqrt{n}} .$$

b Suppose that you have J samples, each of size n, where the jth sample is drawn from an $N(\mu_j, 1)$ distribution. Suppose further that to test homogeneity you decide to run the above two-sample test at level .05 on all pairs of samples. Find the probability that even if all means are equal you will find at least one pair of samples for which you reject equality of means. Do this for $J = 3, 5, 10$. (Use Table VIII, 1, pages 130–135 of the 1966 CRC *Tables for Probability and Statistics*. This is a table of the cumulative distribution function of $X_{(n)} - X_{(1)}$, where $X_{(1)} \leq X_{(2)} \leq \cdots \leq X_{(n)}$ are the order statistics for a sample $X_1, \ldots, X_n$ drawn from an $N(0, 1)$ distribution.)

(Ans. .12, $J = 3$; .28, $J = 5$; .62, $J = 10$.)

The χ^2 Test for Homogeneity

So far, all of the tests discussed for the hypothesis of homogeneity have been for continuous underlying distributions and against shift alternatives. The χ^2 method is designed for discrete data and tests homogeneity against general alternatives.

Consider the case in which both samples $x_1, \ldots, x_n$ and $y_1, \ldots, y_m$ are drawn from distributions having only a finite number of outcomes numbered 1, 2, . . . , J. The problem is to test whether these two distributions are the same. Let $\hat{n}_1, \ldots, \hat{n}_J$ be the number of times we get outcomes $1, \ldots, J$ in the first sample. Then $\hat{n}_1 + \cdots + \hat{n}_J = n$. Let $\hat{m}_1, \ldots, \hat{m}_J$ be the observed frequencies in the second sample, so that

$$\hat{m}_1 + \cdots + \hat{m}_J = m.$$

Denote by $p_1^{(1)}, \ldots, p_J^{(1)}$ the underlying probabilities for the first sample, and by $p_1^{(2)}, \ldots, p_J^{(2)}$ the probabilities for the second sample.

In an ordinary one-sample χ^2 goodness of fit test, we define

$$D = n \sum_1^J \frac{(\hat{p}_j - p_j)^2}{p_j},$$

where $\mathbf{p} = (p_1, \ldots, p_J)$ are the hypothesized probabilities and

$$\hat{p}_j = \hat{n}_j/n, \; j = 1, \ldots, J$$

are the sample maximum likelihood estimates for the underlying probabilities. In other words, D is a sort of weighted squared difference between the hypothesized and estimated probabilities. In the two-sample situation,

$$\hat{p}_j^{(1)} = \frac{\hat{n}_j}{n} \qquad \text{and} \qquad \hat{p}_j^{(2)} = \frac{\hat{m}_j}{m}$$

are the estimates within each sample for their underlying probabilities. To test homogeneity, the basic approach is to see whether the estimates

$$(\hat{p}_1^{(1)}, \hat{p}_2^{(1)}, \ldots, \hat{p}_J^{(1)}) \qquad \text{and} \qquad (\hat{p}_1^{(2)}, \ldots, \hat{p}_J^{(2)})$$

are too far apart. Following the goodness of fit form of D it seems reasonable to try to test homogeneity by something of the form

$$\sum_1^J \frac{(\hat{p}_j^{(1)} - \hat{p}_j^{(2)})^2}{p_j},$$

Now p_j is the probability of the jth outcome under the hypothesis. If the hypothesis of homogeneity holds, then we can use the combined sample to estimate p_j, getting

$$\hat{p}_j = \frac{\hat{m}_j + \hat{n}_j}{m + n}.$$

Therefore, our test should be based on something that looks like

$$\sum_1^J \frac{(\hat{p}_j^{(1)} - \hat{p}_j^{(2)})^2}{\hat{p}_j}.$$

Unfortunately, this heuristic approach, while it gives the right form of the answer, does not give the appropriate normalizing factor, nor do we end up knowing how many degrees of freedom there are.

Here is another way to go at it. Suppose the two samples have the same underlying distribution $\mathbf{p} = (p_1, \ldots, p_J)$. Then

$$\mathsf{D}_1 = n \sum_1^J \frac{(\hat{\mathsf{p}}_j^{(1)} - p_j)^2}{p_j}$$

and

$$\mathsf{D}_2 = m \sum_1^J \frac{(\hat{\mathsf{p}}_j^{(2)} - p_j)^2}{p_j}$$

are independent variables each with a χ^2_{J-1} distribution. Therefore

$$\mathsf{D} = \mathsf{D}_1 + \mathsf{D}_2$$

has a χ^2 distribution with $2J - 2$ degrees of freedom. (Why?) If the data consist of two samples from the distribution $\mathbf{p}$, then the maximum likelihood estimates of p_j are gotten by using the combined sample: that is,

$$\hat{p}_j = \frac{\hat{m}_j + \hat{n}_j}{m + n}.$$

Replace the $J - 1$ unknown parameters $p_1, \ldots, p_J$ appearing in D by their maximum likelihood estimates to get

$$D = n \sum_j \frac{(\hat{p}_j^{(1)} - \hat{p}_j)^2}{\hat{p}_j} + m \sum_j \frac{(\hat{p}_j^{(2)} - \hat{p}_j)^2}{\hat{p}_j}.$$

Now notice that

$$\hat{p}_j = \frac{n}{m + n}\,\hat{p}_j^{(1)} + \frac{m}{m + n}\,\hat{p}_j^{(2)},$$

so

$$\hat{p}_j^{(1)} - \hat{p}_j = \frac{m}{m + n}(\hat{p}_j^{(1)} - \hat{p}_j^{(2)}),$$

$$\hat{p}_j^{(2)} - \hat{p}_j = \frac{n}{m + n}(\hat{p}_j^{(2)} - \hat{p}_j^{(1)}).$$

Then

9.12
$$D = \frac{mn}{m + n} \sum_1^J \frac{(\hat{p}_j^{(1)} - \hat{p}_j^{(2)})^2}{\hat{p}_j}.$$

Furthermore, D has a χ^2 distribution (for large sample size) with

$$2J - 2 - (J - 1) = J - 1$$

degrees of freedom. Finally, we can write D in terms of the observed frequencies as

9.13
$$D = \frac{1}{nm} \sum_{1}^{J} \frac{(m\hat{n}_j - n\hat{m}_j)^2}{\hat{n}_j + \hat{m}_j}.$$

What kind of difference between two probability distributions $\mathbf{p}^{(1)} = (p_1^{(1)}, \ldots, p_J^{(1)})$ and $\mathbf{p}^{(2)} = (p_1^{(2)}, \ldots, p_J^{(2)})$ can we detect at level α with given sample sizes? Look at the case $m = n$. Denote

$$\Delta p_j = p_j^{(1)} - p_j^{(2)}, \qquad p_j = \frac{p_j^{(1)} + p_j^{(2)}}{2}.$$

Define the distance between $\mathbf{p}^{(1)}$ and $\mathbf{p}^{(2)}$ to be

9.14
$$D(\mathbf{p}^{(1)}, \mathbf{p}^{(2)}) = \sum_{1}^{J} \frac{(\Delta p_j)^2}{p_j}.$$

Then, the minimum distance $D_{\min}$ between two distributions that can be detected at level α for n large is

9.15
$$D_{\min} = 2\gamma_K \frac{\sqrt{K}}{n}, \qquad K = J - 1,$$

where γ_K is nearly constant. (See p. 203 for values of γ_K.) Just as in goodness of fit tests, it is surprising how large a sample size is needed to detect moderate differences between distributions.

The same approach works to give a χ^2 test for the homogeneity of more than two populations. Suppose our data consist of K samples of sizes $n_1, \ldots, n_K$. Take $\hat{n}_j^{(k)}$ to denote the number of times outcome j appears in the kth sample. Also denote $\hat{p}_j^{(k)} = \hat{n}_j^{(k)}/n_k$. If each population sampled has the same underlying distribution $\mathbf{p} = (p_1, \ldots, p_J)$ and

$$\mathsf{D}_k = n_k \sum_{j=1}^{J} \frac{(\hat{p}_j^{(k)} - p_j)^2}{p_j},$$

then $\mathsf{D}_1 + \cdots + \mathsf{D}_K$ is χ^2 with $K(J - 1)$ degrees of freedom. Now replace p_j in this sum by its maximum likelihood estimate based on the combined sample

$$\hat{p}_j = \frac{\hat{n}_j^{(1)} + \cdots + \hat{n}_j^{(K)}}{n_1 + \cdots + n_K},$$

getting

9.16
$$D = \sum_{k=1}^{K} \sum_{j=1}^{J} n_k \frac{(\hat{p}_j^{(k)} - \hat{p}_j)^2}{\hat{p}_j}.$$

The variable D is χ^2 with

$$K(J-1)-(J-1)=(K-1)(J-1)$$

degrees of freedom.

We can lay out the observed frequencies in these K samples in a $K \times J$ *contingency table.*

	Sample number 1	2	$\cdots$	K	Totals
Outcome 1	$\hat{n}_1^{(1)}$	$\hat{n}_1^{(2)}$	$\cdots$	$\hat{n}_1^{(K)}$	$\hat{n}_1$
2	$\hat{n}_2^{(1)}$	$\hat{n}_2^{(2)}$	$\cdots$	$\hat{n}_2^{(K)}$	$\hat{n}_2$
3	$\hat{n}_3^{(1)}$	$\hat{n}_3^{(2)}$	$\cdots$	$\hat{n}_3^{(K)}$	$\hat{n}_3$
$\vdots$	$\vdots$	$\vdots$	$\vdots$	$\vdots$	$\vdots$
J	$\hat{n}_J^{(1)}$	$\hat{n}_J^{(2)}$	$\cdots$	$\hat{n}_J^{(K)}$	$\hat{n}_J$
Total	n_1	n_2	$\cdots$	n_K	N

Notice how similar this is to a contingency table layout for testing the independence of two discrete variables X, Y, the first having J outcomes and the second K. One difference is that in a K sample layout, the sum of the observed frequencies for the kth column is the size of the kth sample. This sum is fixed, not random. But in a $K \times J$ contingency table for testing the independence of X and Y, the sum of the kth column is the number of times the **Y** variable equals k, and is random. Regardless of their similarities, the two tables represent different situations. But suppose we ignore this for the moment. Set up the χ^2 test for the above K sample contingency table as though it were the layout for a test of independence. Recall that we estimate the probability

$$p_{jk} = P(X = j, Y = k)$$

by the observed frequency in the (j, k)th cell of the table divided by the total sample size. Hence

$$\hat{p}_{jk} = \frac{\hat{n}_j^{(k)}}{N}.$$

Also, we would estimate $p_j = P(X = j)$ by

$$\hat{p}_j = \frac{\hat{n}_j}{N}$$

and $p'_k = P(Y = k)$ by

$$\hat{p}'_k = \frac{n_k}{N}.$$

Again, ignore the fact that this last estimate is nonrandom. Consider

$$9.17 \qquad N \sum_{j,k} \frac{(\hat{p}_{jk} - \hat{p}_j \hat{p}'_k)^2}{\hat{p}_j \hat{p}'_k} .$$

If the table were the layout for independence, this variable would be χ^2 with $(K - 1)(J - 1)$ degrees of freedom. Simplify the above expression 9.17 by replacing $\hat{p}'_k$ by n_k/N to get

$$N^2 \sum_{j,k} \frac{(\hat{p}_{jk} - n_k \hat{p}_j/N)^2}{n_k \hat{p}_j} = \sum_{j,k} \frac{(\hat{n}_j^{(k)} - n_k \hat{p}_j)^2}{n_k \hat{p}_j} .$$

Factoring n_k^2 out of the numerator, this last expression becomes

$$\sum_{j,k} n_k \frac{(\hat{p}_j^{(k)} - \hat{p}_j)^2}{\hat{p}_j}$$

which we derived above in 9.15 as *a test for homogeneity.*

Thus, once we have laid out the contingency table, whether it be for testing homogeneity of K distributions or for testing independence of two variables, the computation of the test statistic D from the table is exactly the same and the number of degrees of freedom for the resulting variable is $[(K - 1)(J - 1)]$ in both cases.

There is more than an accidental similarity here. Let

$$\mathbf{p}^{(k)} = (p_1^{(k)}, \ldots, p_J^{(k)})$$

denote the conditional distribution of X given that Y $= k$, that is,

$$p_j^{(k)} = P(\mathsf{X} = j \mid \mathsf{Y} = k).$$

Then X and Y are independent if and only if all of the conditional distributions $\mathbf{p}^{(1)}, \ldots, \mathbf{p}^{(K)}$ are the same. So you can look at the test for independence in a contingency table as a test for homogeneity of the conditional distributions.

However, to avoid the confusion that arises out of the formal similarity of the computations, I find it is easier to regard this similarity as accidental and keep a firm fence between the two concepts of homogeneity and independence.

Problem 13 In a recent survey of driving patterns in Los Angeles, a group of drivers kept records of how many distinct trips they made on each day of the week for one week. (The group varied slightly from day to day.) A driver was said to have logged an N-trip day if he made N trips during that day. The data recorded are seen below in Table 9.1.

Table **9.1*** *—Daily Distribution of N-Trip Days in Los Angeles Driving Patterns Survey*

Number of trips per day (N)	Number of N-Trip Days						
	M	T	W	TH	F	SA	SU
0	12	7	7	9	8	28	41
1	4	6	9	11	5	7	5
2	40	27	36	33	30	23	19
3	16	30	16	26	15	18	18
4	34	27	29	25	30	12	19
5	23	19	16	18	23	13	18
6	15	16	15	10	16	19	17
7	8	10	13	8	13	5	10
8	6	11	8	13	5	12	4
9	3	3	3	3	8	12	3
10	3	2	4	7	2	2	1
11	1	1	3	1	3	4	1
12	1	1	1	0	1	3	2
13	2	2	1	1	2	2	0
14	0	0	0	1	2	0	0
15	1	1	0	0	1	0	1
16	0	1	1	0	0	0	0
17	0	0	0	0	1	0	0

* D. H. Kearin and R. L. Lamoureur, *A Survey of Average Driving Patterns in the Los Angeles Urban Area,* (Santa Monica, Calif.: System Development Corporation, 1969). Reprinted by permission of the publisher.

a Explain the sense in which the seven samples above are independent.

b Run a χ^2 test for homogeneity between the distributions for the seven days. Group as necessary.

c We suspect some systematic variation even between weekdays. To detect this variation, begin by running a homogeneity test on the five weekdays, grouping as necessary. Test Friday and Monday for homogeneity. What differences do the data suggest?

Problem 14 The χ^2 test for homogeneity is also used to detect any kind of trend. Suppose one has a sample $x_1, \ldots, x_n$ from a discrete distribution. Divide the sample into K subsamples and run a test of homogeneity to detect any differences in the underlying distributions of the subsamples. For example, to detect the possibility of any seasonal trends in sex distribution of births, the number of births by sex in Sweden in 1935 was given for each month:

	1	2	3	4	5	6
Boys	3743	3550	4017	4173	4117	3944
Girls	3537	3407	3866	3711	3775	3665

	7	8	9	10	11	12
Boys	3964	3797	3712	3512	3392	3761
Girls	3621	3596	3491	3391	3160	3371

How would you use this data to detect any seasonal changes in sex distribution? Run the test; what do you conclude?

As a second example, in the production of a new model television set, the following figures are the number of defective sets produced during the successive daily production runs:

4, 9, 3, 8, 11, 4, 4, 8, 0, 2, 3, 6, 10, 4,
2, 3, 5, 1, 2, 1, 6, 1, 3, 1, 2, 0, 9, 1,
6, 1, 3, 0, 1, 7, 2, 3, 3, 0, 5, 2, 4, 3,
1, 5, 4, 3, 0, 0, 4, 0, 1, 1, 3, 2, 3, 0.

The data seem to indicate a significant downward trend in defectives produced per day. Can you detect this trend using a χ^2 test for homogeneity? Group as necessary.

Kolmogorov and Smirnov Reappear

The companion to the histogram based χ^2 goodness of fit test was the Kolmogorov-Smirnov goodness of fit test based on the cumulative distribution function and designed for continuous distributions. Just as there is a χ^2 test for homogeneity of two populations, there is also a Kolmogorov-Smirnov two sample test. It is designed for continuous underlying distributions. Like χ^2 it tests against general alternatives. Let $\hat{F}_1(x)$ be the sample distribution function computed from the first sample, $\hat{F}_2(x)$ that computed from the second sample. Form

9.18
$$d = \max_x |\hat{F}_1(x) - \hat{F}_2(x)|.$$

Then the test rejects homogeneity if d is too large, and otherwise accepts.

The usefulness of this test is that if the underlying distributions of the two samples are the same, then the distribution of d does not depend on the form of the underlying distribution. In other words, to get the distribution of d under the hypothesis of homogeneity and to get the acceptance region for a level α test, we may as well assume that we are sampling from a uniform distribution. There are some results for large sample size. If n and m are the two sample sizes, accept at

$$\text{level } \alpha = .10 \quad \text{if} \quad d \leq 1.22\sqrt{\frac{1}{n} + \frac{1}{m}}$$

$$\text{level } \alpha = .05 \quad \text{if} \quad d \leq 1.36\sqrt{\frac{1}{n} + \frac{1}{m}},$$

$$\text{level } \alpha = .01 \quad \text{if} \quad d \leq 1.63\sqrt{\frac{1}{n} + \frac{1}{m}}.$$

What differences in the two underlying distributions can the Kolmogorov-Smirnov test detect at level α? Denote by $F_1(x)$ and $F_2(x)$ the two underlying cumulative distribution functions. As with the Kolmogorov-Smirnov single sample test, there is no simple way to exactly characterize detectable differences in terms of a minimum detectable distance. Assume $m = n$. It is known that for $\alpha = .05$, if

$$\max_x |F_1(x) - F_2(x)| \geq \frac{3.1}{\sqrt{n}}, \tag{9.19}$$

then detection is always possible. We can use 9.19 to compute a lower bound for the efficiency of the Kolmogorov-Smirnov test in the normal model. Let P_1 be $N(\mu, \sigma^2)$ and P_2 be $N(\mu + \Delta\mu, \sigma^2)$. For $\Delta\mu$ small,

$$\max_x |F_1(x) - F_2(x)| \simeq \frac{1}{\sqrt{2\pi}} \frac{|\Delta\mu|}{\sigma}.$$

Therefore, using 9.19, we are always certain of detecting the shift of amount $\Delta\mu$ if

$$\frac{1}{\sqrt{2\pi}} \frac{|\Delta\mu|}{\sigma} \geq \frac{3.1}{\sqrt{n}},$$

or if

$$n \geq \frac{60}{(\Delta\mu/\sigma)^2}.$$

Thus, the efficiency of the Kolmogorov-Smirnov test compared with the t test for normal shift alternatives is at least .43. However, the constant 3.1 appearing in 9.19 is an upper bound holding for all distributions. The actual efficiency of the Kolmogorov-Smirnov test as compared with the t test is probably considerably higher.

The fact that the Kolmogorov-Smirnov test combines good efficiency with protection against general alternatives makes it an attractive test. However, if the sample sizes are not equal then $\max_x |\hat{F}_1(x) - \hat{F}_2(x)|$ is difficult to compute. For small sample size, the fastest method is probably graphical. Plot both $\hat{F}_1(x)$ and $\hat{F}_2(x)$ on the same graph and do a visual inspection to determine the maximum of the difference in heights between the two curves. If the sample sizes are equal, then the value of the test statistic can be determined by working from the two ordered samples.

Problem 15 If the two sample sizes are equal, here is a way of computing the maximum difference. Order the two samples:

$$1. \qquad x_{(1)} < x_{(2)} < \cdots < x_{(n)}$$

$$2. \qquad y_{(1)} < y_{(2)} < \cdots < y_{(n)}.$$

To $x_{(j)}$ assign the number

$$r_j = j - \{\# \; y_k < x_{(j)}\}.$$

Then

a Show that

$$\max_x \, [\hat{F}_1(x) - \hat{F}_2(x)] = \frac{1}{n} \max_j r_j.$$

Assign to $y_{(j)}$ the number

$$s_j = j - \{\# \; x_k < y_{(j)}\};$$

then from a. it follows that

$$\max_x \, [\hat{F}_2(x) - \hat{F}_1(x)] = \frac{1}{n} \max_j s_j.$$

b Show that

$$\max_x |\hat{F}_1(x) - \hat{F}_2(x)| = \frac{1}{n} \max\,(\max_j r_j, \max_j s_j).$$

Problem 16 To illustrate the use of the method of Problem 15, look at the two samples given in rows 1 and 4 of Problem 6. Their ordering and associated r_j, s_j are as follows:

rank	1	2	3	4	5	6	7	8	9
$x_{(j)}$	.01	.02	.03	.04	.09	.14	.19	.35	.44
$y_{(j)}$	.21	.27	.29	.30	.33	.36	.50	.51	.54
r_j	1	2	3	4	5	6	7	3	3
s_j	−6	−5	−4	−3	−2	−2	−4	−3	−3

rank	10	11	12	13	14	15	16	17	18
$x_{(j)}$	.48	.49	.52	.55	.59	.76	.87	.92	.94
$y_{(j)}$	.61	.62	.75	.89	.90	1.00	1.06	1.14	1.17
r_j	4	5	4	4	5	3	4	3	4
s_j	−4	−3	−2	−3	−2	−3	−2	−1	0

Therefore

$$\max |\hat{F}_1(x) - \hat{F}_2(x)| = \tfrac{7}{18}.$$

a Does the above test lead to the rejection of homogeneity at the .05 level?

b Repeat the above computation comparing rows 1 and 5 of the same data.

Problem 17 Run a Kolmogorov-Smirnov test on the first and second samples of Problem 11. What are your conclusions?

Problem 18 If P_1 is uniform on [0, 1] and P_2 is the distribution of |X|, where X has an $N(0, 1)$ distribution, then

a Use 9.19 to find an upper bound for the sample size needed to detect the difference at the .05 level if the samples have equal sizes.

b Do an experimental check on the conclusion of **a.** by using samples from Table 10 and Table 11.

Summary

There are a small number of common nonparametric techniques that are available for the three basic problems of this and the previous chapter.

a For testing *symmetry, independence, homogeneity against general alternatives*, both of the basic goodness of fit methods, the χ^2 and the Kolmogorov-Smirnov, provide tests.

b For testing *symmetry and homogeneity against shift alternatives*, we have Student's t test, the median tests, and the Wilcoxon tests. Notice that the sign test for symmetry around zero is a version of the median test.

c For testing *independence and no trend against monotone dependence and trend* we have Kendall's Tau test and Spearman's Rho test. By dividing the sample in two we can apply a median test or a Wilcoxon type test.

All of the above tests are fairly efficient. They range in computational difficulty from the median tests, which involve very little except finding the median, to the t tests and Kolmogorov-Smirnov tests which may involve lengthy computations.

For the two-sample homogeneity problem with continuous underlying distributions, for ease of computation start with the median test. For discrete distributions, use χ^2. For more than two samples, with continuous underlying distributions rank the data and compute the Kruskal-Wallis test. If the data are discrete, use χ^2.

There are many nonparametric tests proposed for these problems and others. The ones we discussed only skim the surface. They are tests that are widely applicable, but for many situations, more specific nonparametric tests should be used. Design your own tests—tailor them to the problem. The only requirement is that you be able to compute the level of significance under the hypothesis.

Problem 19 To test the hypothesis that skill is an element in playing gin rummy, an expert E was seated at one table and a student S at another table, but at the corresponding seat. An identical sequence of hands was dealt at both tables. Records were kept of how much both won or lost on each game. If E won money in the kth hand, he was given a W; if he lost, an L. Similarly for the student. So for each hand we see recorded one of (W, L), (W, W), (L, W), (L, L), where the first entry refers to E and the second to S. On the basis of these records design a simple test for the hypothesis that skill plays no part in the game—that the expert wins or loses just as often as the student. Note that the probability of W or L varies from the kth game to the $(k + 1)$th game.)

10 Multifactor, Multisample Problems

Introduction

In our work up to now with two or more samples, the question we were trying to answer was usually of the form: *Does the change in one factor alter the population distribution*? For example, one group of rats is given a diet containing a certain drug and the weight changes compared with another group of rats given the same diet without the drug. If the two groups were similar originally, then they differ only in the presence or absence of one factor—the drug in their diet.

Or, suppose two batches of steel are tested to breaking point, where one batch contains .02% carbon, the other .05%. When we use the Wilcoxon rank test, say, to determine if there is any significant difference in the breaking points, we are testing to see whether the change in one factor, the carbon content, is effecting the experimental observations.

In the example where a test is administered to three groups of children from different cultural backgrounds, the children are matched as far as possible by age, sex, and other possible relevant factors. Then the only factor that differs among the groups is cultural background. Here, we can test for significant differences by ranking the scores and using the Kruskal-Wallis test.

But suppose we are interested in what happens as a number of factors change. For instance, we might be interested in the breaking strength of an alloy as we vary two or more constituents. Or we may be interested in an experimental situation where both the carbon content and heat treatment of steel are varied. Similarly, in the example of testing we may be interested in a breakdown by sex and age as well as by cultural background.

This chapter is concerned with two methods of tracking down and isolating the effects of varying any number of factors in a multisample situation. The first method is useful when the changes in the various factors can be quantitatively described. It is known as *regression analysis*. The second method applies when the changes are qualitative and is called *analysis of variance*. There are two problems to deal with in these multifactor, multisample situations. One is a hypothesis testing problem.

> *Is the change in this factor or group of factors producing any significant change in the distribution of the sample?*

The second is an estimation, confidence region problem:

> *Estimate the change in the distributions due to the change in some specific factor or factors and give confidence regions for the estimates.*

To illustrate these two approaches we turn to an example.

A One-Factor Example

We propose to investigate the effects of a one-hour heat treatment at a constant temperature x on the tensile strength of steel. We want to find out what happens for various heat treatment temperatures in the range 1000°C to 1200°C.

If we decide on a sample size n then we can heat-treat the n specimens at temperatures $x_1, \ldots, x_n$ that are fairly evenly spaced over the range 1000°–1200°. Denote the associated tensile strengths by $y_1, \ldots, y_n$. Of course, the $y_1, \ldots, y_n$ are not deterministic functions of the heat-treatment temperatures. Even the most carefully selected and matched specimens show some apparently random dispersion in their tensile strength. If there is a systematic change in tensile strength with treatment temperature, then the random fluctuation component superimposed on it tends to hide it. The problem is to detect and estimate any systematic change.

Suppose we make a scatter diagram of the points (x_k, y_k), $k = 1, \ldots, n$ (see Figure 10.1). This diagram leads us to suspect the possibility of an

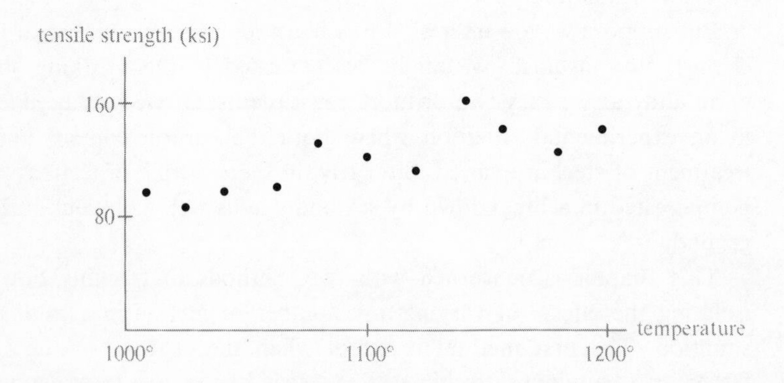

***Figure 10.1** Scatter diagram for strength versus treatment temperature data*

underlying linear relationship between tensile strength and treatment temperature. We take a naive attitude. In a sense, the problem is one of curve fitting. Given a scatter diagram like the above, find the best straight-line fit to the data. This can be done by inspection, or better still, by the method of fitting by least squares. For a straight line $y = a + bx$, the vertical distance from the point (x_k, y_k) to the point on the line with the same x-coordinate is

$$|y_k - a - bx_k|.$$

The sum of the squares of the differences is

$$D(a, b) = \sum_1^n (y_k - a - bx_k)^2.$$

Now pick a and b so as to minimize this sum. (Solving the equations

$$\frac{\partial D}{\partial a} = 0, \qquad \frac{\partial D}{\partial b} = 0$$

will locate the minimizing values of a and b. The exact solution is not terribly important to us right now and is left to you as Problem 1.) The minimizing values of a and b are called the *least squares estimates*. We denote them by $\hat{a}$, $\hat{b}$. The straight line

$$y = \hat{a} + \hat{b}x$$

gives a quantitative description of the dependence of tensile strength on treatment temperature.

Curve fitting this way leads to the following questions:

a How good a fit to the data is $y = \hat{a} + \hat{b}x$?

b Could the data have been described equally well by saying that the tensile strength remained nearly constant; that is, that some horizontal line gave just about as good a fit as $y = \hat{a} + \hat{b}x$?

c Do we get a significantly better fit by using a quadratic expression $y = a + bx + cx^2$, and doing a least squares estimate of a, b, c?

In the presence of random fluctuations, none of these questions make sense or can be answered without a model. While the $x_1, \ldots, x_n$ are fixed constants, the $y_1, \ldots, y_n$ are outcomes of random variables $Y_1, \ldots, Y_n$. Once we select a class of possible underlying distributions for $Y_1, \ldots, Y_n$, we can begin to formulate meaningful questions.

What can we reasonably assume about the $Y_1, \ldots, Y_n$? Since the tensile strength of one specimen should have no effect on that of another, take the variables to be independent. Denote $\mu_k = EY_k$ and write

$$Y_k = \mu_k + Z_k,$$

where Z_k has mean zero. In other words, the tensile strength of a specimen treated at temperature x_k is μ_k plus the random dispersion variable Z_k.

Since the $Y_1, \ldots, Y_n$ are independent, so are the $Z_1, \ldots, Z_n$. Make the further assumption that $Z_1, \ldots, Z_n$ are *normally distributed with the same, but unknown, variance* σ^2. Thus, our model is that $y_1, \ldots, y_n$ are the outcomes of

10.1 $$Y_k = \mu_k + Z_k, \qquad k = 1, \ldots, n$$

where the $Z_1, \ldots, Z_n$ are independent $N(0, \sigma^2)$ variables. The underlying means $\mu_1, \ldots, \mu_n$ are the values we want to determine.

The idea in regression analysis is: *Assume that the* μ_k *have some simple functional relationship to the* x_k *determined by a few parameters.* For instance, in a linear regression model, we assume that

10.2 $$\mu_k = a + bx_k, \qquad k = 1, \ldots, n.$$

We could just as well assume a quadratic relationship

$$\mu_k = a + bx_k + cx_k^2, \qquad k = 1, \ldots, n$$

or an exponential

$$\mu_k = a + be^{-x_k/c}, \qquad k = 1, \ldots, n.$$

Once the functional form for μ_k is selected, the family of underlying distributions for $Y_1, \ldots, Y_k$ is determined by the parameters in the functional relationship between μ_k and x_k and the unknown variance σ^2. After the model is specified, we can use maximum likelihood to estimate all the parameters.

For example, suppose we select a linear regression model

$$Y_k = a + bx_k + Z_k, \qquad k = 1, \ldots, n,$$

to underlie the data. Then the Y_k, $k = 1, \ldots, n$, are independent $N(a + bx_k, \sigma^2)$ variables with joint density

$$f(y_1, \ldots, y_n) = \frac{1}{(\sqrt{2\pi}\sigma)^n} e^{-1/2\sigma^2 \sum_1^n (y_k - a - bx_k)^2}.$$

Taking logarithms gives

$$L = c - n \log \sigma - \frac{1}{2\sigma^2} \sum_1^n (y_k - a - bx_k)^2.$$

To get the maximum likelihood estimates of a, b, and σ we solve the equations

$$\frac{\partial L}{\partial a} = 0, \qquad \frac{\partial L}{\partial b} = 0, \qquad \frac{\partial L}{\partial \sigma} = 0.$$

The first two reduce to

$$\frac{\partial D(a, b)}{\partial a} = 0, \qquad \frac{\partial D(a, b)}{\partial b} = 0,$$

where as before,

$$D(a, b) = \sum_{1}^{n} (y_k - a - bx_k)^2.$$

But these are exactly the equations we got for determining the least squares estimates $\hat{a}$ *and* $\hat{\hat{b}}$. Therefore, in our linear regression model the least squares estimates of a and b are also the maximum likelihood estimates. We will find, in general, that for normally distributed regression models, least squares estimates coincide with maximum likelihood estimates.

Furthermore, let $\hat{D}$ be the minimum value of $D(a, b)$. Then the third maximum likelihood equation gives $\hat{D}/n$ as an estimate for σ^2. An unbiased estimate is given by

$$\hat{\sigma}^2 = \frac{\hat{D}}{n - 2}. \tag{10.3}$$

This also holds true more generally: The minimum value of the sum of the squared deviations, divided by the appropriate factor, gives an unbiased estimate for σ^2.

Assuming that a linear model gives a good fit, we use the data to produce a functional relationship

$$y = \hat{a} + \hat{\hat{b}}x$$

which summarizes the effect of treatment temperature on tensile strength. This is a compact, concise quantitative result. But it was produced at the risk of making the quantitative assumption that *some* linear relationship exists between μ_k and x_k.

Our model has structured the situation and questions can now be given a precise formulation. For instance, the question of whether a linear fit is as good as a quadratic can be translated as:

Given that $\mu_k = a + bx_k + cx_k^2$, test the hypothesis $c = 0$.

There is also the question of accuracy. If, for example, we determine that the mean tensile strength is given by $83.1 + 2.5x_k$ pounds per square inch, what are $100\gamma\%$ confidence intervals for the coefficients? There are other important questions, but in our work in regression analysis we will focus only on the generalization of the above two.

Suppose that we did not want to assume some specific functional relationship between the mean tensile strength and treatment temperature.

Then another way to approach the problem would be to start with the model

$$\mathsf{Y}_k = \mu_k + \mathsf{Z}_k, \qquad k = 1, \ldots, n$$

assuming that the Z_k are $N(0, \sigma^2)$ and try to estimate all of the parameters $\mu_1, \ldots, \mu_n$ and σ^2. This is nonsensical, because for each k, we have only one sample value y_k from the distribution $N(\mu_k, \sigma^2)$. Unless we postulate some connection among the μ_k's, the maximum likelihood estimate for μ_k is y_k. Since the estimates fit the data exactly, the estimate $\hat{\sigma}^2$ of the variance is zero. We wind up with the foolish conclusion that the data have no scatter—whenever we make a measurement at temperature x_k, the tensile strength will be exactly y_k (according to our estimate $\hat{\sigma}^2 = 0$). The difficulty lies in trying to estimate the mean of a distribution from a sample of size one.

The obvious solution is that instead of taking one sample at each temperature x_k, select a smaller number m of temperatures and test $l > 1$ specimens at each temperature treatment. The total sample size is now $n = ml$. For this sampling scheme the scatter diagram for the pairs (x_k, y_k) looks like Figure 10.2. How can we estimate μ_k, $k = 1, \ldots, m$?

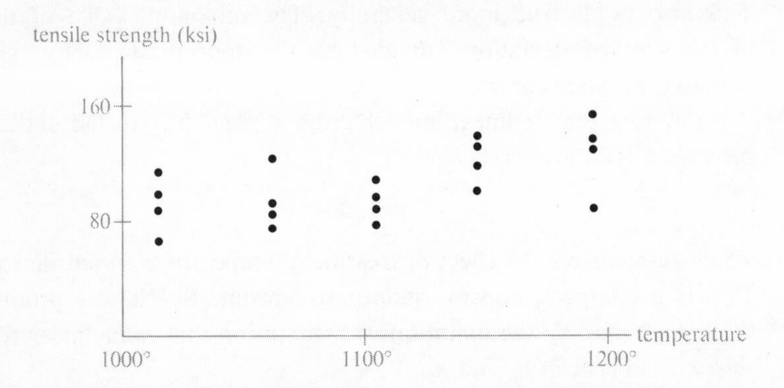

***Figure 10.2** Scatter diagram for alternative sampling scheme in strength versus temperature treatment experiment (m = 5, l = 4)*

The least squares estimates are gotten by taking the mean value of each group of l values. For instance, to get the least squares estimate of μ_1, minimize

$$\sum_1^l (y_k - \mu_1)^2.$$

This gives

$$\mu_1 = \frac{1}{l} \sum_1^l y_k.$$

For this model, if we wrote out the densities of $Y_1, \ldots, Y_n$ and computed the maximum likelihood estimate $\hat{\mu}_k$ of μ_k we would find that it again coincides with the least squares estimate.

Using the sample mean to estimate the mean of a normal distribution is standard. The difference is that the estimate for the variance σ^2 is gotten by combining the whole sample. If $\hat{\sigma}_k^2$ is the usual unbiased estimate for the variance of the kth sample of size l, then we use

$$\hat{\sigma}^2 = \frac{1}{m}(\hat{\sigma}_1^2 + \cdots + \hat{\sigma}_m^2)$$

as the unbiased estimate for σ^2. It is this fact that links the m samples together and makes their analysis different from the analysis of m independent samples with unrelated means and variances. In particular, confidence intervals for the mean when the variance is unknown depend on the estimate $\hat{\sigma}$. Thus, the confidence intervals for $\mu_1, \ldots, \mu_n$ are related in a complex way.

This second model is called a one-way analysis of variance layout. In it, the effect of temperature on tensile strength is analyzed by using the individual estimates of the means $\mu_1, \ldots, \mu_m$ from the m samples and their associated confidence intervals.

To contrast the analysis of variance model with the linear regression model, consider testing the hypothesis that temperature has no effect on tensile strength. In our linear regression model, this translates into, Test the hypothesis $b = 0$. In the analysis of variance model, with $\mu_1, \ldots, \mu_m$ not necessarily related, it translates into the more complicated statement, Test the hypothesis

$$\mu_1 = \mu_2 = \cdots = \mu_m.$$

Denote

$$\mu = \frac{1}{m}\sum_1^m \mu_k$$

and write

$$\mu_k = \mu + \Delta_k, \qquad \sum_1^m \Delta_k = 0.$$

Then the hypothesis can be stated more simply as: $\Delta_k = 0, k = 1, \ldots, m$. In this form the problem for both models falls under the general heading: Test the hypothesis that a parameter or group of parameters is zero.

Now we backtrack for a moment. There is nothing in regression analysis that says we have to spread the points $x_1, \ldots, x_n$ evenly over the range, or, in fact, that they have to be distinct points. Our sampling scheme for a regression analysis could be just the same as for the analysis of variance—pick samples of size l at m temperature points. The basic difference is that in regression analysis we assume some functional relationship between the means and the factors involved (temperature in our example). This functional relationship is specified as one of a finite parameter family of curves, and the first step is to select parameter values to find the best fitting curve. Actually, regression analysis is an outgrowth of curve fitting to scattered data.

In analysis of variance no such functional relationship between means and factors is assumed, and each mean μ_k is separately estimated. Thus, analysis of variance is suitable for use when the factors involved cannot be quantified. For example, suppose we have three samples of test scores where the varying factor is cultural background. How does one assign a numerical value to cultural background? Does it make any sense to assume a linear relationship, say, between mean test score and cultural background? Here, a regression analysis is out of place. But analysis of variance simply computes the mean of each test group individually and proceeds. Analysis of variance can be used in situations where the factors fall into discrete unordered categories, for example, sex, economic class, or blood type. Because of this, we roughly think of regression analysis as applying to quantitative factors, and analysis of variance to qualitative factors.

Problem 1 **a** Solve the equations

$$\frac{\partial D}{\partial A} = 0, \qquad \frac{\partial D}{\partial b} = 0$$

to find expressions for $\hat{a}$, $\hat{b}$.

b Denote

$$\overline{\phi(x, y)} = \frac{1}{n} \sum_{1}^{n} \phi(x_k, y_k);$$

then show that the least squares regression line passes through the center of mass point $(\bar{x}, \bar{y})$.

(Ans. **a**

$$\hat{b} = \frac{\overline{xy} - \bar{x}\bar{y}}{\overline{x^2} - (\bar{x})^2}, \qquad \hat{a} = \bar{y} - \hat{b}\bar{x}.)$$

Problem 2 (Difficult!) Suppose that the mean tensile strength is actually a linear function of the treatment temperature, so that if μ is the mean tensile strength at treatment temperature x, then

$\mu = a + bx$. From a sample $\mathsf{Y}_1, \ldots, \mathsf{Y}_n$ at any n points $x_1, \ldots, x_n$ we get the estimate $\boldsymbol{\mu} = \hat{\mathsf{a}} + \hat{\mathsf{b}}x$. Notice that $\boldsymbol{\mu}$ can be written as

$$\boldsymbol{\mu} = \sum_{1}^{n} \beta_k \mathsf{Y}_k,$$

where β_k depends only on the $x_1, \ldots, x_n$ and x. Show that

a $$E\boldsymbol{\mu} = a + bx,$$

b $$V(\boldsymbol{\mu}) = \frac{\sigma^2}{n}\left(1 + \frac{(x - \bar{x})^2}{\overline{x^2} - (\bar{x})^2}\right).$$

Problem 3 For our linear regression model consider the following two sampling schemes:

a Spread $x_1, \ldots, x_n$ evenly over the range and sample once at every point.

b Assuming n even, $n = 2m$, sample m times at the lower end of the range and m times at the upper end.

If μ is a linear function $ax + b$, then these two schemes can be compared by using b. of Problem 2. Compute $V(\boldsymbol{\mu})$ for each scheme and show that the squared error loss in $\boldsymbol{\mu}$ using the second scheme is always less than that using the first. Why might you still prefer the first scheme to the second?

Problem 4 For values of a factor varying by integer steps from 0 to 20, the following measurements were made on a dependent variable:

0.02, 1.61, 0.39, 0.93, 1.44, 2.04, 2.76,
1.58, 2.90, 1.87, 2.82, 3.89, 4.59, 5.01,
3.59, 4.29, 3.24, 3.34, 4.09, 4.26, 5.23.

Construct the scatter diagram, decide on an appropriate functional relationship, and find the least squares estimates of the parameters.

The General Regression Model

Look at a situation in which we are interested in tracing the effects of J factors on a single variable. We begin by fixing the value of the first factor at $x_1^{(1)}$, the second at $x_1^{(2)}, \ldots,$ the Jth at $x_1^{(J)}$. Carrying out the experiment, we get outcome y_1. Now fix the values of the J factors at $x_2^{(1)}, x_2^{(2)}, \ldots, x_2^{(J)}$. Repeat the experiment, getting outcome y_2. In all, we do n experiments, getting outcomes $y_1, \ldots, y_n$ where y_k is the outcome corresponding to the factor values $x_k^{(1)}, \ldots, x_k^{(J)}$. Notice that in our notation, the superscript always refers to the factor, the subscript to the experiment. Thus $x_k^{(j)}$ is the value of the jth factor in the kth experiment. Take $y_1, \ldots, y_n$ to be outcomes of $Y_1, \ldots, Y_n$ and assume

10.4 **a** *The* $Y_1, \ldots, Y_n$ *are independent normally distributed variables with unknown means* $\mu_1, \ldots, \mu_n$ *and unknown common variance* σ^2.

b *There are unknown constants* $b_1, \ldots, b_J$ *such that*

$$\mu_k = b_1 x_k^{(1)} + b_2 x_k^{(2)} + \cdots + b_J x_k^{(j)}, \qquad k = 1, \ldots, n.$$

These assumptions define what we call a *linear regression model*, which will be our concern for the next three sections. This model is parametric: Each distribution in the class of all possible underlying distributions is specified by the values of the variance σ^2 and the means $\mu_1, \ldots, \mu_n$. But the means are determined by the J parameters $b_1, \ldots, b_J$. Thus the parameter space Θ is $(J + 1)$-dimensional, with σ^2 varying over $[0, \infty)$ and the $b_1, \ldots, b_J$ over $(-\infty, \infty)$.

Taking the means to be linear functions of the J factors seems very restrictive. Why exclude curve fitting by more complicated functions such as parabolas, exponentials, etc.? The answer is that, in fact, many of these possibilities are included. For example, if we have a one-factor situation and want to fit a quadratic $a + bx + cx^2$ to the mean, we use the model of 10.4 with three factors. If the levels (i.e. values) of the single factor are $x_1, \ldots, x_n$, define

$$x_k^{(1)} \equiv 1, \qquad x_k^{(2)} = x_k, \qquad x_k^{(3)} = x_k^2.$$

Then μ_k can be written as either

$$\mu_k = b_1 + b_2 x_k + b_3 x_k^2, \qquad k = 1, \ldots, n,$$

or

$$\mu_k = b_1 x_k^{(1)} + b_2 x_k^{(2)} + b_3 x_k^{(3)}, \qquad k = 1, \ldots, n.$$

Similarly, if we want to fit a linear term plus an exponential, say

$$\mu_k = a + bx_k + ce^{-x_k},$$

we use the three factors

$$x_k^{(1)} \equiv 1, \qquad x_k^{(2)} = x_k, \qquad x_k^{(3)} = e^{-x_k}.$$

The coefficients $b_1, \ldots, b_n$ that appear in 10.4b govern the effect that the variation of each factor will have on the mean. If b_1 is small compared to b_2, say, then a given variation in the first factor will produce a much smaller change in the mean than the same variation in the second factor. To get estimates for these coefficients, think of trying to fit the observations by a linear function $\tilde{y}_k$ of the factor levels,

$$\tilde{y}_k = b_1 x_k^{(1)} + \cdots + b_J x_k^{(J)}.$$

We get the *best least squares fit* by selecting the $b_1, \ldots, b_J$ to minimize the Euclidean squared distance

$$D = \sum_1^n (y_k - \tilde{y}_k)^2.$$

To find the minimizing values of $b_1, \ldots, b_J$, take partial derivatives;

$$\frac{\partial D}{\partial b_j} = 2 \sum_1^n (y_k - \tilde{y}_k) \frac{\partial \tilde{y}_k}{\partial b_j}.$$

Setting these equal to zero, and noting that $\partial \tilde{y}_k / \partial b_j = x_k^{(j)}$, gives the equations

10.5 $$\sum_{k=1}^n x_k^{(j)} y_k = \sum_{k=1}^n x_k^{(j)} \tilde{y}_k$$

$$= \hat{\hat{b}}_1 \sum_{k=1}^n x_k^{(j)} x_k^{(1)} + \cdots + \hat{\hat{b}}_J \sum_{k=1}^n x_k^{(j)} x_k^{(J)},$$

for the minimizing values $\hat{\hat{b}}_1, \ldots, \hat{\hat{b}}_J$. Here j ranges over $1, \ldots, J$. These are J linear equations in the J unknowns $\hat{\hat{b}}_1, \ldots, \hat{\hat{b}}_J$. Assuming there is no linear dependence between equations, they can be solved. The resulting values $\hat{\hat{b}}_1, \ldots, \hat{\hat{b}}_J$ are called the *least squares estimates* for $b_1, \ldots, b_J$. Throughout this chapter we will use a double hat above a symbol to denote a least squares estimate.

To put the solution of the equations 10.5 into a concise form, we need to introduce some vector and matrix notation. Use the vector notation

$$\mathbf{y} = \begin{pmatrix} y_1 \\ \vdots \\ y_n \end{pmatrix}, \quad \hat{\hat{\mathbf{b}}} = \begin{pmatrix} \hat{\hat{b}}_1 \\ \vdots \\ \hat{\hat{b}}_J \end{pmatrix}, \quad \mathbf{x}^{(j)} = \begin{pmatrix} x_1^{(j)} \\ \vdots \\ x_n^{(j)} \end{pmatrix},$$

and the inner product notation

$$s_{jl} = (\mathbf{x}^{(j)}, \mathbf{x}^{(l)}) = \sum_{k=1}^n x_k^{(j)} x_k^{(l)}.$$

Now write 10.5 as

10.6 $$\sum_{k=1}^n x_k^{(j)} y_k = (\mathbf{x}^{(j)}, \mathbf{x}^{(1)}) \hat{\hat{b}}_1 + \cdots + (\mathbf{x}^{(j)}, \mathbf{x}^{(J)}) \hat{\hat{b}}_J$$

$$= \sum_{l=1}^J s_{jl} \hat{\hat{b}}_l.$$

Introduce the $J \times n$ matrix $[X]$ with elements $X_{jk} = x_k^{(j)}$, and the square $J \times J$ matrix $[S]$ with elements s_{jl}. Then we can rewrite 10.6:

10.7 $$[X]\mathbf{y} = [S]\hat{\hat{\mathbf{b}}}.$$

The determinant of the inner product matrix $[S]$ is zero only if the J vectors $\mathbf{x}^{(1)}, \ldots, \mathbf{x}^{(J)}$ are linearly dependent. This implies that one of the factors varies in such a way that its values are always the same linear combination of the remaining factor values. In this case, the means μ_k can be written as a linear function of only $J - 1$ factors. To eliminate this case, we assume that *the vectors* $\mathbf{x}^{(1)}, \ldots, \mathbf{x}^{(J)}$ *are linearly independent.* In other words, there is no way of expressing one of the vectors as a linear combination of the others. The important consequence for us is that $\det[S] \neq 0$. Then $[S]$ has an inverse matrix $[S]^{-1}$ and the least squares estimates are given by

10.8 $$\hat{\hat{\mathbf{b}}} = [S]^{-1}[X]\mathbf{y}.$$

This is the basic equation in linear regression analysis. If we write

$$[A] = [S]^{-1}[X],$$

then $[A]$ is an $J \times n$ matrix with elements a_{jk}, and 10.8 becomes

$$\hat{\hat{b}}_j = \sum_{k=1}^{n} a_{jk} y_k.$$

Notice from this that the estimates $\hat{\hat{b}}_j$ are linear functions of the observations $y_1, \ldots, y_n$. Using random variables, write the above as

10.9 $$\hat{\hat{\mathsf{b}}}_j = \sum_{k=1}^{n} a_{jk} \mathsf{Y}_k.$$

You can see from this that the $\hat{\hat{\mathsf{b}}}_j$ are linear combinations of independent normally distributed variables. Therefore

10.10 ***Proposition*** *Under the assumptions* 10.4 *the estimators* $\hat{\hat{\mathsf{b}}}_1, \ldots, \hat{\hat{\mathsf{b}}}_J$ *have a joint normal distribution.*

The next problem is to determine the means and covariance matrix of the estimators $\hat{\hat{\mathsf{b}}}_1, \ldots, \hat{\hat{\mathsf{b}}}_J$.

10.11 ***Proposition*** *The* $\hat{\hat{\mathsf{b}}}_j$ *are unbiased estimates of* b_j*; that is, if the true parameter values are* $b_1, \ldots, b_J$*, then*

$$E\hat{\hat{\mathsf{b}}}_j = b_j, \qquad j = 1, \ldots, J$$

and their covariance matrix is

$$[\Gamma] = \sigma^2[S]^{-1}.$$

Here is a proof of Proposition 10.11 which is given to illustrate some standard manipulations in linear regression analysis. It is easiest to work with 10.6 written in random variable terms as

10.12 $$\sum_k x_k^{(j)} \mathsf{Y}_k = \sum_l s_{jl} \hat{\hat{\mathsf{b}}}_l.$$

Take expectations of both sides, using

$$E\mathsf{Y}_k = \mu_k = \sum_l x_k^{(l)} b_l$$

to get

$$\sum_{k,l} x_k^{(j)} x_k^{(l)} b_l = \sum_l s_{jl} E\,\hat{\hat{\mathsf{b}}}_l.$$

Summing the left-hand side first over k gives

$$\sum_l s_{jl} b_l = \sum_l s_{jl} E\,\hat{\hat{\mathsf{b}}}_l.$$

Since $\det[S] \neq 0$, this implies that $E\hat{\hat{\mathsf{b}}}_l = b_l$, $l = 1, \ldots, J$. (Why?) Now subtract the means from each term in both sides of 10.12:

$$\sum_k x_k^{(j)} (\mathsf{Y}_k - \mu_k) = \sum_l s_{jl} (\hat{\hat{\mathsf{b}}}_l - b_l).$$

This holds for $j = 1, \ldots, J$ so we have J equations. Multiply together corresponding sides of the ith and jth equations and take expectations:

10.13 $$\sum_{k,m} x_m^{(i)} x_k^{(j)} \Gamma(\mathsf{Y}_m, \mathsf{Y}_k) = \sum_{l,h} s_{ih} s_{jl} \Gamma(\hat{\hat{\mathsf{b}}}_h, \hat{\hat{\mathsf{b}}}_l),$$

where $\Gamma(\ ,\ )$ denotes the covariance of the indicated variables. Since $\Gamma(\mathsf{Y}_m, \mathsf{Y}_k)$ equals σ^2 when $m = k$ and zero otherwise, the left-hand side of 10.13 is $\sigma^2 s_{ij}$. Use the additional fact that $s_{jl} = s_{lj}$ to see that 10.13 can be written in matrix form as

$$\sigma^2[S] = [S][\Gamma][S].$$

Multiplying this equation on the left and the right by $[S]^{-1}$, we conclude that

$$[\Gamma] = \sigma^2[S]^{-1}.$$

Proposition 10.11 indicates that a matrix formulation is essential. Without the concept of the inverse of a matrix, how could we express the covariance of $\hat{\hat{\mathsf{b}}}_j$ and $\hat{\hat{\mathsf{b}}}_l$? In fact, even to get the variance of the estimates b_j, we have to compute the diagonal elements of $[S]^{-1}$.

The form 10.8 of the least squares estimates and the results in 10.11 do not require normality or independence. If you reexamine the derivation, you will see that only the following two assumptions are necessary: First, that the variables $\mathsf{Y}_1, \ldots, \mathsf{Y}_n$ are uncorrelated with the common variance σ^2. Second, that μ_k is a linear function of the factor levels.

However, there is an important result that is not generally true unless the $Y_1, \ldots, Y_n$ are normally distributed.

There is no reason, at this point, to believe that least squares estimates are particularly efficient. Why not use maximum likelihood instead? The density of $Y_1, \ldots, Y_n$ is

$$f(y_1, \ldots, y_n) = \frac{1}{(\sqrt{2\pi}\sigma)^n} e^{-1/2\sigma^2 \sum_1^n (y_k - \mu_k)^2},$$

where μ_k is a function of the parameters $b_1, \ldots, b_J$:

$$\mu_k = b_1 x_k^{(1)} + \cdots + b_J x_k^{(J)}, \qquad k = 1, \ldots, n.$$

We get the maximum likelihood estimates for $b_1, \ldots, b_J$, and σ^2 by maximizing $f(y_1, \ldots, y_n)$ or its logarithm

$$L = -n \log(\sqrt{2\pi}\sigma) - \frac{1}{2\sigma^2} \sum_1^n (y_k - \mu_k)^2.$$

Whatever the value of σ, L is maximized by minimizing

$$D = \sum_1^n (y_k - \mu_k)^2$$

over $b_1, \ldots, b_J$. But this is exactly what we do to get the least squares estimates for $b_1, \ldots, b_J$. Therefore

10.14 ***Proposition*** *The least squares and maximum likelihood estimates for $b_1, \ldots, b_J$ are the same.*

The fact that these two different estimation methods give the same result depends critically on the assumption of normally distributed variables. The least squares estimates will be the same no matter what the distribution of $Y_1, \ldots, Y_n$, as long as their means are assumed to be linear functions of the factor levels, because the values of $b_1, \ldots, b_J$ that minimize $\sum_1^n (y_k - \mu_k)^2$ certainly do not depend on what we assume about the distributions of $Y_1, \ldots, Y_n$. But the appearance of $\sum_1^n (y_k - \mu_k)^2$ in the expression for the joint density of $Y_1, \ldots, Y_n$, and therefore its minimization to obtain the maximum likelihood estimate, is peculiar to the normal distribution.

Problem 5 For the data of Problem 4, assume that the mean is quadratic in the factor level; that is,

$$\mu_k = a + bx_k + cx_k^2.$$

a Compute the matrix S.

b Compute the least squares estimates $\hat{\hat{a}}, \hat{\hat{b}}, \hat{\hat{c}}$.

c Compute the variances of $\hat{\hat{\mathrm{a}}}, \hat{\hat{\mathrm{b}}}, \hat{\hat{\mathrm{c}}}$ in terms of σ^2.

Problem 6 An experimental investigation of the heat evolved during the hardening of Portland cement gave the results shown in Table 10.1. The heat evolved in the curing, given in calories per gram, is denoted by y. The factors $\mathbf{x}^{(1)}$ and $\mathbf{x}^{(2)}$ are the percentages by weight of two chemical components ($3CaO \cdot Al_2O_3$ and $3CaO \cdot SiO_2$) in the clinkers from which the cement was produced. (This example is from the article "Effect of Composition of Portland Cement on Heat Evolved During Hardening" by H. Woods, H. H. Steinour, and H. R. Starke and is given in the book by A. Hald, *Statistical Theory with Engineering Applications*, John Wiley and Sons, 1952.)

Table **10.1**

y	$x^{(1)}$	$x^{(2)}$
78.5	7	26
74.3	1	29
104.3	11	56
87.6	11	31
95.9	7	52
109.2	11	55
102.7	3	71
72.5	1	31
93.1	2	54
115.9	21	47
83.8	1	40
113.3	11	66
109.4	10	68

Estimate the regression coefficients from this data using a constant ($x_k^{(3)} \equiv 1$) as one of the factors. (Ans. $\hat{\hat{b}}_1 = 1.47$, $\hat{\hat{b}}_2 = .662$, $\hat{\hat{b}}_3 = 52.6$.)

Problem 7 (Difficult) Define the distance $\|\mathbf{y} - \tilde{\mathbf{y}}\|$ between any two vectors $(y_1, \ldots, y_n)$ and $(\tilde{y}_1, \ldots, \tilde{y}_n)$ as the ordinary Euclidean distance

$$\|\mathbf{y} - \tilde{\mathbf{y}}\| = \left(\sum_1^n (y_k - \tilde{y}_k)^2\right)^{1/2}.$$

Define the best predictor $\hat{\hat{\mathbf{y}}}$ of $\mathbf{y}$ based on the **J** vectors $\mathbf{x}^{(1)}, \ldots, \mathbf{x}^{(J)}$ to be that linear combination of these **J** vectors which minimizes

$$\|\mathbf{y} - b_1\mathbf{x}^{(1)} - b_2\mathbf{x}^{(2)} - \cdots - b_J\mathbf{x}^{(J)}\|$$

Show that if $\mathbf{x}^{(1)}, \ldots, \mathbf{x}^{(J)}$ are linearly independent, $\hat{\hat{\mathbf{y}}}$ is uniquely defined by the property that $\mathbf{y} - \hat{\hat{\mathbf{y}}}$ is orthogonal to all the vectors $\mathbf{x}^{(1)}, \ldots, \mathbf{x}^{(J)}$; that is,

$$(\mathbf{y} - \hat{\hat{\mathbf{y}}}, \mathbf{x}^{(j)}) = 0, \qquad j = 1, \ldots, J.$$

Put another way, $\hat{\hat{\mathbf{y}}}$ is the projection of $\mathbf{y}$ on the subspace spanned by $\mathbf{x}^{(1)}, \ldots, \mathbf{x}^{(J)}$.

Interaction and Selection of Factor Levels

It is possible that the changes in levels of some of the factors in a linear regression analysis are very closely related, so that they mask the effect of each other on the experimental situation. For example, suppose we are running a two-factor analysis with factor levels $\mathbf{x}^{(1)} = (x_1^{(1)}, \ldots, x_n^{(1)})$ and $\mathbf{x}^{(2)} = (x_1^{(2)}, \ldots, x_n^{(2)})$. Suppose furthermore that $\mathbf{x}^{(2)}$ is nearly a constant multiple of $\mathbf{x}^{(1)}$; that is, $\mathbf{x}^{(2)} \simeq c\mathbf{x}^{(1)}$. We are in the unstable situation where $\det [S] \simeq 0$. By changing $\mathbf{x}^{(1)}$ or $\mathbf{x}^{(2)}$ slightly, we can make $\hat{\hat{b}}_1$ much larger than $\hat{\hat{b}}_2$, or $\hat{\hat{b}}_2$ much larger than $\hat{\hat{b}}_1$, or both $|\hat{\hat{b}}_1|$ and $|\hat{\hat{b}}_2|$ large (see Problem 8).

A more precise statement of the difficulty is that whenever two or more factors have levels that are approximately linearly related, their coefficients are not an accurate reflection of their separate effects. Suppose we treat the factor levels as though they are outcomes of random variables. That is, suppose $x_1^{(1)}, \ldots, x_n^{(1)}$ are n outcomes of a variable $\mathsf{X}^{(1)}$, and so on. The sample mean is defined as

$$\bar{x}^{(1)} = \frac{1}{n} \sum_{k=1}^{n} x_k^{(1)}$$

and the sample variance by

$$\hat{\sigma}_{(1)}^2 = \frac{1}{n} \sum_{k=1}^{n} (x_k^{(1)} - \bar{x}^{(1)})^2.$$

The sample correlation coefficient between $\mathbf{x}^{(j)}$ and $\mathbf{x}^{(k)}$ is

$$\hat{\rho}_{jl} = \frac{1}{\hat{\sigma}_{(j)}\hat{\sigma}_{(l)}} \cdot \frac{1}{n} \sum_{k=1}^{n} (x_k^{(j)} - \bar{x}^{(j)})(x_k^{(l)} - \bar{x}^{(l)}).$$

Just as with the correlation coefficient between random variables we can show that if $|\hat{\rho}_{jl}| = 1$, there are constants a, b such that

10.15 $$\mathbf{x}^{(j)} = a + b\mathbf{x}^{(l)}.$$

Suppose one of the factors we are using is constant; that is, $x_k^{(1)} \equiv 1$, all k. Then 10.15 can be rewritten as

$$\mathbf{x}^{(j)} = a\mathbf{x}^{(1)} + b\mathbf{x}^{(l)}.$$

This is a linear relationship between factors, and implies that det $[S] = 0$. Thus, a large positive or negative correlation between any two factors puts us into the unstable situation det $[S] \simeq 0$ and produces coefficients for the two factors which mask their separate effects. Furthermore, since det $[S] \simeq 0$, the entire regression analysis is thrown out of kilter and the results may be senseless (see Problem 8). For instance, Table 10.2 gives the heat y_k in calories per gram evolved in the drying of cement, together with the percentages by weight of four chemical compounds used in the manufacture of the concrete. The first two, $\mathbf{x}^{(1)}$ and $\mathbf{x}^{(2)}$, are the same as in Table 10.1, Problem 6. The second two, $\mathbf{x}^{(3)}$ and $\mathbf{x}^{(4)}$ are the weights of $4CAO \cdot Al_2O_3$ and $2CaO \cdot S_1O_2$.

Table **10.2**

y	$\mathbf{x}^{(1)}$	$\mathbf{x}^{(2)}$	$\mathbf{x}^{(3)}$	$\mathbf{x}^{(4)}$
78.5	7	26	6	60
74.3	1	29	15	52
104.3	11	56	8	20
87.6	11	31	8	47
95.9	7	52	6	33
109.2	11	55	9	22
102.7	3	71	17	6
72.5	1	31	22	44
93.1	2	54	18	22
115.9	21	47	4	26
83.8	1	40	23	34
113.3	11	66	9	12
109.4	10	68	8	12

If the five regression coefficients (including a constant) are computed, it is found that, based on the results of a standard t test, none of the four coefficients of the chemical compound factors differs significantly from zero. Examining the table you can see that the sum of the four factors is always nearly 100 and that the sum of the first and third factors is nearly constant, as is the sum of the second and fourth factors. Therefore, $\mathbf{x}^{(1)}$, $\mathbf{x}^{(3)}$, and $\mathbf{x}^{(5)}$, the constant vector, are nearly linearly dependent, as are $\mathbf{x}^{(2)}$, $\mathbf{x}^{(4)}$, and $\mathbf{x}^{(5)}$. Even though $\hat{\hat{b}}_1\mathbf{x}^{(1)} + \cdots + \hat{\hat{b}}_5\mathbf{x}^{(5)}$ is a good fit to $\mathbf{y}$, the sizes of the coefficients do not give much information. We will come back to this example later.

The foregoing raises the question of how to choose factor levels if choice is possible. Often choice is not possible. For instance, in a polynomial regression where there is one factor with levels $x_1, \ldots, x_n$, and

$$x_k^{(j)} = (x_k)^j, \qquad j = 0, 1, \ldots, J,$$

then once the levels $x_1, \ldots, x_n$ have been selected, the levels of all the powers of this factor are automatically selected.

But suppose that the choice of levels is at our disposal. How should we select them so as to get good estimates of $b_1, \ldots, b_J$? One approach is based on common sense: To trace the effect of a number of factors, hold all of the factors constant except one; vary that one, and estimate its effect. Now repeat, varying factors one by one, holding the others constant. This does not give as good an estimate for the coefficients as a sampling plan with the same size sample where the n combinations of levels

$$(x_k^{(1)}, \ldots, x_k^{(J)}), \qquad k = 1, \ldots, n$$

are more uniformly spread out over the space of all possible combinations of levels. For instance, consider a two-factor regression where both factors can take only the values 1 or 2 and the sample size is four. Holding the second factor constant, say equal to 1, and varying the first, then repeating with the second factor varying, leads to the sample plan:

I $\qquad (x^{(1)}, x^{(2)})$

$$(1, 1), \quad (2, 1), \quad (1, 1), \quad (1, 2).$$

Using **a.** we get

$$V(\hat{\hat{\mathsf{b}}}_1) = V(\hat{\hat{\mathsf{b}}}_2) = \tfrac{28}{13}\sigma^2$$

(see Problem 10). Sample plan II uses every possible combination of factors:

II $\qquad (x^{(1)}, x^{(2)})$

$$(1, 1), \quad (2, 1), \quad (1, 2), \quad (2, 2).$$

The variances here are

$$V(\hat{\hat{\mathsf{b}}}_1) = V(\hat{\hat{\mathsf{b}}}_2) = \tfrac{36}{19}\sigma^2.$$

Thus II is the better design.

Problem 8 Here are some two-factor, four sample data:

$\mathbf{y}$	$\mathbf{x}^{(1)}$	$\mathbf{x}^{(2)}$
2	5.0	5.3
3	9.3	8.7
0	3.1	1.5
1	7.6	8.8

a Evaluate $\hat{\hat{b}}_1, \hat{\hat{b}}_2$.

Here is the same data with the second levels slightly altered:

$\mathbf{y}$	$\mathbf{x}^{(1)}$	$\mathbf{x}^{(2)}$
2	5.0	4.9
3	9.3	9.2
0	3.1	3.3
1	7.6	7.7

b Evaluate $\hat{\hat{b}}_1, \hat{\hat{b}}_2$ for the altered data.

(Ans. **a.** $\hat{\hat{b}}_1 = .26, \hat{\hat{b}}_2 = .00$, **b.** $\hat{\hat{b}}_1 = 5.8, \hat{\hat{b}}_2 = -5.5$.)

Problem 9 Consider the three-factor data below:

$\mathbf{y}$	$\mathbf{x}^{(1)}$	$\mathbf{x}^{(2)}$	$\mathbf{x}^{(3)}$
6	2	14	$14 + \varepsilon_1$
10	8	7	$7 + \varepsilon_2$
5	4	17	$17 + \varepsilon_3$
13	7	6	$6 + \varepsilon_4$
7	5	8	$8 + \varepsilon_5$

where $\varepsilon_1, \ldots, \varepsilon_5$ are small. If the third factor, which is very close to the second factor, is ignored, then this becomes a reasonable two-factor problem with estimated coefficients $\hat{\hat{b}}_1, \hat{\hat{b}}_2$. Find values for $\varepsilon_1, \ldots, \varepsilon_5$ such that they are small numerically, but that the resulting estimate for $\hat{\hat{b}}_1$ is much different from the estimate obtained from deleting the third factor.

Problem 10 In a two-factor experiment, each factor can have one of three possible values, 1, 2, or 3. Thus, the possible combinations of factor levels are the nine couples (1, 1), (2, 1), . . . , (3, 3). Of the possible four sample experiments, which one minimizes

$$V(\hat{\hat{b}}_1) + V(\hat{\hat{b}}_2)?$$

Problem 11 Outline a practical numerical procedure for evaluating the five regression coefficients (including a constant factor) for the data in Table 10.2.

Estimating σ^2 and Getting Confidence Intervals

The important result 10.8 reduces the computation of estimates for the coefficients $b_1, \ldots, b_J$ to a straightforward procedure: Compute $[S]$, then $[S]^{-1}$, then $[X]\mathbf{y}$, and finally $[S]^{-1}[X]\mathbf{y}$. But as always, coming up with an estimate is not enough. We want confidence intervals for the estimates, or perhaps we want to test hypotheses concerning the coefficients. Therefore we need more detailed information regarding the distribution of the estimates.

Propositions 10.10 and 10.11 give the distribution of the regression coefficient estimates. They are jointly normal and unbiased, with covariance matrix $\sigma^2[S]^{-1}$. Now $[S]^{-1}$ is computable from the known factor levels, but σ^2 is unknown. Hence we have to estimate σ^2 before we can find out how $\hat{\hat{\mathbf{b}}}$ is distributed about the true value $\mathbf{b}$.

If we knew the means $\mu_k = EY_k$, we could use the usual unbiased estimate

10.16
$$\hat{\sigma}^2 = \frac{1}{n}\sum_1^n (y_k - \mu_k)^2.$$

The μ_k are unknown, but since we have estimates of them, namely

$$\hat{\mu}_k = \hat{\hat{b}}_1 x_k^{(1)} + \cdots + \hat{\hat{b}}_J x_k^{(J)},$$

why not estimate σ^2 as

$$\frac{1}{n}\sum_1^n (y_k - \hat{\mu}_k)^2?$$

There is a good reason why we do not. Write the estimate 10.16 as $\hat{\sigma}^2 = D/n$, where

10.17
$$D = \sum_1^n (y_k - b_1 x_k^{(1)} - \cdots - b_J x_k^{(J)})^2,$$

and $b_1, \ldots, b_J$ are the true values of the parameters. We get the estimates $\hat{\hat{b}}_1, \ldots, \hat{\hat{b}}_J$ by selecting parameter values that make 10.17 as small as possible. Denote this minimum value by $\hat{D}$. Then

$$\begin{aligned}\hat{D} &= \sum_1^n (y_k - \hat{\hat{b}}_1 x_k^{(1)} - \cdots - \hat{\hat{b}}_J x_k^{(J)})^2 \\ &= \sum_1^n (y_k - \hat{\mu}_k)^2\end{aligned}$$

will be smaller than

$$D = \sum_1^n (y_k - \mu_k)^2.$$

In fact, we get an unbiased estimate of σ^2 by dividing $\hat{D}$ not by n but by the smaller factor $n - J$. Our estimate for σ^2 will be

$$\hat{\sigma}^2 = \frac{1}{n - J} \sum_1^n (y_k - \hat{\hat{\mu}}_k)^2.$$

What about the distribution of this estimate? The variable Y_k is $N(\mu_k, \sigma^2)$, so $(\mathsf{Y}_k - \mu_k)/\sigma$ is $N(0, 1)$ and

10.18
$$\frac{\mathsf{D}}{\sigma^2} = \sum_1^n \frac{(\mathsf{Y}_k - \mu_k)^2}{\sigma^2}$$

is the sum of the squares of n independent $N(0, 1)$ variables. Thus, it has a χ^2 distribution with n degrees of freedom. We estimate the J parameters $b_1, \ldots, b_J$ using maximum likelihood (equivalently, least squares) and substitute the estimated values into 10.18. As usual, we drop one degree of freedom for every parameter estimated. In other words,

10.19
$$\frac{\hat{\mathsf{D}}}{\sigma^2} = \sum_1^n \frac{(\mathsf{Y}_k - \hat{\hat{\boldsymbol{\mu}}}_k)^2}{\sigma^2}$$

has a χ^2 distribution with $n - J$ degrees of freedom. Summarizing, we have

10.20 ***Proposition*** *Let*

$$\hat{\mathsf{D}} = \sum_1^n (\mathsf{Y}_k - \hat{\hat{\boldsymbol{\mu}}}_k)^2,$$

where

$$\boldsymbol{\mu}_k = \hat{\hat{\mathsf{b}}}_1 x_k^{(1)} + \cdots + \hat{\hat{\mathsf{b}}}_J x_k^{(J)}.$$

Then

$$\hat{\sigma}^2 = \frac{1}{n - J} \hat{\mathsf{D}}$$

is an unbiased estimate of σ^2 and $\hat{D}/\sigma^2$ has a χ^2 distribution with $n - J$ degrees of freedom.

The argument we gave to lead to this result was somewhat informal, but it is not difficult to give rigorous proof (see Problem 14 or Theorem 10.30).

Now we can get some confidence intervals for our estimates. Denote the elements of $[S]^{-1}$ by r_{ij}. We know that the estimate $\hat{\hat{\mathsf{b}}}_j$ is normally distributed with mean b_j and variance $r_{jj}\sigma^2$. The estimated variance of $\hat{\hat{\mathsf{b}}}_j$ is

$$\hat{\sigma}_j^2 = r_{jj}\hat{\sigma}^2.$$

To get a confidence interval, look at the variable

10.21 $$\mathsf{S} = \frac{\hat{\hat{b}}_j - b_j}{\hat{\sigma}_j},$$

This has a Student's t distribution with $n - J$ degrees of freedom. Therefore we have the following:

10.22 ***Proposition*** *Take t so that*

$$P(|\mathsf{S}| \leq t) = \gamma,$$

where S *has a t distribution with $n - J$ degrees of freedom. A $100\gamma\%$ confidence interval for b_j is given by*

$$J_j = [\hat{\hat{b}}_j - t\hat{\sigma}_j, \hat{\hat{b}}_j + t\hat{\sigma}_j].$$

We have to be careful here. Suppose we compute 95% intervals $J_1, \ldots, J_J$ for every coefficient $b_1, \ldots, b_J$. Does this imply that the probability that no one of the coefficients is outside of its associated interval is $\leq .05$? Certainly not. Recall the discussion on page 106, Chapter 4. The repeated use of 10.22 to get confidence intervals for a number of different coefficients does not give simultaneous 100% confidence intervals and therefore can be misleading.

Suppose we are interested in the first six factors. What we would like are *simultaneous* confidence intervals for $b_1, \ldots, b_6$. This means we want intervals $\mathsf{J}_1, \ldots, \mathsf{J}_6$ such that

10.23 $$P(b_1 \in \mathsf{J}_1, \ldots, b_6 \in \mathsf{J}_6) = \gamma.$$

One way of getting such intervals is the *S*-method (*S* standing for Scheffé), which computes intervals that satisfy 10.23 by using an F distribution (see Chapter 2, page 42).

Suppose that we want to get simultaneous confidence intervals on M of the coefficients, $M \leq J$. Let F have an F distribution with M and $n - J$ degrees of freedom. From the $F_{M,n-J}$ tables find f such that

$$P(\mathsf{F} \leq f) = \gamma.$$

10.24 ***Theorem*** *The probability that all of the M coefficients b_j satisfy*

$$\hat{\hat{\mathsf{b}}}_j - \sqrt{Mf}\hat{\sigma}_j \leq b_j \leq \hat{\hat{\mathsf{b}}}_j + \sqrt{Mf}\hat{\sigma}_j.$$

is greater than or equal to γ.

If $M = 1$, then $f = t^2$ where t is computed from a t_{n-J} distribution, and the interval in 10.24 reduces to the confidence interval in 10.22 based on Student's t distribution. For any $M \geq 1$, if the sample size is large and $J \ll n$, the variable F has approximately the distribution of a χ_M^2 variable divided by M. Denote this by C/M. Then

10.25 $$P(\mathsf{F} \leq f) \simeq P(\mathsf{C} \leq Mf),$$

so that $Mf \simeq \delta$, where δ is gotten from a χ_M^2 table as the value that satisfies

$$P(\mathsf{C} \leq \delta) = \gamma.$$

The conclusion is that for $n - J$ large, simultaneous confidence intervals for M of the coefficients are given by

10.26 $$\hat{\hat{b}}_j - \sqrt{\delta}\,\hat{\sigma}_j \leq b_j \leq \hat{\hat{b}}_j + \sqrt{\delta}\,\hat{\sigma}_j.$$

To give some idea of how $\sqrt{\delta}$ behaves as M increases, here is a brief table.

Table **10.3**

	$\sqrt{\delta}$	
M	$\gamma = .9$	$\gamma = .95$
1	1.64	1.96
2	2.15	2.45
3	2.50	2.97
4	2.79	2.97
5	3.04	3.33
6	3.26	3.55
7	3.46	3.74
8	3.66	3.94
9	3.83	3.99
10	4.00	4.28
25	5.86	6.14
100	11.16	11.39

The lengths of the simultaneous confidence intervals increase slowly with M. For $M = 10$, they are slightly more than twice as long as for $M = 1$. For M large they are close to $\sqrt{M}$.

Actually, we can make a much stronger statement than Theorem 10.24. For ease of notation, suppose the M coefficients in question are $b_1, \ldots, b_M$. If we want to compare the sizes of these coefficients, we would look at the differences

$$b_k - b_j$$

and get confidence intervals for these. More generally, consider the problem of computing simultaneous confidence intervals for all linear combinations of the form

10.27 $$\theta = a_1 b_1 + \cdots + a_M b_M,$$

where $a_1, \ldots, a_M$ are specified constants. Denote

$$\hat{\hat{\theta}} = a_1\hat{\hat{b}}_1 + \cdots + a_M\hat{\hat{b}}_M$$

The variance of θ is

$$E(\hat{\hat{\boldsymbol{\theta}}} - \theta)^2 = \sum_{j,k} a_j a_k \Gamma(\hat{\hat{\mathrm{b}}}_j, \hat{\hat{\mathrm{b}}}_k) = \sigma^2 \sum_{j,k} a_j a_k r_{jk}.$$

We get an estimate $\hat{\sigma}^2_{\hat{\hat{\theta}}}$ of the variance of $\hat{\hat{\boldsymbol{\theta}}}$ by replacing σ^2 by $\hat{\sigma}^2$ in the above expression. Theorem 10.24 extends to

10.28 ***Theorem*** *The probability that all linear combinations θ of the form* 10.27 *simultaneously satisfy*

10.29 $$\theta \in \hat{\hat{\theta}} \pm \sqrt{Mf}\,\hat{\sigma}_{\hat{\hat{\theta}}}$$

is γ.

Thus with probability γ not only do the individual coefficients all fall into the confidence intervals given by Theorem 10.24, but at the same time, all linear combinations fall into the intervals given by 10.29.

This result has some useful applications. It allows unrestricted computation of confidence intervals for any and all linear combinations of $b_1, \ldots, b_M$ without worrying about their interaction. As an application, look at the model where the mean is linear in the factor level,

$$\mu_k = b_1 + b_2 x_k.$$

The assumption in this model is that the mean value at any level x is given by

$$\mu = b_1 + b_2 x.$$

Call this the *true regression line.* The estimated regression (or mean value) line is

$$\hat{\hat{\mu}} = \hat{\hat{b}}_1 + \hat{\hat{b}}_2 x.$$

Fix the value of x. Then $\hat{\hat{\mu}}$ is a linear combination of $\hat{\hat{b}}_1$ and $\hat{\hat{b}}_2$. To compute a confidence interval for μ we first need to compute the variance of $\hat{\hat{\mu}}$. In this example

$$[S] = n\begin{pmatrix} 1 & \bar{x} \\ \bar{x} & \overline{x^2} \end{pmatrix}.$$

Let $d = (\overline{x^2}) - (\bar{x})^2$. By an easy inversion

$$[S]^{-1} = \frac{1}{nd}\begin{pmatrix} \overline{x^2} & -\bar{x} \\ -\bar{x} & 1 \end{pmatrix},$$

so

$$\hat{\sigma}^2_{\hat{\hat{\mu}}} = \hat{\sigma}^2(r_{11} + 2r_{12}x + r_{22}x^2)$$

$$= \frac{\hat{\sigma}^2}{nd}(\overline{x^2} - 2\bar{x}x + x^2)$$

$$= \frac{\hat{\sigma}^2}{n} + \frac{\hat{\sigma}^2}{nd}(x - \bar{x})^2.$$

Take $M = J = 2$. Then applying Theorem 10.24, we conclude that the inclusions

10.30 $$b_1 + b_2x \in \hat{\hat{b}}_1 + \hat{\hat{b}}_2x \pm c\hat{\sigma}\sqrt{1 + \frac{1}{d}(x - \bar{x})^2},$$

where $c = \sqrt{2f/n}$, hold simultaneously for all values of x with probability at least γ.

There is a simple and neat geometric interpretation of 10.30. On a single sheet of graph paper, graph the two functions

$$f^+(x) = \hat{\hat{b}}_1 + \hat{\hat{b}}_2x + c\hat{\sigma}\sqrt{1 + \frac{1}{d}(x - \bar{x})^2},$$

$$f^-(x) = \hat{\hat{b}}_1 + \hat{\hat{b}}_2x - c\hat{\sigma}\sqrt{1 - \frac{1}{d}(x - \bar{x})^2}.$$

With probability at least γ, the true regression line $y = b_1 + b_2x$ will fall between the two graphed curves. Therefore, in a manner reminiscent of the Kolmogorov-Smirnov work, the band of values within the two curves forms a $100\gamma\%$ confidence band for the true regression line.

The graphs of both $f^+(x)$ and $f^-(x)$ are parabolic for x near $\bar{x}$ and taper off to straight line asymptotes as $|x - \bar{x}|$ increases. Their graphs typically look like Figure 10.3 below.

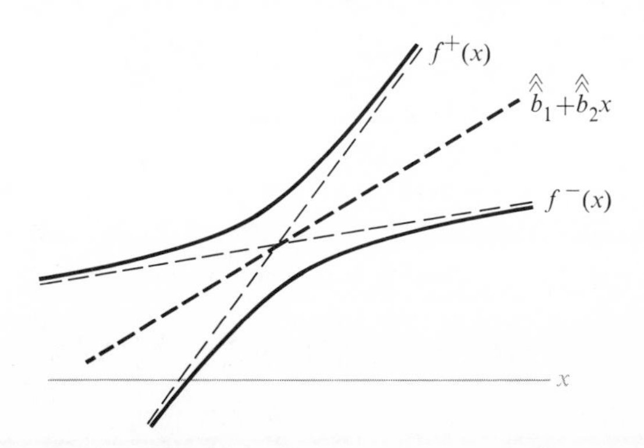

Figure 10.3 *The $100\gamma\%$ confidence band for the true regression line*

Problem 12 Estimate σ^2 for the data of Problem 4, assuming that

$$\mu_k = b_1 + b_2 x_k.$$

Find 95% simultaneous confidence intervals for b_1, b_2. Graph the 95% confidence band for the regression line.

Problem 13 Find 95% simultaneous confidence intervals for the coefficients of the three factors (including a constant) in a linear regression analysis of the data in Table 10.1. Compare the simultaneous intervals with the 95% individual intervals computed according to Proposition 10.22.

Problem 14 Here is a proof that the distribution of $\hat{\mathsf{D}}$ is χ^2_{n-J}. Assume that $\mathbf{x}^{(1)}, \ldots, \mathbf{x}^{(J)}$ are linearly independent. Now take $\mathbf{e}^{(1)}, \ldots, \mathbf{e}^{(J)}$ to be n-dimensional orthogonal unit vectors, such that $\mathbf{x}^{(1)}, \ldots, \mathbf{x}^{(J)}$ are linear combinations of $\mathbf{e}^{(1)}, \ldots, \mathbf{e}^{(J)}$. We can find n-dimensional unit vectors $\mathbf{e}^{(J+1)}, \ldots, \mathbf{e}^{(n)}$ which together with $\mathbf{e}^{(1)}, \ldots, \mathbf{e}^{(J)}$ form a set of n orthogonal vectors in n-dimensional space. Every n-dimensional vector can be written as a linear combination of these unit vectors. In particular, we can write

$$\mathbf{y} = \sum_{1}^{n} z_k \mathbf{e}^{(k)}.$$

Take the inner product of both sides with $\mathbf{e}^{(j)}$. Since $(\mathbf{e}^{(k)}, \mathbf{e}^{(j)}) = 0$, $j \neq k$, $(\mathbf{e}^{(k)}, \mathbf{e}^{(j)}) = 1$ if $j = k$,

$$(\mathbf{y}, \mathbf{e}^{(j)}) = z_j.$$

To minimize

$$\|\mathbf{y} - \tilde{\mathbf{y}}\|^2 = \sum_{1}^{n} (y_k - \tilde{y}_k)^2$$

over all possible linear combinations $\tilde{\mathbf{y}}$ of $\mathbf{x}^{(1)}, \ldots, \mathbf{x}^{(J)}$ is equivalent to minimizing when $\tilde{\mathbf{y}}$ ranges over all linear combinations $\sum_1^J a_k \mathbf{e}^{(k)}$ of the vectors $\mathbf{e}^{(1)}, \ldots, \mathbf{e}^{(J)}$. Note that

$$\|\mathbf{y} - \tilde{\mathbf{y}}\|^2 = \left\| \sum_{1}^{n} z_k \mathbf{e}^{(k)} - \sum_{1}^{J} a_k \mathbf{e}^{(k)} \right\|^2$$

$$= \sum_{J+1}^{n} z_k^2 + \sum_{1}^{J} (z_k - a_k)^2.$$

To minimize $\|\mathbf{y} - \tilde{\mathbf{y}}\|^2$ it is easy to see that we take $a_k = z_k$, $k = 1, \ldots, J$. The minimum value $\hat{D}$ is then

$$\hat{D} = \sum_{J+1}^{n} z_k^2.$$

As a random variable z_k is the outcome of the variable $\mathsf{Z}_k = (\mathbf{Y}, \mathbf{e}^{(k)})$, where $\mathbf{Y} = (\mathsf{Y}_1, \ldots, \mathsf{Y}_n)$. Let the true mean $\boldsymbol{\mu} = (\mu_1, \ldots, \mu_n)$. By assumption $\boldsymbol{\mu}$ is some linear combination of $\mathbf{x}^{(1)}, \ldots, \mathbf{x}^{(J)}$, and hence of $\mathbf{e}^{(1)}, \ldots, \mathbf{e}^{(J)}$. Therefore $(\boldsymbol{\mu}, \mathbf{e}^{(k)}) = 0$, $k \geq J + 1$, and we can write for $k \geq J + 1$,

$$\begin{aligned} \mathsf{Z}_k &= (\mathbf{Y}, \mathbf{e}^{(k)}) - (\boldsymbol{\mu}, \mathbf{e}^{(k)}) \\ &= (\mathbf{Y} - \boldsymbol{\mu}, \mathbf{e}^{(k)}) \\ &= (\mathbf{X}, \mathbf{e}^{(k)}), \end{aligned}$$

where $\mathbf{X} = \mathbf{Y} - \boldsymbol{\mu}$. Now $\mathbf{X} = (\mathsf{Y}_1 - \mu_1, \ldots, \mathsf{Y}_n - \mu_n)$ has independent $N(0, \sigma^2)$ components. We leave it to the student to show that

a the variables $\mathsf{Z}_k = (\mathbf{X}, \mathbf{e}^{(k)})$, $k = 1, \ldots, n$ are independent $N(0, \sigma^2)$ variables; (see PSP, pp. 237–238)

b in consequence $\hat{\mathsf{D}}/\sigma^2$ has a χ^2_{n-J} distribution;

c $\hat{\mathsf{D}}/(n - J)$ is an unbiased estimate of σ^2.

Testing Hypotheses on Regression Coefficients

The most frequent sort of hypothesis in regression analysis is that some coefficient or group of coefficients are zero. If $\mathbf{x}^{(1)}, \ldots, \mathbf{x}^{(J)}$ all correspond to different factors, then the hypothesis $b_J = 0$ is the statement that the Jth factor has no effect on the mean value of the observed variable. The hypothesis that $b_{J-M+1} = 0, \ldots, b_J = 0$ is the translation of the statement that the last group of M factors have no effect on the mean value of the observed variable.

Now suppose we are going to do a polynomial regression with $x_k^{(1)} \equiv 1$, $x_k^{(2)} = x_k$, $x_k^{(3)} = x_k^2$. Then $b_3 = 0$ is the statement that the observed data are fit as well by a linear function of x_k as they are by a quadratic function. More generally, if $x_k^{(j)} = (x_k)^{j-1}, j = 1, 2, \ldots, J$, then $b_{J-M+1} = 0, \ldots, b_J = 0$ is the statement that the data are fit as well by a $(J - M - 1)$th degree polynomial as by a $(J - 1)$st degree polynomial.

To understand the test that is used, look at

10.31 $$D = \sum_{1}^{n} (y_k - b_1 x_k^{(1)} - \cdots - b_J x_k^{(J)})^2,$$

for *any* choice of the parameters $b_1, \ldots, b_J$. The size of D is a measure of how well that particular set of parameter values fits the data. The smaller D is, the better the fit. The best fit is obtained by using the least squares estimates $\hat{\hat{b}}_1, \ldots, \hat{\hat{b}}_J$, giving D its minimum value $\hat{D}$.

Now suppose we restrict the selection of $b_1, \ldots, b_J$ to those parameter values such that $b_{J-M+1} = 0, \ldots, b_J = 0$. In other words, see how well the data can be fitted by linear combinations of the levels of the first $J - M$ factors only. The best fit is gotten by using the parameter values that minimize

$$D_1 = \sum_{1}^{n} (y_k - b_1 x_k^{(1)} - \cdots - b_{J-M} x_k^{(J-M)})^2.$$

Denote the minimum value of D_1 by $\hat{D}_1$. As long as $M \geq 1$, then $\hat{D}_1 \geq \hat{D}$, because the fit with some of the parameter values restricted to zero will certainly be no better than the best unrestricted fit. The difference

$$\hat{D}_1 - \hat{D}$$

is a measure of how much worse the fit gets if we delete the last M factors. But this difference also depends on σ^2. Whether or not the hypothesis is true, $\hat{D}/(n - J)$ is a good estimator for σ^2. If the hypothesis is true, then $\hat{D}_1/(n - J + M)$ is a good estimator for σ^2, so

$$\hat{D}_1 \sim (n - J + M)\sigma^2,$$
$$\hat{D}_1 - \hat{D} \sim M\sigma^2,$$

and $(\hat{D}_1 - \hat{D})/M$ should be of magnitude σ^2. If the hypothesis is not true, the fit should be much worse, and $(\hat{D}_1 - \hat{D})/M$ considerably larger than σ^2. Logically, the thing to do is to compare $(\hat{D}_1 - \hat{D})/M$ with an estimate of σ^2. Since $\hat{\sigma}^2 = \hat{D}/(n - J)$ is a good estimator of σ^2, look at the ratio

10.32 $$\mathsf{R} = \frac{n - J}{M}\left(\frac{\hat{\mathsf{D}}_1 - \hat{\mathsf{D}}}{\hat{\mathsf{D}}}\right).$$

If we knew the distribution of R under the hypothesis, we could test the hypothesis in the usual way: select c such that under the hypothesis

$$P(\mathsf{R} \leq c) = 1 - \alpha.$$

We get a level α test by accepting the hypothesis whenever $R \leq c$ and rejecting otherwise. The distribution of R is settled by

10.33 ***Proposition*** *R has an $F_{M,n-J}$ distribution.*

Actually, this proposition is a special case of a theorem which has been hinted at by one of the problems and which is at the heart of most of the distribution results in both regression analysis and analysis of variance. We state it this way: Suppose $y_1, \ldots, y_n$ are the outcomes of independent $N(\mu_k, \sigma^2)$ variables $\mathsf{Y}_1, \ldots, \mathsf{Y}_n$ with σ^2 and $\boldsymbol{\mu} = (\mu_1, \ldots, \mu_n)$ unknown.

10.34 ***Theorem*** **a** *Let*

$$\hat{D} = \min_{\tilde{\mathbf{y}} \in \mathscr{L}_J} \|\mathbf{y} - \tilde{\mathbf{y}}\|^2$$

where $\mathscr{L}_J$ is a J-dimensional subspace of the space of n-vectors. If the true mean $\boldsymbol{\mu} \in \mathscr{L}_J$, then

$$\hat{\sigma}^2 = \frac{\hat{\mathsf{D}}}{n - J}$$

is an unbiased estimator for σ^2 and $\hat{\mathsf{D}}/\sigma^2$ has a χ^2_{n-J} distribution.

b *Let*

$$\hat{D}_1 = \min_{\tilde{\mathbf{y}}_1 \in \mathscr{L}_{J-M}} \|\mathbf{y} - \tilde{\mathbf{y}}_1\|^2$$

where $\tilde{\mathbf{y}}_1$ ranges over $\mathscr{L}_{J-M}$, a $(J - M)$-dimensional subspace of $\mathscr{L}_J$. Then, under the hypothesis that

$$\boldsymbol{\mu} \in \mathscr{L}_{J-M},$$

the ratio

$$\frac{n - J}{M} \frac{\hat{\mathsf{D}}_1 - \hat{\mathsf{D}}}{\hat{\mathsf{D}}}$$

has an $F_{M,n-J}$ distribution.

Taking $\mathscr{L}_J$ to consist of all linear combinations of $\mathbf{x}^{(1)}, \ldots, \mathbf{x}^{(J)}$ and $\mathscr{L}_{J-M}$ of all linear combinations of $\mathbf{x}^{(1)}, \ldots, \mathbf{x}^{(J-M)}$, you can see how part b. of this theorem gives Proposition 10.33.

We use this theorem to test the hypothesis $\boldsymbol{\mu} \in \mathscr{L}_{J-M}$ by accepting if

$$\frac{n - J}{M} \frac{\hat{D}_1 - \hat{D}}{\hat{D}} \leq f,$$

where f is selected from the $F_{M,n-J}$ table so as to give the desired level. If we rewrite

$$\frac{n - J}{M} \frac{\hat{D}_1 - \hat{D}}{\hat{D}} = \frac{(\hat{D}_1 - D)/M}{\hat{D}/(n - J)},$$

we see that the denominator $\hat{\mathsf{D}}/(n - J)$, has the same distribution whether the hypothesis is true or false. It essentially acts as a factor which divides out the size of the variance. If $\mathbf{Y}$ has mean $\boldsymbol{\mu}$, then its values are centered around $\boldsymbol{\mu}$. Therefore, when the hypothesis is true and $\boldsymbol{\mu} \in \mathscr{L}_{J-M}$, then $\hat{D}_1$ is generally not much larger than $\hat{D}$. But as $\boldsymbol{\mu}$ moves outside of $\mathscr{L}_{J-M}$, $\hat{D}_1$ becomes larger. Hence the numerator $\hat{D}_1 - \hat{D}$ is distributed around its smallest values when the hypothesis is true and becomes generally larger when the hypothesis is false.

Actually, it is easier to describe the detection capability of this test in the more general form given by the above theorem. Let $d(\boldsymbol{\mu})$ be the squared distance of $\boldsymbol{\mu}$ from the subspace $\mathscr{L}_{J-M}$. That is,

$$d(\boldsymbol{\mu}) = \min_{\boldsymbol{\mu}_1 \in \mathscr{L}_{J-M}} \|\boldsymbol{\mu} - \boldsymbol{\mu}_1\|^2.$$

10.35 ***Theorem*** *$\boldsymbol{\mu}$ can be detected at level α if*

$$d(\boldsymbol{\mu}) \geq k\sigma^2,$$

where k depends only on M, $n - J$, and α.

For $n - J$ moderately large here is a short table of k values:

Table **10.4**

	α	
M	.01	.05
1	24	13
2	28	15
3	30	17
4	32	18
5	33	19
6	34	20
7	36	21

While the F test described above seems reasonable, it could still have much higher error probabilities than the best test. It is comforting to know that the F test is the likelihood ratio test for the hypothesis $\boldsymbol{\mu} \in \mathscr{L}_{J-M}$ (see Problem 18). It may be even more comforting to know that under the model assumptions, it is in an appropriate sense, the best test of the hypothesis.

The following proof of Theorem 10.34 is optional reading. Both parts are proved together. The proof is an interesting application of some simple vector concepts together with the basic fact that any orthogonal transformation of independent $N(0, \sigma^2)$ variables produces independent $N(0, \sigma^2)$ variables. (This was proved in PSP, pp. 237–238, for $\sigma^2 = 1$. Just multiply all equations by σ to get the general result.) Without some background in vector spaces you may find the proof rough going, but it does illuminate the theorem.

First of all, select orthogonal unit vectors $\mathbf{e}^{(1)}, \ldots, \mathbf{e}^{(n)}$ in n-dimensional space such that

a $\mathscr{L}_{J-M}$ consists of all linear combinations of $\mathbf{e}^{(1)}, \ldots, \mathbf{e}^{(J-M)}$;

b $\mathscr{L}_J$ consists of all linear combinations of $\mathbf{e}^{(1)}, \ldots, \mathbf{e}^{(J)}$.

We can always do this since $\mathscr{L}_{J-M}$ is a $(J - M)$-dimensional linear space *contained* in $\mathscr{L}_J$. Then $\hat{D}$ is computed by minimizing $\|\mathbf{y} - \tilde{\mathbf{y}}\|^2$, where $\tilde{\mathbf{y}}$ ranges over all linear combinations of $\mathbf{e}^{(1)}, \ldots, \mathbf{e}^{(J)}$, and $\hat{D}_1$, by minimizing $\|\mathbf{y} - \tilde{\mathbf{y}}_1\|^2$ where $\tilde{\mathbf{y}}_1$ ranges over all linear combinations of $\mathbf{e}^{(1)}, \ldots, \mathbf{e}^{(J-M)}$.

Since every n-dimensional vector can be written as a linear combination of $\mathbf{e}^{(1)}, \ldots, \mathbf{e}^{(n)}$, we have that

$$\mathbf{y} = \sum_{1}^{n} z_k \mathbf{e}^{(k)},$$

where the coefficients are given by

$$z_k = (\mathbf{y}, \mathbf{e}^{(k)}).$$

Write

$$\tilde{\mathbf{y}} = a_1 \mathbf{e}^{(1)} + \cdots + a_J \mathbf{e}^{(J)},$$

so

$$\mathbf{y} - \tilde{\mathbf{y}} = \sum_{1}^{J} (z_k - a_k)\mathbf{e}^{(k)} + \sum_{J+1}^{n} z_k \mathbf{e}^{(k)}.$$

Now $\|\mathbf{y} - \tilde{\mathbf{y}}\|^2$ is the ordinary squared Euclidean length of the vector $\mathbf{y} - \tilde{\mathbf{y}}$. Since the $\mathbf{e}^{(1)}, \ldots, \mathbf{e}^{(n)}$ are orthogonal unit vectors,

$$\|\mathbf{y} - \tilde{\mathbf{y}}\|^2 = \sum_{1}^{J} (z_k - a_k)^2 + \sum_{J+1}^{n} z_k^2.$$

What selection of $a_1, \ldots, a_J$ minimizes this expression? It is easy to see that the selection $a_1 = z_1, \ldots, a_J = z_J$ makes the first sum zero; otherwise it is positive. We arrive at

10.36a
$$\hat{D} = \sum_{J+1}^{n} z_k^2.$$

A similar argument, expressing $\tilde{\mathbf{y}}_1 = c_1 \mathbf{e}^{(1)} + \cdots + c_{J-M} \mathbf{e}^{(J-M)}$ gives

10.36b
$$\hat{D}_1 = \sum_{J-M+1}^{n} z_k^2.$$

Hence

10.37
$$\hat{D}_1 - \hat{D} = \sum_{J-M+1}^{J} z_k^2.$$

If $z_{J+1}, \ldots, z_n$ were outcomes of independent $N(0, \sigma^2)$ variables, then the first part of the theorem would follow immediately from 10.36a.

If $z_{J-M+1}, \ldots, z_n$ were outcomes of independent $N(0, \sigma^2)$ variables, then 10.36 and 10.37 give the result we want; for we can use them to write

$$\frac{\hat{D}_1 - \hat{D}}{\hat{D}} = \frac{\sum_{J-M+1}^{J} (z_k/\sigma)^2}{\sum_{J+1}^{n} (z_k/\sigma)^2},$$

which makes it clear that $(_1 - \hat{\mathsf{D}})\hat{\mathsf{D}}/\hat{\mathsf{D}}$ is the ratio of the sum of the squares of M independent $N(0, 1)$ variables divided by the sum of the squares of $n - J$ other $N(0, 1)$ variables, independent of each other and of the M variables in the numerator.

Write

$$\mathbf{Y} = \mathbf{V} + \boldsymbol{\mu},$$

so that $\mathbf{V}$ has independent $N(0, \sigma^2)$ components. Then

$$\mathsf{Z}_k = (\mathbf{V}, \mathbf{e}^{(k)}) + (\boldsymbol{\mu}, \mathbf{e}^{(k)}).$$

Look at the transformation of the components of $\mathbf{V}$ given by

$$\mathsf{U}_k = (\mathbf{V}, \mathbf{e}^{(k)}) = \sum_{j=1}^{n} e_j^{(k)} \mathsf{V}_j.$$

The variables $\mathsf{U}_1, \ldots, \mathsf{U}_n$ are gotten from an orthogonal transformation on $\mathsf{V}_1, \ldots, \mathsf{V}_n$. Hence they are independent $N(0, \sigma^2)$ distributed variables. Now write

$$\mathsf{Z}_k = \mathsf{U}_k + (\boldsymbol{\mu}, \mathbf{e}^{(k)}).$$

If $\boldsymbol{\mu}$ is in $\mathscr{L}_J$, $\boldsymbol{\mu}$ is a linear combination of $\mathbf{e}^{(1)}, \ldots, \mathbf{e}^{(J)}$, so that $(\boldsymbol{\mu}, \mathbf{e}^{(k)}) = 0$ for $k \geq J + 1$. Therefore, $\mathsf{Z}_k = \mathsf{U}_k$ for $k \geq J + 1$; $\mathsf{Z}_{J+1}, \ldots, \mathsf{Z}_n$ are independent $N(0, \sigma^2)$ variables and part a. of the theorem is proved.

If $\boldsymbol{\mu}$ is in $\mathscr{L}_{J-M}$, then exactly the same argument gives $(\boldsymbol{\mu}, \mathbf{e}^{(k)}) = 0$ for $k \geq J - M + 1$. In consequence, $\mathsf{Z}_{J-M+1}, \ldots, \mathsf{Z}_n$ are independent $N(0, \sigma^2)$ variables, and this proves part b.

Problem 15 Consider the data of Problem 4.

a Which of the two models $\mu = a + bx$ and $\mu = a + b\sqrt{x}$ provides the better fit?

b Assume a model of the form

$$\mu = a + b\sqrt{x} + cx$$

and test the hypothesis $a = c = 0$.

Problem 16 Scheffé ("Analysis of Variance," John Wiley and Sons, 1959) reproduces the following data where x is the percentage of natural rubber in samples of vulcanizates and $y = 1 + \log_{10} r$, where r is the ratio of transmittances at two selected wave lengths. The purpose is to get a calibration curve to allow the use of y to determine the value of x.

Table **10.5**

x	0	20	40	60	80	100
y	0.727	0.884	1.073	1.194	1.350	1.442
	0.721	.0880	1.050	1.184	1.291	1.369
	0 742	0.885	1.045	1.205	1.291	1.458
	0.746	0.890	1.033	1.180	1.323	1.459

The following questions were also borrowed from Scheffé:

a Plot the 24 data points.

b At each x-value plot the average of the four y-values and circle these.

c Fit a line by least squares to the data, and plot it on the graph.

d Fit a second degree polynomial by least squares and plot it on the graph. Visually, do you judge that there is a great improvement over the line?

e Test to see whether the coefficient of x^2 in **d.** differs significantly from zero.

Problem 17 For the regression model with

$$\mu_k = a + bx_k,$$

let the hypothesis be $b = 0$. Simplify the expression for $d(\boldsymbol{\mu})$ and find the range of values of b that can be detected at the .05 level for sample size $n \gg 1$, using the test given in the preceding section. (Ans.

$$|b| \geq \frac{3.6}{\sqrt{n}}\left(\frac{\sigma}{\sigma_x}\right),$$

where

$$\sigma_x^2 = \overline{x^2} - (\bar{x})^2.)$$

Problem 18 Show that the likelihood ratio test for $\boldsymbol{\mu} \in \mathscr{L}_{J-M}$ in the model of Theorem 3.10 is the F test. You may use the following outline:

a Let $\hat{\sigma}_1^2$, $\hat{\sigma}^2$ be the maximum likelihood estimates of σ^2 for $\boldsymbol{\mu}$ ranging over $\mathscr{L}_{J-M}$ and over $\mathscr{L}_J$, respectively. Show that the likelihood ratio test can be put into the form: Accept if

$$\frac{\hat{\sigma}_1^2 - \hat{\sigma}^2}{\hat{\sigma}^2} \leq c'$$

(see p. 154, Chapter 5 for a similar procedure);

b show that $\hat{\sigma}_1^2 = n\hat{D}_1$, and $\hat{\sigma}^2 = n\hat{D}$.

Does a Regression Model Fit?

The linear regression model is based on many assumptions. The least squares estimates for $b_1, \ldots, b_J$ will usually be fairly accurate as long as the mean is actually a linear function of the factor levels, even if the observations are not normally distributed with a common variance. But we do need normality and common variance to compute the confidence intervals around the estimates of the coefficients and the acceptance regions for testing whether various groups of coefficients are zero. What happens if the distributions are not normal or if the variances are not the same? Even before we can answer this, there are other questions we must consider: How can we tell if the observations are normally distributed, if they have a common variance, if the means are linear in the factor levels? This is, how can we tell if our model fits?

There is one rough answer. If the model fits, then for large sample size $n \gg 1$, $\hat{\sigma} \simeq \sigma$ and $\hat{\hat{b}}_j \simeq b_j$, so that $\hat{\hat{\mu}}_k \simeq \mu_k$. Therefore the sequence of numbers

$$z_k = \frac{y_k - \hat{\hat{\mu}}_k}{\hat{\sigma}}, \qquad k = 1, \ldots, n$$

should look like a sample of size n drawn independently from an $N(0, 1)$ distribution. By looking at $z_1, \ldots, z_n$ we may be able to spot departures from this distribution. If normality is suspect, we may want to run a goodness of fit test on the $z_1, \ldots, z_n$ to see whether the $N(0, 1)$ is a reasonable fit to their underlying distribution. If we suspect that the variance changes with the magnitude of the factor levels, this will show up in systematic increases or decreases in $|z_1|, |z_2|, \ldots, |z_n|$. Similarly, if the mean is not a linear function of the given factor levels, this may show up as systematic variations in the sizes of the z_k as the various factor levels are changed. Suppose that all the factors are actually functions of one factor, such as in a polynomial fit. Then, assuming that $z_1, \ldots, z_n$ correspond to ordered factor levels $x_1 \leq \cdots \leq x_n$, probably the best way to start is to plot the z_k as a function of k and examine the graph carefully.

Putting questions of normality or constant variance aside for the moment, a more basic question is: Do we have any systematic way of telling whether the assumption that the means are linear functions of the factor levels gives a reasonable fit to the data? This is really a fairly general goodness of fit question. For example, look at the data tabled in 10.6 and graphed in Figure 10.4.

Table **10.6**

x	1	2	3	4	5	6	7	8
y	.395	1.059	.886	1.147	1.557	1.770	.945	.976

x	9	10	11	12	13	14	15	16
y	1.909	1.59	1.255	1.681	1.249	1.799	1.773	1.722

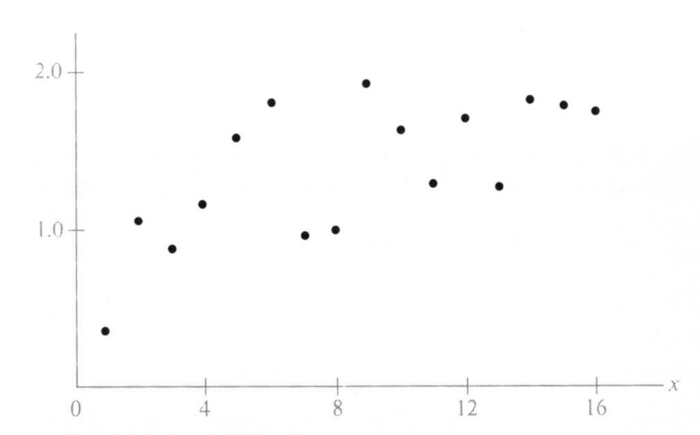

Figure 10.4 *Data of Table 10.6*

For these data we might be tempted to try a fit by setting

$$\mu_k = b_1 + b_2 \log x_k.$$

This is a linear model with factors $x_k^{(1)} = 1$, $x_k^{(2)} = \log x_k$. Or, we could try a fit using

$$\mu_k = b_1 + b_2 x_k + b_3 x_k^2.$$

These two fits could be compared by seeing which function of x_k leads to the smaller $\hat{D}$. But this does not settle the question of whether either of them is really a good fit. One way to approach this might be to assume the model

$$\mu_k = b_1 + b_2 x_k + b_3 x_k^2 + b_4 x_k^3$$

and test for $b_4 = 0$. Another is to assume

$$\mu_k = b_1 + b_2 x_k + \cdots + b_{11} x_k^{10}$$

and test the hypothesis $b_4 = b_5 = \cdots = b_{11} = 0$. The latter is a grand computational problem, while the former assumes to begin with that a cubic in x_k is a good fit to the data. Furthermore, neither of these procedures could be easily extended to check whether $b_1 + b_2 \log x_k$ gives a good fit.

One goodness of fit check is based on the following idea: Suppose there were an "intrinsic" way of getting an estimate of σ^2. Here "intrinsic" means that the estimate does not depend on any assumptions regarding the functional dependence of the means μ_k on the factor levels. Denote this estimate by $\hat{\sigma}_0^2$. The estimate we get using our linear model is $\hat{\sigma}^2 = \hat{D}/(n - J)$. If linearity is a reasonable assumption, then the *residual variance estimate* $\hat{\sigma}^2$ should be comparable in size to the intrinsic estimate $\hat{\sigma}_0^2$. If the fit is bad, then $\hat{\sigma}^2$ will be much larger.

To illustrate, suppose there is only one factor with m different levels $x_1, \ldots, x_m$. At every level x_k, we make N observations. The total sample size is $n = mN$, but the observations occur in m groups of size N

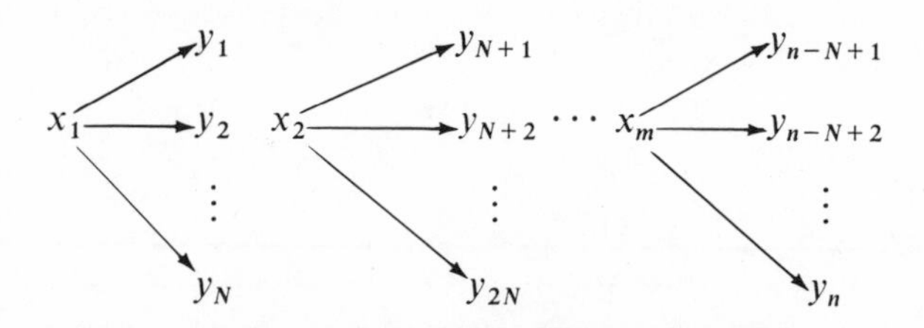

Assume normality and common variance σ^2. The observations in the first group are from an $N(\mu_1, \sigma^2)$ distribution, in the second group, from an $N(\mu_2, \sigma^2)$ distribution, and in the mth group, from an $N(\mu_m, \sigma^2)$ distribution. We get estimates for $\mu_1, \ldots, \mu_m$ from the sample means

$$\hat{\mu}_1 = \frac{1}{N} \sum_{1}^{N} y_k, \ldots, \qquad \hat{\mu}_m = \frac{1}{N} \sum_{n-N+1}^{n} y_k.$$

From each group, we get unbiased estimates for σ^2:

$$\hat{\sigma}_1^2 = \frac{1}{N-1} \sum_{1}^{N} (y_k - \hat{\mu}_1)^2, \ldots, \hat{\sigma}_m^2 = \frac{1}{N-1} \sum_{n-N+1}^{n} (y_k - \hat{\mu}_m)^2.$$

Pool these estimates to get the aggregate intrinsic unbiased estimator for σ^2,

$$\hat{\sigma}_0^2 = \frac{1}{m}(\hat{\sigma}_1^2 + \cdots + \hat{\sigma}_m^2).$$

Suppose now that we want to check the linear regression model

10.38 $$\mu_k = b_1 x_k^{(1)} + \cdots + b_J x_k^{(J)}, \qquad k = 1, \ldots, m.$$

Get the least squares fit by minimizing

$$D = \sum_1^N (y_k - b_1 x_1^{(1)} - \cdots - b_J x_1^{(J)})^2 + \cdots$$

$$+ \sum_{n-N+1}^{n} (y_k - b_1 x_m^{(1)} - \cdots - b_J x_m^{(J)})^2.$$

If the model fits, the minimum value $\hat{D}$ divided by $n - J$ is a good estimate $\hat{\sigma}^2$ for σ^2. A reasonable test for goodness of fit would be to look at the ratio

10.39 $$\frac{\hat{\sigma}^2}{\hat{\sigma}_0^2}$$

and reject the linear model (10.38) if this ratio is too large. To get a level α test we must know the distribution of the ratio (10.39) under the hypothesis that (10.38) holds. If we know this distribution, we then select c such that under the hypothesis,

$$P(\hat{\sigma}^2/\hat{\sigma}_0^2 \leq c) = 1 - \alpha$$

and reject whenever $\hat{\sigma}^2/\hat{\sigma}_0^2 > c$.

To get a convenient distribution, compute

10.40 $$w = \frac{n - J}{m - J} \cdot \left(\frac{\hat{\sigma}^2}{\hat{\sigma}_0^2}\right) - \frac{n - m}{m - J}$$

and use the equivalent test that rejects whenever the value of w is too large. The reason we use the linear function w of the ratio instead of the ratio itself is the following.

10.41 ***Proposition*** *The underlying distribution of* w *when the means are linear functions of the factor levels* (10.38) *is* $F_{m-J,n-m}$.

Thus, we have a simple procedure for a goodness of fit test: Compute f such that $P(\mathsf{F} \leq f) = 1 - \alpha$ where F has an $F_{m-J,n-m}$ distribution. Reject the fit of the linear model at level α if $w > f$.

The proof of Proposition 10.41 uses some familiar techniques. Let

$$D_0 = \sum_{1}^{N} (y_k - \mu_1)^2 + \cdots + \sum_{n-N+1}^{n} (y_k - \mu_m)^2,$$

so $\hat{\sigma}_0^2 = \hat{D}_0/m(N-1)$. Substitute this expression for $\hat{\sigma}_0^2$ and $\hat{\sigma}^2 = \hat{D}/(n-J)$ into the expression for w getting

$$\text{10.42} \qquad w = \frac{n-m}{m-J} \cdot \frac{\hat{D}}{\hat{D}_0} - \frac{n-m}{m-J} = \frac{n-m}{m-J} \cdot \frac{\hat{D} - \hat{D}_0}{\hat{D}_0}.$$

Now $\hat{D}_0$ equals the minimum over $\alpha_1, \ldots, \alpha_m$ of the expression

$$D_0 = \sum_{1}^{N} (y_k - \alpha_1)^2 + \cdots + \sum_{n-N+1}^{n} (y_k - \alpha_m)^2,$$

so $\hat{D}_0$ is a measure of the best fit when we assume *only* that each group of observations has a common mean, the kth group having mean μ_k. The value $\hat{D}$ is a measure of the fit when we assume, in addition, that $\mu_1, \ldots, \mu_m$ are linear functions of the factor levels. Look at it this way—$\hat{D}_0$ is the minimum of

$$\|\mathbf{y} - \tilde{\mathbf{y}}_0\|^2$$

when $\tilde{\mathbf{y}}_0$ ranges over all n-dimensional vectors $(\tilde{y}_1, \ldots, \tilde{y}_n)$ such that the first N coordinates $\tilde{y}_1, \ldots, \tilde{y}_N$ are equal, the second N coordinates $\tilde{y}_{N+1}, \ldots, \tilde{y}_{2N}$ are equal, and so on. $\hat{D}$ is the minimum of $\|\mathbf{y} - \tilde{\mathbf{y}}_1\|^2$ where $\tilde{\mathbf{y}}_1$ ranges over all vectors $(\tilde{y}_1, \ldots, \tilde{y}_n)$ where not only do successive groups of N coordinates have a common value, but also each of these common values is given by a linear combination of the factor levels. In brief, the proof is a direct application of the second part of Theorem 10.34. That is, $\tilde{\mathbf{y}}_0$ ranges over an m-dimensional linear vector space and $\tilde{\mathbf{y}}_1$ ranges over a J-dimensional linear subspace of the $\tilde{\mathbf{y}}_0$ space. This completes the proof.

The usual situation in regression analysis is that we do not have groups of repeated observations at the same factor levels. How can we test goodness of fit in this situation? A procedure which gives a good approximation is to group together observations for which the factor levels are *nearly* equal and assume that these have the same means. For example, for the data in Table 10.6, group the first and second observations together, the third and fourth, . . . , the fifteenth and sixteenth. Assuming the first and second observations have the same mean and variance, the unbiased estimate for σ^2 is

$$\hat{\sigma}_1^2 = (y_1 - \bar{y})^2 + (y_2 - \bar{y})^2,$$

where

$$\bar{y} = \frac{y_1 + y_2}{2}.$$

Simplifying gives the expression

$$\hat{\sigma}_1^2 = \tfrac{1}{2}(y_1 - y_2)^2.$$

Hence, an approximate intrinsic estimate for the variance of the data is given by

$$\begin{aligned}\hat{\sigma}_0^2 &= \tfrac{1}{8} \cdot \left(\tfrac{1}{2}(y_2 - y_1)^2 + \tfrac{1}{2}(y_4 - y_3)^2 + \cdots + \tfrac{1}{2}(y_{16} - y_{15})^2\right) \\ &= \frac{1}{16} \sum_{k=1}^{8} (y_{2k} - y_{2k-1})^2.\end{aligned}$$

This is simply the normalized sum of the squares of the successive differences. If we are trying to check goodness of fit of the model $\mu_k = a + bx_k$, then $J = 2$, and, of course, $m = 8$, $N = 2$. Now compute $\hat{\sigma}^2$ as $\hat{D}/16$ where $\hat{D}$ is the minimum of

$$D = \sum_{1}^{16} (y_k - a - bx_k)^2.$$

By 10.40,

$$w = \frac{14}{6}\left(\frac{\hat{\sigma}_2}{\hat{\sigma}_0^2}\right) - \frac{8}{6}$$

and we check this value against f computed from an $F_{6,8}$ table.

This kind of grouping will work in all problems provided that the difference in the means of the grouped observations is small compared to the standard deviation of the data. In multi-factor problems, try to group together those observations for which *all* factor levels are nearly the same. Grouping in pairs and estimating σ^2 by

$$\hat{\sigma}_0^2 = \frac{1}{n} \sum_{1}^{n/2} (y_{2k} - y_{2k-1})^2$$

is safest in the sense that the larger the groups, the more diversity there will be in means between groups.

The usual cautionary words need to be added here. If you accept the adequacy of your model, for instance if you accept the fit of a linear model

$$\mu = a + bx_k,$$

this does not mean that μ is a linear function of x, but only that the linear function is among the great many possible functional expressions for the mean that give acceptable fits.

Sometimes the factor levels are such that the best grouping of data does not give equal numbers of observations in every group. For example, three sets of factor levels may be fairly close to each other while relatively distant from the other sets of factor levels. If we decide to group into pairs, then one of these observations is left out, with factor levels unmatchable with any other available readings. Or one set of factor levels may be relatively distant from the other sets. The appropriate procedure for this situation is set up in Problem 21, and applied in Problem 22.

Problem 19 For the data of Problem 16, check goodness of fit to the linear model

$$\mu = a + bx.$$

Problem 20 For the data of Table 10.6, check the fit of the models $\mu = a + b \log x$ and $\mu = a + bx$.

Problem 21 Consider a situation where N_1 of the observations are at one fixed set of factor values, N_2 at the next, and N_m at the last set. Denote by $\mu_1, \ldots, \mu_m$ the true means of the observations in the m groups, and let $n = N_1 + \cdots + N_m$. The groups need not have more than one observation; all we demand is $N_k \geq 1$. The least squares estimates for $\mu_1, \ldots, \mu_m$ are those values of $\alpha_1, \ldots, \alpha_m$ that minimize

$$D_0 = \sum_{1}^{N_1} (y_k - \alpha_1)^2 + \cdots + \sum_{n-N_m+1}^{n} (y_k - \alpha_m)^2.$$

A simple computation shows that these minimizing values are the sample means

$$\hat{\mu}_1 = \frac{1}{N_1} \sum_{1}^{N_1} y_k, \ \ldots, \ \hat{\mu}_m = \frac{1}{N_m} \sum_{n-N_m+1}^{n} y_k.$$

In particular, if $N_k = 1$, notice that $\hat{\mu}_k = y_k$. Let

$$\hat{D}_0 = \sum_{1}^{N_1} (y_k - \hat{\mu}_1)^2 + \cdots + \sum_{n-N_m+1}^{n} (y_k - \hat{\mu}_k)^2.$$

An intrinsic unbiased estimate for σ^2 is given by

$$\hat{\sigma}_0^2 = \frac{\hat{D}_0}{(N_1 - 1) + \cdots + (N_m - 1)} = \frac{\hat{D}_0}{n - m}.$$

If a linear model

$$\mu_k = b_1 x_k^{(1)} + \cdots + b_J x_k^{(J)}, \qquad k = 1, \ldots, m,$$

is assumed, show that the argument of this section applies in the sense that putting $\hat{\sigma}^2 = \hat{D}/(n - J)$, the expression

$$w = \frac{n - J}{m - J}\left(\frac{\hat{\sigma}^2}{\hat{\sigma}_0^2}\right) - \frac{n - m}{m - J}$$

is the outcome of a variable with an $F_{m-J,n-m}$ distribution.

Problem 22 Using the results of Problem 21, select a reasonable grouping for the data in Table 10.1, Problem 6, and test for linearity.

Problem 23 Use the method of this section to see whether the data of Table 10.5 could have come from a sequence of independent identically distributed normal variables. That is, is it possible that the visually observed upward trend in the data can be reasonably interpreted as a random occurrence in data actually selected from the same underlying distribution? What other tests are available for checking this?

Least Squares Estimates in Nonstandard Situations

Usually, we know very little about the underlying distribution of the observations $y_1, \ldots, y_n$. We may hope that they are normal or nearly normal but without any real evidence. For normal distributions least squares estimates is a good estimation method. But is it possible for least squares to be badly thrown off by heavily skewed or discontinuous underlying distributions?

There is evidence that if the variances of the observations remain nearly constant, then least squares estimates are fairly good no matter what the underlying distributions. This evidence is partly provided by an interesting result known as the Gauss-Markov theorem.

Least squares estimates have two properties that we want to focus on.

a The estimates $\hat{\hat{b}}_1, \ldots, \hat{\hat{b}}_J$ are linear functions of the observations. That is, each one is of the form $\sum_k a_k y_k$.

b The estimates are unbiased:

$$E\hat{\hat{b}}_j = b_j, \quad j = 1, \ldots, J. \tag{10.43}$$

Look at a model for linear regression where we drop the assumption of normality, but keep all others: independence, common variance σ^2, and means linear in the factor levels. We want to estimate $b_1, \ldots, b_J$ by linear functions of the observations. Furthermore, we will look only at unbiased estimates. Now the problem is: *Among all linear unbiased estimates for* $b_1, \ldots, b_J$, *find the best,* (best meaning lowest expected squared error loss). Since the estimates we are looking at are all unbiased, squared error risk is equal to variance. So the problem becomes: *Among all linear unbiased estimators, find the one with minimum variance.* The answer is:

10.44 ***Theorem*** *The minimum variance linear unbiased estimates of* $b_1, \ldots, b_J$ *are the least squares estimates.*

We illustrate this in Problem 24.

What happens if the distributions are not normal and the variances are not constant? For example, take $\mathsf{Y}_1, \ldots, \mathsf{Y}_n$ to be exponentially distributed with means $\mu_k = bx_k$. The densities are

$$f_k(y) = \frac{1}{\mu_k} e^{-y/\mu_k}, \qquad y \geq 0,$$

and the variances are $V(\mathsf{Y}_k) = \mu_k^2$. The exponential distributions have very little resemblance to the normal distributions. So comparing the maximum likelihood and least squares estimates of b will give a rigorous test of least squares methods.

We get the maximum likelihood estimate by maximizing

$$L = \sum_1^n \log f_k(y_k) = -\sum_1^n \left(\frac{y_k}{\mu_k} + \log \mu_k\right).$$

The maximizing equation is

$$\frac{\partial L}{\partial b} = \sum_1^n \left(\frac{y_k}{b^2 x_k} - \frac{1}{b}\right) = 0.$$

Therefore

$$\hat{b} = \frac{1}{n} \sum_1^n \frac{y_k}{x_k}.$$

The variance is given by

$$V(\hat{\mathsf{b}}) = \frac{1}{n^2} V\left(\sum_1^n \frac{\mathsf{Y}_k}{x_k}\right) = \frac{1}{n^2} \sum_1^n \frac{1}{x_k^2} V(\mathsf{Y}_k) = \frac{b^2}{n}.$$

The least squares estimate minimizes

$$D = \sum_1^n (y_k - bx_k)^2.$$

A quick computation gives

$$\hat{\hat{b}} = \frac{\sum_1^n y_k x_k}{\sum_1^n x_k^2}.$$

Use the notations

$$\overline{x^2} = \frac{1}{n} \sum_1^n x_k^2$$

$$\overline{x^4} = \frac{1}{n} \sum_1^n x_k^4.$$

Then

$$V(\hat{\hat{\mathrm{b}}}) = \frac{1}{n^2(\overline{x^2})^2} \sum_1^n V(x_k \mathrm{Y}_k) = \frac{b^2}{n} \frac{\overline{x^4}}{(\overline{x^2})^2}.$$

Both estimates are unbiased, so their variances are also their squared error loss. The efficiency of $\hat{\hat{\mathrm{b}}}$ as compared to $\hat{\mathrm{b}}$ is $(\overline{x^2})^2/\overline{x^4}$. No matter what the values $x_1, \ldots, x_n$ are, this ratio is always ≤ 1. That is, maximum likelihood is always at least as good as least squares. In general, the more spread out the x_k are, the smaller the efficiency of least squares. This is not completely unexpected. Since $V(\mathrm{Y}_k) = b^2 x_k^2$, the more diverse the factor levels, the more unequal the variances. For example, if the factor levels are proportional to $1, 2, \ldots, n$, then the asymptotic efficiency of $\hat{\hat{\mathrm{b}}}$ compared to $\hat{\mathrm{b}}$ is .56. If the factor levels are proportional to the squares $1, 4, 9, \ldots, n^2$, then the efficiency is .36. If the levels are proportional to the cubes $1, 8, 27, \ldots, n^3$, the efficiency drops to .27.

Problem 24 In a linear regression model, suppose there is only one factor with levels $x_1, \ldots, x_n$ and $\mu_k = b x_k$.

a Show that the least squares estimate is

$$\hat{\hat{b}} = \frac{(\mathbf{x}, \mathbf{y})}{(\mathbf{x}, \mathbf{x})}.$$

b Let $\hat{b}$ be any linear estimate

$$\hat{b} = \sum a_k y_k = (\mathbf{a}, \mathbf{y}).$$

Show that $\hat{\mathrm{b}}$ is unbiased if and only if $(\mathbf{a}, \mathbf{x}) = 1$.

c Show that the variance of $\hat{\mathrm{b}}$ is

$$V(\hat{\mathrm{b}}) = \sigma^2 \sum a_k^2 = \sigma^2(\mathbf{a}, \mathbf{a}).$$

d For any $\mathbf{a}$, write $\mathbf{a}$ as

$$\mathbf{a} = \beta \mathbf{x} + \mathbf{z}$$

where $\mathbf{z}$ is perpendicular to $\mathbf{x}$. That is, decompose $\mathbf{a}$ into its component parallel to $\mathbf{x}$ and its component $\mathbf{z}$ such that $(\mathbf{x}, \mathbf{z}) = 0$. Use this to show that for all $\mathbf{a}$ such that $(\mathbf{a}, \mathbf{x}) = 1$,

$$(\mathbf{a}, \mathbf{a}) = \frac{1}{(\mathbf{x}, \mathbf{x})} + (\mathbf{z}, \mathbf{z}).$$

e Use **d.** to show that the vector $\mathbf{a}$ that minimizes $(\mathbf{a}, \mathbf{a})$ under the restraint $(\mathbf{a}, \mathbf{x}) = 1$ is

$$\mathbf{a} = \frac{\mathbf{x}}{(\mathbf{x}, \mathbf{x})}.$$

Conclude that the best unbiased linear estimator is

$$\hat{b} = \frac{(\mathbf{x}, \mathbf{y})}{(\mathbf{x}, \mathbf{x})}.$$

The Analysis of Variance

The analysis of variance model is similar to the regression analysis model. We pointed out earlier that the main difference was that in analysis of variance the factor levels are not necessarily numerical values. Another less essential difference is that often in analysis of variance, a number of observations are made at each set of factor levels.

The goals of regression analysis and analysis of variance are the same: to find what factors significantly effect the size of the observed variable and to estimate their effects. But the techniques are different. Regression analysis first estimates the coefficients $b_1, \ldots, b_J$ and then analyzes the effects of the various factors by working with these coefficients and the values of $\hat{D}$. In analysis of variance there are no such coefficients and no linear expression for the means in terms of numerical factor levels. The means are estimated directly and the analysis proceeds by working with these direct estimates.

As an example, consider the following fictitious experiment with data gotten from various operations on a random number table. The problem is to trace the effects of x ray radiation on the appearance of genetic irregularities in two generations of flies. Four groups of equal numbers of flies are set up. Of these, groups 3 and 4 are the offspring of flies that have been exposed to x ray radiation. Groups 1 and 2 have unexposed parent generations. Groups 2 and 4 now receive x ray radiation, and groups 1 and 3 are used as unexposed controls. The number of genetic irregularities is then counted. In twelve repetitions of this experiment, the numbers of genetic irregularities are:

Table **10.7**

Group			
1	**2**	**3**	**4**
34	81	43	95
53	50	118	58
73	58	42	68
33	110	78	84
31	54	55	101
10	71	60	44
6	117	25	72
11	35	76	109
69	56	90	120
52	119	26	79
50	123	93	139
41	72	36	66

If we wanted to, we could run a Kruskal-Wallis test on these four sets of observations. The result at the .05 level is that the observations come from different underlying distributions, so the radiation is having some effect. But two factors are operating here. One is the hereditary effect from the parent generation, and the other is the effect of the x ray radiation on the experimental groups. The real question we want to answer is: Which of these factors is making a significant contribution? It could be that the hereditary effect is negligible. At any rate, we can exhibit the data in a square layout (Table 10.8) which makes the interdependence of the factors more apparent.

Table **10.8**

		Factor **2**	
		Unexposed parent generation	**Exposed parent generation**
Factor **1**	**No x ray treatment**	34 ⋮ 41	43 ⋮ 36
	x ray treatment	81 ⋮ 72	95 ⋮ 66

The first factor is present generation x ray treatment. It has two levels

level 1: no x ray treatment;
level 2: x ray treatment.

The hereditary effect is the second factor with two levels

level 1: unexposed parent generation;
level 2: exposed parent generation.

Each of the four sub-squares in Table 10.8 contains observations taken at the same set of factor levels. We will refer to any such subsquare as a *cell*; equivalently, a group of observations taken at the same factor levels forms a cell. The model begins with the assumption:

10.45 *The observations are outcomes of independent normally distributed variables with common but unknown variance σ^2 and having the same mean within each cell.*

If we refer to the i, jth cell as that cell for which factor 1 has level i and factor 2 has level j, where $i = 1,2$, $j = 1,2$ then all the observations in the i, jth cell are the outcomes of independent $N(\mu_{ij}, \sigma^2)$ variables, where we use μ_{ij} to denote the common mean for the i, jth cell. The cell means μ_{ij} are unknown, as is σ^2, but notice that σ^2 is assumed to be the same for all cells.

We want to trace the effect of each factor on the means μ_{ij}. First, let's analyze what happens when the first factor changes levels. If the second factor is at level 1, changing the second factor from level 1 to level 2 increases the mean by

$$\Delta_1 = \mu_{21} - \mu_{11}.$$

For the second factor at its second level, changing the levels of the first factor increases the means by

$$\Delta_2 = \mu_{22} - \mu_{12}.$$

The layout of means looks like this:

		Factor **2**	
		1	2
Factor **1**	1	μ_{11}	μ_{12}
	2	$\mu_{11} + \Delta_1$	$\mu_{12} + \Delta_2$

The average increase in mean due to change in the levels of factor 1 is

$$\Delta = \frac{\Delta_1 + \Delta_2}{2}$$

and this quantity is called *the main effect due to factor 1.* Similarly, as factor 2 changes levels, we get the increases

$$\delta_1 = \mu_{12} - \mu_{11}$$
$$\delta_2 = \mu_{22} - \mu_{21}$$

for factor 1 at fixed levels 1 and 2, respectively. The layout of means is:

	Factor 1: 1	2
Factor 2: 1	μ_{11}	$\mu_{11} + \delta_1$
2	μ_{21}	$\mu_{21} + \delta_2$

The average increase

$$\delta = \frac{\delta_1 + \delta_2}{2}$$

is called *the main effect due to factor 2.*

Consider the notion of *interaction* between the two factors. It seems difficult to define exactly what interaction is and how much of an effect it has. So instead, we will define what no interaction is. If the presence or absence of the second factor has no effect on the operation of the first factor, then the amount by which the first factor raises the mean should be the same, regardless of the level the second factor is, and similarly for the second factor. This translates into

$$\Delta_1 = \Delta_2 = \Delta$$
$$\delta_1 = \delta_2 = \delta;$$

the increases due to changes in the level of one factor are the same for each level of the other factor. For no interaction, the means are

	1	2
1	μ_{11}	μ_{12}
2	$\mu_{11} + \Delta$	$\mu_{12} + \Delta$

or

	1	2
1	μ_{11}	$\mu_{11} + \delta$
2	μ_{21}	$\mu_{21} + \delta$

If in the first layout we substitute $\mu_{12} = \mu_{11} + \delta$, or in the second layout, $\mu_{21} = \mu_{11} + \Delta$, we get

	1	2
1	μ_{11}	$\mu_{11} + \delta$
2	$\mu_{11} + \Delta$	$\mu_{11} + \Delta + \delta$

No interaction between factors implies that they operate additively. If a change in factor 1 by itself produces a change Δ in the means, and a change in factor 2 by itself produces a change δ in the means, then changing both factors produces the change $\Delta + \delta$ in means, because, as you can see from the above layout,

$$\mu_{22} = \mu_{11} + \Delta + \delta.$$

Hence, *no interaction* means there is *additivity of effects.*

Roughly, analysis of variance is concerned with estimating and testing hypotheses concerning main effects and interactions between factors. For example, the hypothesis that exposure of the parent generation has no effect on the number of genetic irregularities can be translated as $\delta = 0$. The hypothesis of no interaction can be translated as

$$\mu_{22} = \mu_{11} + \delta_1 + \Delta_1.$$

Suppose you did not know any analysis of variance and you wanted to estimate the effects of the above two factors. The classical "scientific" approach is to keep all factors but one at their normal levels and vary that one, then repeat with the other factors. Suppose we carried out this approach on the above example using the same sample size ($n = 48$). Take three groups of the same size: the first one a control, the second one with parent generation exposed, and the third one with unexposed parent generation that we subject to x rays. Repeat this 16 times. Then we get a 3 column listing:

Experiment Number	Group 1	Group 2	Group 3
1	·	·	·
2	·	·	·
⋮	⋮	⋮	⋮
16	·	·	·

How does this compare to analysis of variance? Let the three group means be μ_1, μ_2, μ_3. The maximum likelihood estimates for Δ_1 and δ_1 are

$$\hat{\Delta}_1 = \hat{\mu}_3 - \hat{\mu}_1$$
$$\hat{\delta}_1 = \hat{\mu}_2 - \hat{\mu}_1,$$

where $\hat{\mu}_1, \hat{\mu}_2, \hat{\mu}_3$ are sample means. Thus, the estimate for $\hat{\Delta}_1$ is based on the first and third columns and utilizes only two-thirds of the data. Similarly for the estimate of δ_1. Now, if we wanted to estimate Δ_1 and δ_1 in the analysis of variance layout, we would take

$$\hat{\Delta}_1 = \hat{\mu}_{21} - \hat{\mu}_{11}$$
$$\hat{\delta}_1 = \hat{\mu}_{12} - \hat{\mu}_{11}.$$

These estimates are based on the difference of sample means of 2 groups of 12 each with variance σ^2. Naturally they will not be as accurate as the former estimates which are based on the difference of sample means of 2 groups of 16 each, with the same variance σ^2. But if we estimate instead the main effect Δ, we have

$$\hat{\Delta} = \frac{\hat{\Delta}_1 + \hat{\Delta}_2}{2}$$
$$= \tfrac{1}{2}(\hat{\mu}_{21} - \hat{\mu}_{11} + \hat{\mu}_{22} - \hat{\mu}_{12}).$$

The estimate $\hat{\Delta}$ uses *all* the data and involves the difference of means of 2 groups of 24 each. If there is no interaction, then $\Delta = \Delta_1$, and the estimate $\hat{\Delta}$ above is more accurate (there is smaller squared error loss) than the estimate based on the classical "hold all factors but one constant" approach. Furthermore, the classical approach gives no information about the possible interaction between factors. Analysis of variance (and regression analysis) were the tools involved in the recognition that, in designing an experiment, more accurate information can be obtained by varying all factor levels simultaneously. But it is not simply that this is a preferred way of designing an experiment. The major reason for the development of analysis of variance was the need to analyze experiments (mainly agricultural) in which it was simply impossible to hold all but one factor level constant.

Problem 25 A 2×2 layout like the example of this section with no interaction can be translated into a linear regression model by taking

$$\mathbf{x}^{(1)} = (0, 0, 1, 1), \qquad \mathbf{x}^{(2)} = (0, 1, 0, 1), \qquad \mathbf{x}^{(3)} = (1, 1, 1, 1)$$

as factor levels. In fact, linearity is equivalent to no interaction. Consider the two statements:

a There is no interaction if there are numbers Δ, δ such that

$$\begin{aligned} \mu_{12} &= \delta + \mu_{11} \\ \mu_{21} &= \Delta + \mu_{11} \\ \mu_{22} &= \delta + \Delta + \mu_{11}. \end{aligned}$$

b A linear regression model with the above factor vectors applies if there are numbers b_1, b_2, b_3, b_4 such that the four means are given by

$$\begin{aligned} \mu_{11} &= b_1 x_1^{(1)} + b_2 x_1^{(2)} + b_3 x_1^{(3)} \\ \mu_{12} &= b_1 x_2^{(1)} + b_2 x_2^{(2)} + b_3 x_2^{(3)} \\ \mu_{21} &= b_1 x_3^{(1)} + b_2 x_3^{(2)} + b_3 x_3^{(3)} \\ \mu_{22} &= b_1 x_4^{(1)} + b_2 x_4^{(2)} + b_3 x_4^{(3)}. \end{aligned}$$

To see the equivalence of **a.** and **b.**, substitute the given values of $x^{(j)}$ into the regression equations, setting

$$\begin{aligned} \mu_{11} &= b_3 \\ \mu_{12} &= b_2 + b_3 \\ \mu_{21} &= b_1 + b_3 \\ \mu_{22} &= b_1 + b_2 + b_3. \end{aligned}$$

Suppose there are numbers b_1, b_2, b_3 for which these last equations hold. Then put $\Delta = b_1$, $\delta = b_2$ to see that we get the no interaction expressions for the means. Conversely, if the latter hold, then put $b_1 = \Delta$, $b_2 = \delta$, $b_3 = \mu_{11}$ to get the linear regression equations.

a′ Using the above remarks, test the data of Table 10.7 for no effect due to factor 2 at the .05 level (assuming no interaction).

b′ Since linearity of the means in the factor vectors $\mathbf{x}^{(1)}$, $\mathbf{x}^{(2)}$, $\mathbf{x}^{(3)}$ is equivalent to no interaction, test the hypothesis of zero interaction at the .05 level by testing goodness of fit of the linear model.

The Two-Factor Layout

Suppose that we have factor 1 with I levels, factor 2 with J levels, and m observations per cell. How do we estimate and test for main effects and interaction? Let μ_{ij} be the mean for the variables in the i, jth cell. The overall mean μ is the average of the cell means

$$\mu = \frac{1}{IJ} \sum_{i,j} \mu_{ij}.$$

Define the mean $\mu_i^{(1)}$ for the ith level of the first factor as the average over all cells where factor 1 is at the ith level:

$$\mu_i^{(1)} = \frac{1}{J} \sum_j \mu_{ij}.$$

Similarly, define the mean for the jth level of factor 2 by

$$\mu_j^{(2)} = \frac{1}{I} \sum_i \mu_{ij}.$$

Now, define the *main effect* $v_i^{(1)}$ *of the ith level of factor* 1 as

$$v_i^{(1)} = \mu_i^{(1)} - \mu.$$

This is the excess of the mean for the ith level over the overall level. Similarly, define the *main effect* $v_j^{(2)}$ *of the jth level of factor* 2 by

$$v_j^{(2)} = \mu_j^{(2)} - \mu.$$

Notice the identities

$$\sum_i v_i^{(1)} = 0, \qquad \sum_j v_j^{(2)} = 0.$$

Write the mean as

10.46
$$\mu_{ij} = \mu + v_i^{(1)} + v_j^{(2)} + \gamma_{ij}.$$

This decomposes the mean of the i, jth cell into the sum of:

a the overall mean μ;

b the main effect due to the ith level of factor 1;

c the main effect due to the jth level of factor 2;

d the residual term γ_{ij} called the interaction between the ith level of factor 1 and the jth level of factor 2.

In the last section we discussed the idea of no interaction between factors and defined it as follows: There is no interaction if a change in the level of the first factor from i to l produces the same increase in mean regardless of the level of the second factor. That is, the differences

10.47
$$\mu_{lj} - \mu_{ij}$$

are the same for all values $j = 1, \ldots, J$. The analogous condition holds for the second factor—the differences

10.48
$$\mu_{ih} - \mu_{ij}$$

are the same for all values $i = 1, \ldots, I$.

If there is no interaction in this sense, then it is easy to show that

10.49 ***Proposition*** *No interaction is equivalent to* $\gamma_{ij} = 0$ *for all i, j, or*

$$\mu_{ij} = \mu + v_i^{(1)} + v_j^{(2)}.$$

Thus the cell means are the sum of the overall mean and the main effects. The effect of the ith level of factor 1 and the jth level of factor 2 simply add together, and the factor effects are said to be additive.

Consider the situation where there are an equal number of observations per cell, say N. With IJ cells, the total sample size is $n = NIJ$. Two problems come first:

a Assuming there is no interaction, test the hypothesis that factor 1, say, has no effect.

b Test the hypothesis of no interaction.

Tests for these can be constructed by using methods we have already developed. There are n observations $y_1, \ldots, y_n$. Of these, the first N are in the 1,1 cell, the second N in the 1,2 cell, and so on. Let the notation $k \in C_{i,j}$ stand for all subscripts k such that y_k is an observation in the i, jth cell. First of all, we want to get estimates for μ, $v_i^{(1)}$, and $v_j^{(2)}$. To get least squares (equivalently, maximum likelihood) estimates write

10.50
$$D = \sum_{i,j} \sum_{k \in C_{ij}} (y_k - \mu - v_i^{(1)} - v_j^{(2)})^2$$

and minimize over the parameters μ, $v_i^{(1)}$, and $v_j^{(2)}$. Taking partial derivatives with respect to each parameter subject to the restraints that $v_i^{(1)}$ and $v_2^{(2)}$ sum to zero, and setting the partials equal to zero gives the not unexpected results

10.51
$$\hat{\mu} = \frac{1}{n} \sum_{1}^{n} y_k$$

$$\hat{v}_i^{(1)} = \frac{1}{NJ} \sum_{j} \sum_{k \in C_{ij}} (y_k - \hat{\mu})$$

$$\hat{v}_j^{(2)} = \frac{1}{NI} \sum_{i} \sum_{k \in C_{ij}} (y_k - \hat{\mu}).$$

Notice that $\hat{v}_i^{(1)}$ is the sample mean of $y_k - \hat{\mu}$ for all observations y_k made at the ith level of factor 1, and analogously for $\hat{v}_j^{(2)}$. Denote, as before, the minimum value of D by

10.52
$$\hat{D} = \sum_{i,j} \sum_{k \in C_{ij}} (y_k - \hat{\mu} - \hat{v}_i^{(1)} - \hat{v}_j^{(2)})^2.$$

Now, let the numbers μ, $v_i^{(1)}$, $v_j^{(2)}$, $i = 1, \ldots, I$, $j = 1, \ldots, J$, range over all possible values subject only to the two restrictions $\sum_i v_i^{(1)} = 0$, $\sum_j v_j^{(2)} = 0$. As they do this, the vector $\tilde{\mathbf{y}}$ with coordinates $(\tilde{y}_1, \ldots, \tilde{y}_n)$ such that for all $k \in C_{i,j}$

$$\tilde{y}_k = \mu + v_i^{(1)} + v_j^{(2)}$$

ranges over a linear space $\mathscr{L}_{I+J-1}$ of dimension $I + J - 1$. Roughly, μ gives 1 dimension, the $v_1^{(1)}, \ldots, v_I^{(1)}$ under one restriction give $I - 1$ dimensions, and the $v_1^{(2)}, \ldots, v_J^{(2)}$ give $J - 1$ dimensions. We get $\hat{D}$ by minimizing

$$D = \|\mathbf{y} - \tilde{\mathbf{y}}\|$$

as $\tilde{\mathbf{y}}$ ranges over $\mathscr{L}_{I+J-1}$. Therefore, applying Theorem 10.34 a. gives

10.53 ***Proposition*** *If there is no interaction*

$$\hat{\sigma}^2 = \frac{\hat{D}}{n - I - J + 1}$$

is an unbiased estimator for σ^2.

What does it mean to say factor 1 has no effect? In terms of the means μ_{ij} it translates as—the means μ_{ij} are not effected by a change in i. That is,

10.54 $$\mu_{1j} = \mu_{2j} = \cdots = \mu_{Ij}, \qquad j = 1, \ldots, J.$$

With a little elementary algebraic manipulation, we find that 10.54 is equivalent to

10.55 $$v_i^{(1)} = 0, \qquad i = 1, \ldots, I.$$

No effect from factor 1 translates into the hypothesis that all the main effects $v_1^{(1)}, v_2^{(1)}, \ldots, v_I^{(1)}$ are zero. The measure of the fit under this hypothesis is the minimum value of

$$D_1 = \sum_{i,j} \sum_{k \in C_{ij}} (y_k - \mu - v_j^{(2)})^2$$

over all possible values of μ, $v_j^{(2)}$ subject to the one restraint $\sum_j v_j^{(2)} = 0$. The least squares estimates for μ and $v_j^{(2)}$ are the same as before. If $\hat{D}_1$ denotes the minimum value of D_1, then $\hat{D}_1$ is the minimum of $\|\mathbf{y} - \tilde{\mathbf{y}}\|^2$ where $\tilde{\mathbf{y}}_1$ ranges over a J-dimensional linear subspace of $\mathscr{L}_{I+J-1}$. Therefore

10.56 ***Proposition***

$$\frac{n - I - J + 1}{I - 1}\left(\frac{\hat{D}_1 - \hat{D}}{\hat{D}}\right)$$

has an $F_{I-1, n-I-J+1}$ *distribution.*

We test for no effect by rejecting if the above ratio is too large. The way to develop a test for no effect of factor 2 should now be clear, and we leave it to the reader.

If $N \geq 2$, we can get an intrinsic estimate of the variance by using the observations within each cell to get an estimate of σ^2 and then averaging these estimates over all cells. Since we can get an intrinsic estimate of σ^2, this implies that we can test how well the assumption of no interaction fits.

Without the assumption of no interaction, the "residual error" is given by the minimum of

$$D_0 = \sum_{i,j} \sum_{k \in C_{ij}} (y_k - \mu_{ij})^2$$

over all values of μ_{ij}. The least squares estimates for μ_{ij} are simply the cell means

$$\hat{\mu}_{ij} = \frac{1}{N} \sum_{k \in C_{ij}} y_k.$$

Then

$$\hat{D}_0 = \sum_{i,j} \sum_{k \in C_{ij}} (y_k - \hat{\mu}_{ij})^2,$$

so that $\hat{D}_0$ is, except for a factor of $(N - 1)IJ$, just the average over all cells of the unbiased variance estimates

$$\hat{\sigma}_{ij}^2 = \frac{1}{N - 1} \sum_{k \in C_{ij}} (y_k - \hat{\mu}_{ij})^2,$$

computed within each cell.

In vector language, $\hat{D}_0$ is the minimum of $\|\mathbf{y} - \hat{\mathbf{y}}_0\|$, where $\tilde{\mathbf{y}}_0$ ranges over all n-vectors such that all coordinates with subscript k falling into the same cell are equal. Thus, $\tilde{\mathbf{y}}_0$ ranges over an IJ-dimensional vector space. Under the hypothesis of no interaction, $\hat{D}$ is the minimum of $\|\mathbf{y} - \tilde{\mathbf{y}}\|^2$ where $\tilde{\mathbf{y}}$ ranges over an $(I + J - 1)$-dimensional subspace of the $\hat{\mathbf{y}}_0$ space. Using Theorem 10.34 again we get

10.57 ***Proposition*** *Under the hypothesis of no interaction,*

$$\frac{(N - 1)IJ}{(I - 1)(J - 1)} \frac{\hat{D} - \hat{D}_0}{\hat{D}_0}$$

has an $F_{(I-1)(J-1),(N-1)IJ}$ *distribution.*

The coefficient of the ratio came from

$$n - IJ = (N - 1)IJ,$$
$$IJ - I - J + 1 = (I - 1)(J - 1).$$

Unfortunately, Proposition 10.57 cannot be used if there is only one observation per cell. In this case there is another, more complicated, test that can be used, or, if the change in mean from cell to cell is not too large, cells can be combined into larger cells.

Problem 26 A. Hald in *Statistical Theory with Engineering Applications* gives data consisting of the logarithms of the observed permeability of 81 sheets of building material, selected over 9 days from 3 machines, 3 samples from each machine per day.
(John Wiley and Sons, 1952)

Table **10.9**

Day	Machine	log of Permeability		
1	1	1.404	1.346	1.618
	2	1.306	1.628	1.410
	3	1.932	1.674	1.399
2	1	1.447	1.569	1.820
	2	1.241	1.185	1.516
	3	1.426	1.768	1.859
3	1	1.914	1.477	1.894
	2	1.506	1.575	1.644
	3	1.382	1.690	1.361
4	1	1.887	1.485	1.392
	2	1.673	1.372	1.114
	3	1.721	1.528	1.371
5	1	1.772	1.728	1.545
	2	1.227	1.397	1.531
	3	1.320	1.489	1.336
6	1	1.665	1.539	1.680
	2	1.404	1.452	1.627
	3	1.633	1.612	1.359
7	1	1.918	1.931	2.129
	2	1.229	1.508	1.436
	3	1.328	1.802	1.385
8	1	1.845	1.790	2.042
	2	1.583	1.627	1.282
	3	1.689	2.248	1.795
9	1	1.540	1.428	1.704
	2	1.636	1.067	1.384
	3	1.703	1.370	1.839

Find out if

a there is a significant difference in log permeability between machines;

b there is a significant difference in log permeability from day to day.

In setting up **a.** and **b.** assume no interaction.

c Now test for zero interaction. Set the levels yourself.

Decomposing the Sum of Squares

There is another approach and notation in analysis of variance. Let a dot as a subscript mean the average over the missing subscript. For instance $\hat{\hat{\mu}}_{.j}$ is the sample mean of all observations at the jth level of factor 2:

$$\hat{\hat{\mu}}_{.j} = \frac{1}{NI} \sum_i \sum_{k \in C_{ij}} y_k .$$

It is also called the mean of the jth column. Similarly

$$\hat{\hat{\mu}}_{i.} = \frac{1}{NJ} \sum_j \sum_{k \in C_{ij}} y_k$$

is the sample mean of all observations at the ith level of factor 1, and is called the mean of the ith row. In terms of our old notation,

$$\hat{\hat{v}}_i^{(1)} = \hat{\hat{\mu}}_{i.} - \hat{\hat{\mu}}$$
$$\hat{\hat{v}}_j^{(2)} = \hat{\hat{\mu}}_{.j} - \hat{\hat{\mu}} .$$

That is, the main effect of the ith level is estimated as the ith row mean minus the overall mean, and similarly for the main effect of the jth level of factor 2.

Our test of the hypothesis of no factor 1 effect, also called no row effect, is based on the ratio of $\hat{D}_1 - \hat{D}$ to $\hat{D}$, where

$$\hat{D} = \sum_{i,j} \sum_{k \in C_{ij}} (y_k - \hat{\hat{\mu}} - \hat{\hat{v}}_i^{(1)} - \hat{\hat{v}}_j^{(2)})^2$$

and

$$\hat{D}_1 = \sum_{i,j} \sum_{k \in C_{ij}} (y_k - \hat{\hat{\mu}} - \hat{\hat{v}}_j^{(2)})^2.$$

We can simplify $\hat{D}_1 - \hat{D}$ by writing

$$\hat{D}_1 = \sum_{i,j} \sum_{k \in C_{ij}} (y_k - \hat{\hat{\mu}} - \hat{\hat{v}}_i^{(1)} - \hat{\hat{v}}_j^{(2)} + \hat{\hat{v}}_i^{(1)})^2.$$

Expand the square into:

$$(y_k - \hat{\hat{\mu}} - \hat{\hat{v}}_i^{(1)} - \hat{\hat{v}}_j^{(2)})^2 + 2(y_k - \hat{\hat{\mu}} - \hat{\hat{v}}_i^{(1)} - \hat{\hat{v}}_j^{(2)})\hat{\hat{v}}_i^{(1)} + (\hat{\hat{v}}_i^{(1)})^2.$$

Summing gives

$$\hat{D}_1 = \hat{D} + \sum_{i,j} \sum_{k \in C_{ij}} (\hat{\hat{v}}_i^{(1)})^2$$
$$+ 2 \sum_{i,j} \sum_{k \in C_{ij}} (y_k - \hat{\hat{\mu}} - \hat{\hat{v}}_i^{(1)} - \hat{\hat{v}}_j^{(2)})\hat{\hat{v}}_i^{(1)}.$$

A bit of manipulation will show that the third term is zero. The second term equals

$$NJ \cdot \sum_i (\hat{\nu}_i^{(1)})^2.$$

In the dot notation, then,

10.58 $$\hat{D}_1 - \hat{D} = NJ \sum_i (\hat{\mu}_{i.} - \hat{\mu})^2.$$

The same sort of computation carried out for no factor 2 effect, or no column effect, gives

10.59 $$\hat{D}_2 - \hat{D} = NI \sum_j (\hat{\mu}_{.j} - \hat{\mu})^2,$$

where

$$\hat{D}_2 = \sum_{i,j} \sum_{k \in C_{ij}} (y_k - \hat{\mu} - \hat{\nu}_i^{(1)})^2.$$

The *total sum of squares* of the differences of the observations from the overall mean is

$$\sum_k (y_k - \hat{\mu})^2.$$

By subtracting and adding the same thing, write

$$(y_k - \hat{\mu})^2 = [(y_k - \hat{\mu}_{i.} - \hat{\mu}_{.j} + \hat{\mu}) + (\hat{\mu}_{i.} - \hat{\mu}) + (\hat{\mu}_{.j} - \hat{\mu})]^2$$

Expanding out this square, summing, and noticing that cross-product terms vanish gives

10.60 $$\sum_k (y_k - \hat{\mu})^2 = NJ \sum_i (\hat{\mu}_{i.} - \hat{\mu})^2 + NI \sum_j (\hat{\mu}_{.j} - \hat{\mu})^2 + \sum_{i,j} \sum_{k \in C_{ij}} (y_k - \hat{\mu}_{i.} - \hat{\mu}_{.j} + \hat{\mu})^2.$$

This is the fundamental decomposition of the total sum of squares. It exhibits the sum of squares as the sum of

a NJ times the sum of the squares of the differences of the row means from the overall means:

$$NJ \sum_i (\hat{\mu}_{i.} - \hat{\mu})^2;$$

b NI times the sum of the squares of the differences of the column means from the overall means:

$$NI \sum_j (\hat{\mu}_{.j} - \hat{\mu})^2;$$

c the residual sum of squares:

$$\sum_{i,j} \sum_{k \in C_{ij}} (y_k - \hat{\mu}_{i.} - \hat{\mu}_{.j} + \hat{\mu})^2.$$

Notice that

$$\hat{\mu}_{i.} + \hat{\mu}_{.j} - \hat{\mu} = \hat{v}_i^{(1)} + \hat{v}_j^{(2)} + \hat{\mu},$$

so the residual sum of squares is actually $\hat{D}$.

Suppose we came across this decomposition of the total sum of squares and wanted to test for no row effect or for no column effect. If there is no row effect then the row means $\hat{\mu}_{i.}$ should be fairly close to $\hat{\mu}$, so the contribution to the total sum of squares due to the sum of the squares

$$NJ \sum_i (\hat{\mu}_{i.} - \hat{\mu})^2$$

"between rows" should be small compared, say, to the residual sum of the squares. Therefore, it seems reasonable to base a test on the ratio of a. to c. But this is exactly the test developed in the last section, except for the constant $(N - 1)IJ/(I - 1)(J - 1)$ needed to give the ratio an F distribution.

Thus, an alternative approach to testing for the various effects is to decompose the total sum of squares into the part due to the row effect, the part due to the column effect, and the residual. This approach works with many factors. You can decompose the total sum of squares into the effects due to each factor separately plus residual and possibly interaction terms. The size of the contribution to the sum of the squares due to some factor is then a measure of the effect of that factor.

This alternative approach to analysis of variance has a more intuitive feeling. It makes it clear why the method is called analysis of variance, since the approach is to take the total sum of squares, which is essentially the sample variance (multiplied by $n - 1$) and analyze it into the contributions to the sample variance due to the various factors and their interactions. I happen to prefer the approach of least squares estimation and Theorem 10.34, but this approach lacks the immediate appeal of the decomposition of the total sum of squares.

At any rate, this section gives some important computational results. To do tests for no effect in a two-way layout with equal numbers of observations in each cell, compute the row and column means $\hat{\mu}_{i.}$, $\hat{\mu}_{.j}$, and the overall mean $\hat{\mu}$. Then compute the sum of squares between rows, between columns, and the residual sum of squares.

Simultaneous Confidence Intervals

If you have tested an analysis of variance layout and, for instance, rejected zero effect for factor 1, then the next step is to estimate the effects of factor 1 and put confidence intervals around the estimates. For example, we might want to put confidence intervals around the overall mean μ

and the main effects $v_1^{(1)}, \ldots, v_I^{(1)}$. Or, we might want to put confidence intervals around the means $\mu_1^{(1)}, \ldots, \mu_I^{(1)}$. Or perhaps only three of four of the levels are of interest. Or perhaps we want to put confidence intervals around the various differences of the means, for example, around $\mu_3^{(1)} - \mu_2^{(1)}$, to see how a change in level increases the mean.

We are interested not only in the parameters of main effect and mean but also in various linear combinations of these. For instance,

$$\mu_i^{(1)} = \mu + v_i^{(1)}, \qquad i = 1, 2, \ldots, I$$

are sums of the factor 1 main effects and the overall mean μ. The differences in means $\mu_3^{(1)} - \mu_2^{(1)}$ equals the differences in the main effects $v_3^{(1)} - v_2^{(1)}$. In general, we suppose that the parameters we are interested in consist of some selected subset of

$$\mu, v_1^{(1)}, \ldots, v_I^{(1)}, v_1^{(2)}, \ldots, v_J^{(2)}$$

together with various linear combinations of the selected parameters.

We get least squares estimates for these combinations in the obvious way: if θ is any linear combination, e.g., if

$$\theta = 5\mu + 3v_1^{(1)} + \cdots - 7v_J^{(2)},$$

then

$$\hat{\hat{\theta}} = 5\hat{\mu} + 3\hat{v}_1^{(1)} + \cdots - 7\hat{v}_J^{(2)}.$$

We also want to estimate the variances of such $\hat{\hat{\theta}}$. The estimates for all the main effects and overall mean are linear functions of the y_k. For instance, $\hat{\mu} = (\sum_k y_k)/n$. Therefore, since $\hat{\hat{\theta}}$ is a linear combination of the main effect and mean estimates, it has the form

$$\hat{\hat{\theta}} = \sum a_k y_k.$$

The y_k are outcomes of independent $N(\mu_{ij}, \sigma^2)$ variables; hence the variance of the random variable $\hat{\hat{\boldsymbol{\theta}}} = \sum_k a_k Y_k$ is

$$V(\hat{\hat{\boldsymbol{\theta}}}) = \sigma^2 \sum_k a_k^2.$$

We know or can compute the a_k, and, assuming no interaction,

$$\hat{\sigma}^2 = \frac{\hat{D}}{n - I - J + 1}$$

is an unbiased estimate of σ^2. Thus, an unbiased estimate for the variance of $\hat{\hat{\boldsymbol{\theta}}}$ is

$$\hat{\sigma}^2_{\hat{\hat{\theta}}} = \hat{\sigma}^2 \sum_n a_k^2.$$

Now the problem is to get simultaneous confidence intervals for the selected main effects and mean parameters as well as the various linear combinations of these. One solution is provided by the S-method in essentially the same way as in regression.

To begin with, because of the two restrictions $\sum_i v_i^{(1)} = 0$ and $\sum_j v_j^{(2)} = 0$, the group of main effect and mean parameters we selected for study may not be linearly independent. For instance, the group of $I + 1$ parameters $\mu, v_1^{(1)}, \ldots, v_I^{(1)}$ contains only I linearly independent parameters. We assume that the group selected contains exactly M linearly independent parameters.

Now, for preset γ, select f to satisfy

$$P(\mathsf{F} \leq f) = \gamma,$$

where F has an $F_{M,n-J-I+1}$ distribution. Then

10.61 ***Theorem*** *The intervals*

$$J(\hat{\hat{\theta}}) = [\hat{\hat{\theta}} - \sqrt{Mf}\,\hat{\sigma}_{\hat{\hat{\theta}}}, \hat{\hat{\theta}} + \sqrt{Mf}\,\hat{\sigma}_{\hat{\hat{\theta}}}]$$

form a set of $100\gamma\%$ *simultaneous confidence intervals for all linear combinations* θ *of the selected* M *parameters.*

For example, suppose $I = 2$ and we want to study the parameters $\mu, v_1^{(1)}, v_2^{(1)}$. Then $M = 2$ and we use the $F_{2,n-J-1}$ distribution to get δ. We can get simultaneous $100\gamma\%$ confidence intervals on $v_1^{(1)}, v_2^{(1)}$ from Theorem 10.61. But the theorem says more than this. It also says that the probability is γ that the confidence intervals for all linear combinations of $\mu, v_1^{(1)}, v_2^{(1)}$ will cover the true parameter value. Thus, for instance, the probability is γ that simultaneously

$$\begin{gathered}
\mu \in J(\hat{\hat{\mu}}) \\
v_1^{(1)} \in J(\hat{\hat{v}}_1^{(1)}) \\
v_2^{(1)} \in J(\hat{\hat{v}}_2^{(1)}) \\
\mu_1^{(1)} \in J(\hat{\hat{\mu}}_1^{(1)}) \\
\mu_2^{(1)} \in J(\hat{\hat{\mu}}_2^{(1)}) \\
\mu_2^{(1)} - \mu_1^{(1)} \in J(\hat{\hat{\mu}}_2^{(1)} - \hat{\hat{\mu}}_1^{(1)})
\end{gathered}$$

and so on.

Suppose that one cannot reject the hypothesis of no interaction and wants to estimate the interaction γ_{ij}. From the definition, γ_{ij} satisfies

$$\sum_i \gamma_{ij} = 0, \qquad j = 1, \ldots, J$$

$$\sum_j \gamma_{ij} = 0, \qquad i = 1, \ldots, I.$$

There are thus $IJ - I - J$ linearly independent parameters in the set γ_{ij}. Assume $N \geq 2$. The least squares estimate for γ_{ij} is

$$\hat{\hat{\gamma}}_{ij} = \hat{\hat{\mu}}_{ij} - \hat{\hat{\mu}} - \hat{\hat{v}}_i^{(1)} - \hat{\hat{v}}_j^{(2)},$$

where

$$\hat{\hat{\mu}}_{ij} = \frac{1}{N} \sum_{k \in C_{ij}} y_k.$$

The appropriate estimate for σ^2 is

10.62 $$\hat{\sigma}^2 = \frac{1}{(N-1)IJ} \sum_{i,j} \sum_{k \in C_{ij}} (y_k - \hat{\hat{\mu}}_{ij})^2.$$

Using this estimate, we can compute the estimate $\hat{\sigma}^2_{\hat{\hat{\gamma}}_{ij}}$ for the variance of $\hat{\hat{\gamma}}_{ij}$. Now the S-method gives the simultaneous intervals

$$J(\hat{\hat{\gamma}}_{ij}) = [\hat{\hat{\gamma}}_{ij} - \sqrt{Mf}\,\hat{\sigma}_{\hat{\hat{\gamma}}_{ij}}, \hat{\hat{\gamma}}_{ij} + \sqrt{Mf}\,\hat{\sigma}_{\hat{\hat{\gamma}}_{ij}}]$$

where $M = IJ - I - J$; and f is computed from an $F_{M,n-IJ}$ distribution.

***Problem** 27* Find simultaneous 95% confidence intervals for the overall mean, main effects, and increase in mean due to x ray treatment for the data of Table 10.7, p. 355.

Other Analysis of Variance Layouts

There is a multitude of other analysis of variance layouts. But the methods that we used for analyzing the two-factor layout extend to all the other layouts.

For example, suppose we have a three-factor layout with levels

first factor: $1, \ldots, I$
second factor: $1, \ldots, J$
third factor: $1, \ldots, L$

and N observations per cell. The cell mean μ_{ijl} can be decomposed:

10.63 $$\mu_{ijl} = \mu + v_i^{(1)} + v_j^{(2)} + v_l^{(3)} + \gamma_{ij}^{(12)} + \gamma_{il}^{(13)} + \gamma_{jl}^{(23)} + \gamma_{ijl}$$

(see Problem 28). The three next to last terms are the two-factor interactions and γ_{ijl} is the three-factor interaction. You should be able to figure out how to test for no effect of the third factor, assuming no two- or three-factor interactions. Or, how would you test for no interaction between the first and third factors, assuming no three-factor interaction? Use least squares concepts and check the dimensionality of the vector spaces you are minimizing over (see Problem 29).

Once you see that the same method works on the three-factor layout, you realize that there is no more trouble, other than the formidable computations, in analyzing the four-factor or five-factor layout, and so on. This type of layout, in which equal numbers of observations are made at every possible combination of the various factor levels, is called a *factorial design.*

In factorial designs the number of cells and observations increases rapidly as the number of factors increases. To do a factorial layout on three factors each having six levels would require $216 = 6^3$ cells—a fairly extensive experiment. Layouts have been developed to cut down the number of cells required for multifactor experiments. To give you the flavor of these, we will discuss a layout for three factors all having the same number of levels. This layout goes by the name of *Latin square.*

Suppose each factor has six levels. Draw a two-way layout for factors 1 and 2, say,

		Factor 2					
		1	2	3	4	5	6
	1						
	2						
Factor 1	3						
	4						
	5						
	6						

In the cells of the first row write 1, 2, 3, 4, 5, 6. In the second row, begin with 2 and write 2, 3, 4, 5, 6, 1. The next row is 3, 4, 5, 6, 1, 2, and so on. We get the following layout:

		Factor 2					
		1	2	3	4	5	6
	1	1	2	3	4	5	6
	2	2	3	4	5	6	1
Factor 1	3	3	4	5	6	1	2
	4	4	5	6	1	2	3
	5	5	6	1	2	3	4
	6	6	1	2	3	4	5

Notice that each number appears once in each row and once in each column. We use the square as follows:

> *To level i of factor 1 and level j of factor 2 associate the level l of factor 3 where l is the number in the i, jth cell.*

Now make your observations at the levels i, j, l. Instead of 216 cells, use of a Latin square as above (each number appearing once in every row and column) reduces the number of cells to 36. Do we still have enough different combinations of factors to analyze main effects and interactions? Look at it this way: Each level of each factor shows up in six different cells. Main effects can be estimated by averaging over these six cells. But level 4 of factor 1, say, and level 5 of factor 2 only occur together with level 2 of factor 3. Hence, there is no way of estimating, without further assumptions, the two-factor interaction $\gamma_{4,5}^{(12)}$. Similarly, since a large number of combinations i, j, l simply do not occur, there is no way, in general, of estimating the three-factor interactions. In a Latin square layout with equal numbers of observations in each cell we generally assume that there are no interactions, either double or triple, so that

$$\mu_{ijl} = \mu + v_i^{(1)} + v_j^{(2)} + v_l^{(3)}.$$

If there is more than one observation per cell, then we can get an intrinsic estimate for σ^2 and check the goodness of fit of the no interaction assumption.

But, again, the methods you already know for testing hypotheses and constructing confidence intervals carry over to Latin square designs without modification. The only additional pieces of information necessary are the dimensions of the vector spaces you are minimizing over (see Problem 30).

Problem 28 Given cell means μ_{ijl}, $i = 1, \ldots, I, j = 1, \ldots, J$, $l = 1, \ldots, L$, define the overall mean μ as

$$\mu = \frac{1}{IJL} \sum_{i,j,l} \mu_{ijl},$$

and the excess of the cell mean over the overall mean by

$$v_{ijl} = \mu_{ijl} - \mu.$$

The main effects are defined by

$$v_i^{(1)} = \frac{1}{JL} \sum_{j,l} v_{ijl}$$

$$v_j^{(2)} = \frac{1}{IL} \sum_{i,l} v_{ijl}$$

$$v_l^{(3)} = \frac{1}{IJ} \sum_{i,j} v_{ijl}.$$

The second order interactions are defined as follows: Let

$$v_{ij}^{(12)} = \frac{1}{L} \sum_{l} v_{ijl},$$

then

$$\gamma_{ij}^{(12)} = v_{ij}^{(12)} - v_i^{(1)} - v_j^{(2)},$$

and similarly for the others. Finally, γ_{ijk} is defined as the difference between v_{ijl} and the sum of the main effects and second order interactions.

a Show that 10.63 holds.

b Show that for either the second or third order interactions, the sum over any one subscript is zero. For instance, show that

$$\sum_j v_{ij}^{(12)} = 0.$$

Problem 29 Look at a three-factor factorial design with N observations per cell C_{ijl}.

a Assuming no two- or three-factor interaction, discuss completely the form of the test for no factor 1 effect and its distribution.

b Assuming no three-factor interaction, what are the least squares estimates for the two-factor interactions? Write down the test for no interaction between factors 1 and 2, and give its distribution.

Problem 30 In a Latin square design where each factor has six levels and there are N observations per cell, set up and find the distribution of the test for no factor 1 effect.

Robustness

In the analysis of variance and regression analysis models, there are three bothersome assumptions about the variables $Y_1, \ldots, Y_n$:

a normality,

b equal variances,

c independence.

What effect do departures from these three assumptions have on the various test levels and confidence intervals we have derived in this chapter?

For a. and b. the answer is, fortunately, very little. Non-normality and inequality of variances, as long as they are not too drastic, do not have a marked effect on the distribution of the ratios that have an F distribution under the hypotheses a. and b. But correlation between the observations can have much more disastrous effects. Therefore, watch out for dependence.

If there are enough observations in each cell, then these can be tested for serial independence. Furthermore, tests can be constructed to test whether the variances change from cell to cell. But since we have robustness with respect to unequal variances, these tests should be applied only if serious inequalities among the variances are suspected.

In general, if there are only a few observations per cell but the sample size is fairly large, we can look at the difference x_k between the observation y_k and its estimated mean. If the model is a good fit, then the $x_1, \ldots, x_n$ should be approximately the outcomes of independent $N(0, \hat{\sigma}^2)$ variables where $\hat{\sigma}^2$ is the unbiased estimate of σ^2 for the model. If the model gives a poor description of the data, this may be clear from a visual inspection of the x_k.

Summary

The things to remember are

First: The definitions of the linear regression and analysis of variance models.

Second: How to test hypotheses in these models.

Third: How to construct simultaneous confidence intervals for a number of parameters.

Both models make the assumption that the observations $y_1, \ldots, y_n$ are the outcomes of independent $N(\mu_k, \sigma^2)$ variables with common unknown variance σ^2 and unknown means μ_k. The models differ in their assumptions on how the factor levels and means are linked together.

Regression analysis assumes that the factor levels have numerical values assigned and if $x_k^{(1)}, \ldots, x_k^{(J)}$ are the J factor levels for the kth observation, then the means are linear functions of the factor levels. That is, there are (unknown) coefficients $b_1, \ldots, b_J$ such that

$$\mu_k = b_1 x_k^{(1)} + \cdots + b_J x_k^{(J)}, \qquad k = 1, \ldots, n.$$

Analysis of variance does not assume that the levels of the various factors have numerical values assigned to them. Instead, the usual assumption is that higher order interactions between the effects of the factors are zero. For instance, with two factors, the no interaction assumption is that

$$\mu_{ij} = \mu + v_i^{(1)} + v_j^{(2)},$$

where μ_{ij} is the mean of the ith level of factor 1 and jth level of factor 2, μ is the overall mean, and $v_i^{(1)}$, $v_j^{(2)}$ are the main effects of the two factors. Equivalently, no interaction means that factor effects are additive.

The procedures for construction of estimates, tests, and confidence intervals in both models are similar:

Estimation. The possible means vectors $\boldsymbol{\mu} = (\mu_1, \ldots, \mu_n)$ range over a subspace $\mathscr{L}_L$ of dimension L. We get least squares estimates of the parameters on which $\boldsymbol{\mu}$ depends by minimizing

$$D = \|\mathbf{y} - \tilde{\mathbf{y}}\|^2$$

where $\tilde{\mathbf{y}}$ ranges over $\mathscr{L}_L$. This is curve-fitting in the sense that we are getting the best fit to the data $\mathbf{y}$ by a vector in $\mathscr{L}_L$. The least squares estimates of the parameters are their maximum likelihood estimates.

Testing. Denote the minimum of D by $\hat{D}$. Then $\hat{D}/(n - L)$ is an unbiased estimate of σ^2, and $\hat{D}/\sigma^2$ has a χ^2_{n-L} distribution. Under the hypotheses we have looked at, $\boldsymbol{\mu}$ is confined to a subspace $\mathscr{L}_{L_1}$ of $\mathscr{L}_L$ of dimension L_1. Denote

$$\hat{D}_1 = \min \|\mathbf{y} - \tilde{\mathbf{y}}_1\|^2$$

where $\tilde{\mathbf{y}}_1$ ranges over $\mathscr{L}_{L_1}$. Then we reject the hypothesis if

$$\frac{n - L}{L - L_1} \frac{\hat{D}_1 - \hat{D}}{\hat{D}} \geq f$$

The value of f is gotten by using the fact that if the hypothesis is true, the ratio has an $F_{L-L_1, n-L}$ distribution.

This idea was used to construct tests of the following hypotheses:

a a certain set of the regression coefficients are zero;

b one of the factors in an analysis of variance model has no effect on the observations;

c there is goodness of fit in a linear regression model;

d there is zero interaction in analysis of variance models.

Confidence Intervals. The S-method is used to construct simultaneous $100\gamma\%$ confidence intervals for M of the regression coefficients or for M linearly independent main effect or mean parameters and their linear combinations.

There are various words of warning interspersed throughout the chapter. Beware of linear regression situations where the various factor levels $\mathbf{x}^{(1)}, \ldots, \mathbf{x}^{(J)}$ are nearly linearly dependent. Beware of concluding on the basis of a modest sample size that your model gives a uniquely good fit. On the other hand, for equal numbers of observations per cell, both models are robust with respect to inequality of variances and departures from normality. The test levels and confidence intervals hold up fairly well, but beware of dependence between observations.

Table 1—Table of the Cumulative Distribution Function for the Normal N(0, 1) Distribution

$$P(Z < z) = \int_{-\infty}^{z} \frac{1}{\sqrt{2\pi}} e^{-t^2/2}\, dt$$

z	.00	.01	.02	.03	.04	.05	.06	.07	0.8	.09
.0	.5000	.5040	.5080	.5210	.5160	.5199	.5239	.5279	.5319	.5359
.1	.5398	.5438	.5478	.5517	.5557	.5596	.5636	.5675	.5714	.5753
.2	.5793	.5832	.5871	.5910	.5948	.5987	.6026	.6064	.6103	.6141
.3	.6179	.6217	.6255	.6293	.6331	.6368	.6406	.6443	.6480	.6517
.4	.6554	.6591	.6628	.6664	.6700	.6736	.6772	.6808	.6844	.6879
.5	.6915	.6950	.6985	.7019	.7054	.7088	.7123	.7157	.7190	.7224
.6	.7257	.7291	.7324	.7357	.7389	.7422	.7454	.7486	.7517	.7549
.7	.7580	.7611	.7642	.7673	.7704	.7734	.7764	.7794	.7823	.7852
.8	.7881	.7910	.7939	.7967	.7995	.8023	.8051	.8078	.8106	.8133
.9	.8159	.8186	.8212	.8238	.8264	.8289	.8315	.8340	.8365	.8389
1.0	.8413	.8438	.8461	.8485	.8508	.8531	.8554	.8577	.8599	.8621
1.1	.8643	.8665	.8686	.8708	.8729	.8749	.8770	.8790	.8810	.8830
1.2	.8849	.8869	.8888	.8907	.8925	.8944	.8962	.8980	.8897	.9015
1.3	.9032	.9049	.9066	.9082	.9099	.9115	.9131	.9147	.9162	.9177
1.4	.9192	.9207	.9222	.9236	.9251	.9265	.9279	.9292	.9306	.9319
1.5	.9332	.9345	.9357	.9370	.9382	.9394	.9406	.9418	.9429	.9441
1.6	.9452	.9463	.9474	.9484	.9495	.9505	.9515	.9525	.9535	.9545
1.7	.9554	.9564	.9573	.9582	.9591	.9599	.9608	.9616	.9625	.9633
1.8	.9641	.9649	.9656	.9664	.9671	.9678	.9686	.9693	.9699	.9706
1.9	.9713	.9719	.9726	.9732	.9738	.9744	.9750	.9756	.9761	.9767
2.0	.9772	.9778	.9783	.9788	.9793	.9798	.9803	.9808	.9812	.9817
2.1	.9821	.9826	.9830	.9834	.9838	.9842	.9846	.9850	.9854	.9857
2.2	.9861	.9864	.9868	.9871	.9875	.9878	.9881	.9884	.9887	.9890
2.3	.9893	.9896	.9898	.9901	.9904	.9906	.9909	.9911	.9913	.9916
2.4	.9918	.9920	.9922	.9925	.9927	.9929	.9931	.9932	.9934	.9936
2.5	.9938	.9940	.9941	.9943	.9945	.9946	.9948	.9949	.9951	.9952
2.6	.9953	.9955	.9956	.9957	.9959	.9960	.9961	.9962	.9963	.9964
2.7	.9965	.9966	.9967	.9968	.9969	.9970	.9971	.9972	.9973	.9974
2.8	.9974	.9975	.9976	.9977	.9977	.9978	.9979	.9979	.9980	.9981
2.9	.9981	.9982	.9982	.9983	.9984	.9984	.9985	.9985	.9986	.9986
3.0	.9987	.9987	.9987	.9988	.9988	.9989	.9989	.9989	.9990	.9990
3.1	.9990	.9991	.9991	.9991	.9992	.9992	.9992	.9992	.9993	.9993
3.2	.9993	.9993	.9994	.9994	.9994	.9994	.9994	.9995	.9995	.9995
3.3	.9995	.9995	.9995	.9996	.9996	.9996	.9996	.9996	.9996	.9997
3.4	.9997	.9997	.9997	.9997	.9997	.9997	.9997	.9997	.9997	.9998

Table 2—Chi-Squared Distribution

Entries c are defined in terms of P, n by equation

$$p = P(C_n < c)$$

n \ p	.005	.010	.025	.050	.100	.250	.500	.750	.900	.950	.975	.990	.995
1	.0000393	.000157	.000982	.00393	.0158	.102	.455	1.32	2.71	3.84	5.02	6.63	7.88
2	.0100	.0201	.0506	.103	.211	.575	1.39	2.77	4.61	5.99	7.38	9.21	10.6
3	.0717	.115	.216	.352	.584	1.21	2.37	4.11	6.25	7.81	9.35	11.3	12.8
4	.207	.297	.484	.711	1.06	1.92	3.36	5.39	7.78	9.49	11.1	13.3	14.9
5	.412	.554	.831	1.15	1.61	2.67	4.35	6.63	9.24	11.1	12.8	15.1	16.7
6	.676	.872	1.24	1.64	2.20	3.45	5.35	7.84	10.6	12.6	14.4	16.8	18.5
7	.989	1.24	1.69	2.17	2.83	4.25	6.35	9.04	12.0	14.1	16.0	18.5	20.3
8	1.34	1.65	2.18	2.73	3.49	5.07	7.34	10.2	13.4	15.5	17.5	20.1	22.0
9	1.73	2.09	2.70	3.33	4.17	5.90	8.34	11.4	14.7	16.9	19.0	21.7	23.6
10	2.16	2.56	3.25	3.94	4.87	6.74	9.34	12.5	16.0	18.3	20.5	23.2	25.2
11	2.60	3.05	3.82	4.57	5.58	7.58	10.3	13.7	17.3	19.7	21.9	24.7	26.8
12	3.07	3.57	4.40	5.23	6.30	8.44	11.3	14.8	18.5	21.0	23.3	26.2	28.3
13	3.57	4.11	5.01	5.89	7.04	9.30	12.3	16.0	19.8	22.4	24.7	27.7	29.8
14	4.07	4.66	5.63	6.57	7.79	10.2	13.3	17.1	21.1	23.7	26.1	29.1	31.3
15	4.60	5.23	6.26	7.26	8.55	11.0	14.3	18.2	22.3	25.0	27.5	30.6	32.8
16	5.14	5.81	6.91	7.96	9.31	11.9	15.3	19.4	23.5	26.3	28.8	32.0	34.3
17	5.70	6.41	7.56	8.67	10.1	12.8	16.3	20.5	24.8	27.6	30.2	33.4	35.7
18	6.26	7.01	8.23	9.39	10.9	13.7	17.3	21.6	26.0	28.9	31.5	34.8	37.2
19	6.84	7.63	8.91	10.1	11.7	14.6	18.3	22.7	27.2	30.1	32.9	36.2	38.6
20	7.43	8.26	9.59	10.9	12.4	15.5	19.3	23.8	28.4	31.4	34.2	37.6	40.0
21	8.03	8.90	10.3	11.6	13.2	16.3	20.3	24.9	29.6	32.7	35.5	38.9	41.4
22	8.64	9.54	11.0	12.3	14.0	17.2	21.3	26.0	30.8	33.9	36.8	40.3	42.8
23	9.26	10.2	11.7	13.1	14.8	18.1	22.3	27.1	32.0	35.2	38.1	41.6	44.2
24	9.89	10.9	12.4	13.8	15.7	19.0	23.3	28.2	33.2	36.4	39.4	43.0	45.6
25	10.5	11.5	13.1	14.6	16.5	19.9	24.3	29.3	34.4	37.7	40.6	44.3	46.9
26	11.2	12.2	13.8	15.4	17.3	20.8	25.3	30.4	35.6	38.9	41.9	45.6	48.3
27	11.8	12.9	14.6	16.2	18.1	21.7	26.3	31.5	36.7	40.1	43.2	47.0	49.6
28	12.5	13.6	15.3	16.9	18.9	22.7	27.3	32.6	37.9	41.3	44.5	48.3	51.0
29	13.1	14.3	16.0	17.7	19.8	23.6	28.3	33.7	39.1	42.6	45.7	49.6	52.3
30	13.8	15.0	16.8	18.5	20.6	24.5	29.3	34.8	40.3	43.8	47.0	50.9	53.7

SOURCE: E. S. Pearson and H. O. Hartley, *Biometrika Tables for Statisticians*, vol. 1 (London: Cambridge University Press, 1962). Reprinted by permission of the publisher.

Table 3—Student's t-Distribution

Entries t are determined from p, n by the equation

$$P(T_n < t) = p$$

n \ p	.60	.75	.90	.95	.975	.99	.995
1	.325	1.000	3.078	6.314	12.706	31.821	63.657
2	.289	.816	1.886	2.920	4.303	6.965	9.925
3	.277	.765	1.638	2.353	3.182	4.541	5.841
4	.271	.741	1.533	2.132	2.776	3.747	4.604
5	.267	.727	1.476	2.015	2.571	3.365	4.032
6	.265	.718	1.440	1.943	2.447	3.143	3.707
7	.263	.711	1.415	1.895	2.365	2.998	3.499
8	.262	.706	2.397	1.860	2.306	2.896	3.355
9	.261	.703	1.383	1.833	2.262	2.821	3.250
10	.260	.700	1.372	1.812	2.228	2.764	3.169
11	.260	.697	1.363	1.796	2.201	2.718	3.106
12	.259	.695	1.356	1.782	2.179	2.681	3.055
13	.259	.694	1.350	1.771	2.160	2.650	3.012
14	.258	.692	1.345	1.761	2.145	2.624	2.977
15	.258	.691	1.341	1.753	2.131	2.602	2.947
16	.258	.690	1.337	1.746	2.120	2.583	2.921
17	.257	.689	1.333	1.740	2.110	2.567	2.898
18	.257	.688	1.330	1.734	2.101	2.552	2.878
19	.257	.688	1.328	1.729	2.093	2.539	2.861
20	.257	.687	1.325	1.725	2.086	2.528	2.845
21	.257	.686	1.323	1.721	2.080	2.518	2.831
22	.256	.686	1.321	1.717	2.074	2.508	2.819
23	.256	.685	1.319	1.714	2.069	2.500	2.807
24	.256	.685	1.318	1.711	2.064	2.492	2.797
25	.256	.684	1.316	1.708	2.060	2.485	2.787
26	.256	.684	1.315	1.706	2.056	2.479	2.779
27	.256	.684	1.314	1.703	2.052	2.473	2.771
28	.256	.683	1.313	1.701	2.048	2.467	2.763
29	.256	.683	1.311	1.699	2.045	2.462	2.756
30	.256	.683	1.310	1.697	2.042	2.457	2.750
40	.255	.681	1.303	1.684	2.021	2.423	2.704
60	.254	.679	1.296	1.671	2.000	2.390	2.660
120	.254	.677	1.289	1.658	1.980	2.358	2.617
∞	.253	.674	1.282	1.645	1.960	2.326	2.576

SOURCE: Table 3 is taken from Table IV.1 of Fisher and Yates: *Statistical Tables for Biological, Agricultural and Medical Research*, published by Oliver and Boyd, Edinburgh, and by permission of the authors and publishers.

Table 4(a)—F-Distribution, 90% Points

Entries f are determined from the equation

$$P(F_{m,n} < f) = .90$$

n \ m	1	2	3	4	5	6	7	8	9	10	12	15	20	24	30	40	60	120	∞
1	39.86	49.50	53.59	55.83	57.24	58.20	58.91	59.44	59.86	60.19	60.71	61.22	61.74	62.00	62.26	62.53	62.79	63.06	63.33
2	8.53	9.00	9.16	9.24	9.29	9.33	9.35	9.37	9.38	9.39	9.41	9.42	9.44	9.45	9.46	9.47	9.47	9.48	9.49
3	5.54	5.46	5.39	5.34	5.31	5.28	5.27	5.25	5.24	5.23	5.22	5.20	5.18	5.18	5.17	5.16	5.15	5.14	5.13
4	4.54	4.32	4.19	4.11	4.05	4.01	3.98	3.95	3.94	3.92	3.90	3.87	3.84	3.83	3.82	3.80	3.79	3.78	3.76
5	4.06	3.78	3.62	3.52	3.45	3.40	3.37	3.34	3.32	3.30	3.27	3.24	3.21	3.19	3.17	3.16	3.14	3.12	3.10
6	3.78	3.46	3.29	3.18	3.11	3.05	3.01	2.98	2.96	2.94	2.90	2.87	2.84	2.82	2.80	2.78	2.76	2.74	2.72
7	3.59	3.26	3.07	2.96	2.88	2.83	2.78	2.75	2.72	2.70	2.67	2.63	2.59	2.58	2.56	2.54	2.51	2.49	2.47
8	3.46	3.11	2.92	2.81	2.73	2.67	2.62	2.59	2.56	2.54	2.50	2.46	2.42	2.40	2.38	2.36	2.34	2.32	2.29
9	3.36	3.01	2.81	2.69	2.61	2.55	2.51	2.47	2.44	2.42	2.38	2.34	2.30	2.28	2.25	2.23	2.21	2.18	2.16
10	3.29	2.92	2.73	2.61	2.52	2.46	2.41	2.38	2.35	2.32	2.28	2.24	2.20	2.18	2.16	2.13	2.11	2.08	2.06
11	3.23	2.86	2.66	2.54	2.45	2.39	2.34	2.30	2.27	2.25	2.21	2.17	2.12	2.10	2.08	2.05	2.03	2.00	1.97
12	3.18	2.81	2.61	2.48	2.39	2.33	2.28	2.24	2.21	2.19	2.15	2.10	2.06	2.04	2.01	1.99	1.96	1.93	1.90
13	3.14	2.76	2.56	2.43	2.35	2.28	2.23	2.20	2.16	2.14	2.10	2.05	2.01	1.98	1.96	1.93	1.90	1.88	1.85
14	3.10	2.73	2.52	2.39	2.31	2.24	2.19	2.15	2.12	2.10	2.05	2.01	1.96	1.94	1.91	1.89	1.86	1.83	1.80
15	3.07	2.70	2.49	2.36	2.27	2.21	2.16	2.12	2.09	2.06	2.02	1.97	1.92	1.90	1.87	1.85	1.82	1.79	1.76
16	3.05	2.67	2.46	2.33	2.24	2.18	2.13	2.09	2.06	2.03	1.99	1.94	1.89	1.87	1.84	1.81	1.78	1.75	1.72
17	3.03	2.64	2.44	2.31	2.22	2.15	2.10	2.06	2.03	2.00	1.96	1.91	1.86	1.84	1.81	1.78	1.75	1.72	1.69
18	3.01	2.62	2.42	2.29	2.20	2.13	2.08	2.04	2.00	1.98	1.93	1.89	1.84	1.81	1.78	1.75	1.72	1.69	1.66
19	2.99	2.61	2.40	2.27	2.18	2.11	2.06	2.02	1.98	1.96	1.91	1.86	1.81	1.79	1.76	1.73	1.70	1.67	1.63
20	2.97	2.59	2.38	2.25	2.16	2.09	2.04	2.00	1.96	1.94	1.89	1.84	1.79	1.77	1.74	1.71	1.68	1.64	1.61
21	2.96	2.57	2.36	2.23	2.14	2.08	2.02	1.98	1.95	1.92	1.87	1.83	1.78	1.75	1.72	1.69	1.66	1.62	1.59
22	2.95	2.56	2.35	2.22	2.13	2.06	2.01	1.97	1.93	1.90	1.86	1.81	1.76	1.73	1.70	1.67	1.64	1.60	1.57
23	2.94	2.55	2.34	2.21	2.11	2.05	1.99	1.95	1.92	1.89	1.84	1.80	1.74	1.72	1.69	1.66	1.62	1.59	1.55
24	2.93	2.54	2.33	2.19	2.10	2.04	1.98	1.94	1.91	1.88	1.83	1.78	1.73	1.70	1.67	1.64	1.61	1.57	1.53
25	2.92	2.53	2.32	2.18	2.09	2.02	1.97	1.83	1.89	1.87	1.82	1.77	1.72	1.69	1.66	1.63	1.59	1.56	1.52
26	2.91	2.52	2.31	2.17	2.08	2.01	1.96	1.92	1.88	1.86	1.81	1.76	1.71	1.68	1.65	1.61	1.58	1.54	1.50
27	2.90	2.51	2.30	2.17	2.07	2.00	1.95	1.91	1.87	1.85	1.80	1.75	1.70	1.67	1.64	1.60	1.57	1.53	1.49
28	2.89	2.50	2.29	2.16	2.06	2.00	1.94	1.90	1.87	1.84	1.79	1.74	1.69	1.66	1.63	1.59	1.56	1.52	1.48
29	2.89	2.50	2.28	2.15	2.06	1.99	1.93	1.89	1.86	1.83	1.78	1.73	1.68	1.65	1.62	1.58	1.55	1.51	1.47
30	2.88	2.49	2.28	2.14	2.05	1.98	1.93	1.88	1.85	1.82	1.77	1.72	1.67	1.64	1.61	1.57	1.54	1.50	1.46
40	2.84	2.44	2.23	2.09	2.00	1.93	1.87	1.83	1.79	1.76	1.71	1.66	1.61	1.57	1.54	1.51	1.47	1.42	1.38
60	2.79	2.39	2.18	2.04	1.95	1.87	1.82	1.77	1.74	1.71	1.66	1.60	1.54	1.51	1.48	1.44	1.40	1.35	1.29
120	2.75	2.35	2.13	1.99	1.90	1.82	1.77	1.72	1.68	1.65	1.60	1.55	1.48	1.45	1.41	1.37	1.32	1.26	1.19
∞	2.71	2.30	2.08	1.94	1.85	1.77	1.72	1.67	1.63	1.60	1.55	1.49	1.42	1.38	1.34	1.30	1.24	1.17	1.00

SOURCE: E. S. Pearson and H. O. Hartley *Biometrika Tables for Statisticians*, vol. 1 (London: Cambridge University Press, 1962). Reprinted by permission of the publisher.

Table 4(b)—F-Distribution, 95% Points

Entries f are determined from the equation

$$P(F_{m,n} < f) = .95$$

n \ m	1	2	3	4	5	6	7	8	9	10	12	15	20	24	30	40	60	120	∞
1	161.4	199.5	215.7	224.6	230.2	234.0	236.8	238.9	240.5	241.9	243.9	245.9	248.0	249.1	250.1	251.1	252.2	253.3	254.3
1	18.51	19.00	19.16	19.25	19.30	19.33	19.35	19.37	19.38	19.40	19.41	19.43	19.45	19.45	19.46	19.47	19.48	19.49	19.50
3	10.13	9.55	9.28	9.12	9.01	8.94	8.89	8.85	8.81	8.79	8.74	8.70	8.66	8.64	8.62	8.59	8.57	8.55	8.53
4	7.71	6.94	6.59	6.39	6.26	6.16	6.09	6.04	6.00	5.96	5.91	5.86	5.80	5.77	5.75	5.72	5.69	5.66	5.63
5	6.61	5.79	5.41	5.19	5.05	4.95	4.88	4.82	4.77	4.74	4.68	4.62	4.56	4.53	4.50	4.46	4.43	4.40	4.36
6	5.99	5.14	4.76	4.53	4.39	4.28	4.21	4.15	4.10	4.06	4.00	3.94	3.87	3.84	3.81	3.77	3.74	3.70	3.67
7	5.59	4.74	4.35	4.12	3.97	3.87	3.79	3.73	3.68	3.64	3.57	3.51	3.44	3.41	3.38	3.34	3.30	3.27	3.23
8	5.32	4.46	4.07	3.84	3.69	3.58	3.50	3.44	3.39	3.35	3.28	3.22	3.15	3.12	3.08	3.04	3.01	2.97	2.93
9	5.12	4.26	3.86	3.63	3.48	3.37	3.29	3.23	3.18	3.14	3.07	3.01	2.94	2.90	2.86	2.83	2.79	2.75	2.71
10	4.96	4.10	3.71	3.48	3.33	3.22	3.14	3.07	3.02	2.98	2.91	2.85	2.77	2.74	2.70	2.66	2.62	2.58	2.54
11	4.84	3.98	3.59	3.36	3.20	3.09	3.01	2.95	2.90	2.85	2.79	2.72	2.65	2.61	2.57	2.53	2.49	2.45	2.40
12	4.75	3.89	3.49	3.26	3.11	3.00	2.91	2.85	2.80	2.75	2.69	2.62	2.54	2.51	2.47	2.43	2.38	2.34	2.30
13	4.67	3.81	3.41	3.18	3.03	2.92	2.83	2.77	2.71	2.67	2.60	2.53	2.46	2.42	2.38	2.34	2.30	2.25	2.21
14	4.60	3.74	3.34	3.11	2.96	2.85	2.76	2.70	2.65	2.60	2.53	2.46	2.39	2.35	2.31	2.27	2.22	2.18	2.13
15	4.54	3.68	3.29	3.06	2.90	2.79	2.71	2.64	2.59	2.54	2.48	2.40	2.33	2.29	2.25	2.20	2.16	2.11	2.07
16	4.49	3.63	3.24	3.01	2.85	2.74	2.66	2.59	2.54	2.49	2.42	2.35	2.28	2.24	2.19	2.15	2.11	2.06	2.01
17	4.45	3.59	3.20	2.96	2.81	2.70	2.61	2.55	2.49	2.45	2.38	2.31	2.23	2.19	2.15	2.10	2.06	2.01	1.96
18	4.41	3.55	3.16	2.93	2.77	2.66	2.58	2.51	2.46	2.41	2.34	2.27	2.19	2.15	2.11	2.06	2.02	1.97	1.92
19	4.38	3.52	3.13	2.90	2.74	2.63	2.54	2.48	2.42	2.38	2.31	2.23	2.16	2.11	2.07	2.03	1.98	1.93	1.88
20	4.35	3.49	3.10	2.87	2.71	2.60	2.51	2.45	2.39	2.35	2.28	2.20	2.12	2.08	2.04	1.99	1.95	1.90	1.84
21	4.32	3.47	3.07	2.84	2.68	2.57	2.49	2.42	2.37	2.32	2.25	2.18	2.10	2.05	2.01	1.96	1.92	1.87	1.81
22	4.30	3.44	3.05	2.82	2.66	2.55	2.46	2.40	2.34	2.30	2.23	2.15	2.07	2.03	1.98	1.94	1.89	1.84	1.78
23	4.28	3.42	3.03	2.80	2.64	2.53	2.44	2.37	2.32	2.27	2.20	2.13	2.05	2.01	1.96	1.91	1.86	1.81	1.76
24	4.26	3.40	3.01	2.78	2.62	2.51	2.42	2.36	2.30	2.25	2.18	2.11	2.03	1.98	1.94	1.89	1.84	1.79	1.73
25	4.24	3.39	2.99	2.76	2.60	2.49	2.40	2.34	2.28	2.24	2.16	2.09	2.01	1.96	1.92	1.87	1.82	1.77	1.71
26	4.23	3.37	2.98	2.74	2.59	2.47	2.39	2.32	2.27	2.22	2.15	2.07	1.99	1.95	1.90	1.85	1.80	1.75	1.69
27	4.21	3.35	2.96	2.73	2.57	2.46	2.37	2.31	2.25	2.20	2.13	2.06	1.97	1.93	1.88	1.84	1.79	1.73	1.67
28	4.20	3.34	2.95	2.71	2.56	2.45	2.36	2.29	2.24	2.19	2.12	2.04	1.96	1.91	1.87	1.82	1.77	1.71	1.65
29	4.18	3.33	2.93	2.70	2.55	2.43	2.35	2.28	2.22	2.18	2.10	2.03	1.94	1.90	1.85	1.81	1.75	1.70	1.64
30	4.17	3.32	2.92	2.69	2.53	2.42	2.33	2.27	2.21	2.16	2.09	2.01	1.93	1.89	1.84	1.79	1.74	1.68	1.62
40	4.08	3.23	2.84	2.61	2.45	2.34	2.25	2.18	2.12	2.08	2.00	1.92	1.84	1.79	1.74	1.69	1.64	1.58	1.51
60	4.00	3.15	2.76	2.53	2.37	2.25	2.17	2.10	2.04	1.99	1.92	1.84	1.75	1.70	1.65	1.59	1.53	1.47	1.39
120	3.92	3.07	2.68	2.45	2.29	2.17	2.09	2.02	1.96	1.91	1.83	1.75	1.66	1.61	1.55	1.50	1.43	1.35	1.25
∞	3.84	3.00	2.60	2.37	2.21	2.10	2.01	1.94	1.88	1.83	1.75	1.67	1.57	1.52	1.46	1.39	1.32	1.22	1.00

SOURCE: E. S. Pearson and H. O. Hartley *Biometrika Tables for Statisticians*, vol. 1 (London: Cambridge University Press, 1962). Reprinted by permission of the publisher.

Table 4(c)—F-Distribution, 99% Points

Entries f are determined from the equation

$$P(F_{m,n} < f) = .99$$

n \ m	1	2	3	4	5	6	7	8	9	10	12	15	20	24	30	40	60	120	∞
1	4052	4999.5	5403	5625	5764	5859	5928	5982	6022	6056	6106	6157	6209	6235	6261	6287	6313	6339	6366
2	98.50	99.00	99.17	99.25	99.30	99.33	99.36	99.37	99.39	99.40	99.42	99.43	99.45	99.46	99.47	99.47	99.48	99.49	99.50
3	34.12	30.82	29.46	28.71	28.24	27.91	27.67	27.49	27.35	27.23	27.05	26.87	26.69	26.60	26.50	26.41	26.32	26.22	26.13
4	21.20	18.00	16.69	15.98	15.52	15.21	14.98	14.80	14.66	14.55	14.37	14.20	14.02	13.93	13.84	13.75	13.65	13.56	13.46
5	16.26	13.27	12.06	11.39	10.97	10.67	10.46	10.29	10.16	10.05	9.89	9.72	9.55	9.47	9.38	9.29	9.20	9.11	9.02
6	13.75	10.92	9.78	9.15	8.75	8.47	8.26	8.10	7.98	7.87	7.72	7.56	7.40	7.31	7.23	7.14	7.06	6.97	6.88
7	12.25	9.55	8.45	7.85	7.46	7.19	6.99	6.84	6.72	6.62	6.47	6.31	6.16	6.07	5.99	5.91	5.82	5.74	5.65
8	11.26	8.65	7.59	7.01	6.63	6.37	6.18	6.03	5.91	5.81	5.67	5.52	5.36	5.28	5.20	5.12	5.03	4.95	4.86
9	10.56	8.02	6.99	6.42	6.06	5.80	5.61	5.47	5.35	5.26	5.11	4.96	4.81	4.73	4.65	4.57	4.48	4.40	4.31
10	10.04	7.56	6.55	5.99	5.64	5.39	5.20	5.06	4.94	4.85	4.71	4.56	4.41	4.33	4.25	4.17	4.08	4.00	3.91
11	9.65	7.21	6.22	5.67	5.32	5.07	4.89	4.74	4.63	4.54	4.40	4.25	4.10	4.02	3.94	3.86	3.78	3.69	3.60
12	9.33	6.93	5.95	5.41	5.06	4.82	4.64	4.50	4.39	4.30	4.16	4.01	3.86	3.78	3.70	3.62	3.54	3.45	3.36
13	9.07	6.70	5.74	5.21	4.86	4.62	4.44	4.30	4.19	4.10	3.96	3.82	3.66	3.59	3.51	3.43	3.34	3.25	3.17
14	8.86	6.51	5.56	5.04	4.69	4.46	4.28	4.14	4.03	3.94	3.80	3.66	3.51	3.43	3.35	3.27	3.18	3.09	3.00
15	8.68	6.36	5.42	4.89	4.56	4.32	4.14	4.00	3.89	3.80	3.67	3.52	3.37	3.29	3.21	3.13	3.05	2.96	2.87
16	8.53	6.23	5.29	4.77	4.44	4.20	4.03	3.89	3.78	3.69	3.55	3.41	3.26	3.18	3.10	3.02	2.93	2.84	2.75
17	8.40	6.11	5.18	4.67	4.34	4.10	3.93	3.79	3.68	3.59	3.46	3.31	3.16	3.08	3.00	2.92	2.83	2.75	2.65
18	8.29	6.01	5.09	4.58	4.25	4.01	3.84	3.71	3.60	3.51	3.37	3.23	3.08	3.00	2.92	2.84	2.75	2.66	2.57
19	8.18	5.93	5.01	4.50	4.17	3.94	3.77	3.63	3.52	3.43	3.30	3.15	3.00	2.92	2.84	2.76	2.67	2.58	2.49
20	8.10	5.85	4.94	4.43	4.10	3.87	3.70	3.56	3.46	3.37	3.23	3.09	2.94	2.86	2.78	2.69	2.61	2.52	2.42
21	8.02	5.78	4.87	4.37	4.04	3.81	3.64	3.51	3.40	3.31	3.17	3.03	2.88	2.80	2.72	2.64	2.55	2.46	2.36
22	7.95	5.72	4.82	4.31	3.99	3.76	3.59	3.45	3.35	3.26	3.12	2.98	2.83	2.75	2.67	2.58	2.50	2.40	2.31
23	7.88	5.66	4.76	4.26	3.94	3.71	3.54	3.41	3.30	3.21	3.07	2.93	2.78	2.70	2.62	2.54	2.45	2.35	2.26
24	7.82	5.61	4.72	4.22	3.90	3.67	3.50	3.36	3.26	3.17	3.03	2.89	2.74	2.66	2.58	2.49	2.40	2.31	2.21
25	7.77	5.57	4.68	4.18	3.85	3.63	3.46	3.32	3.22	3.13	2.99	2.85	2.70	2.62	2.54	2.45	2.36	2.27	2.17
26	7.72	5.53	4.64	4.14	3.82	3.59	3.42	3.29	3.18	3.09	2.96	2.81	2.66	2.58	2.50	2.42	2.33	2.23	2.13
27	7.68	5.49	4.60	4.11	3.78	3.56	3.39	3.26	3.15	3.06	2.93	2.78	2.63	2.55	2.47	2.38	2.29	2.20	2.10
28	7.64	5.45	4.57	4.07	3.75	3.53	3.36	3.23	3.12	3.03	2.90	2.75	2.60	2.52	2.44	2.35	2.26	2.17	2.06
29	7.60	5.42	4.54	4.04	3.73	3.50	3.33	3.20	3.09	3.00	2.87	2.73	2.57	2.49	2.41	2.33	2.23	2.14	2.03
30	7.56	5.39	4.51	4.02	3.70	3.47	3.30	3.17	3.07	2.98	2.84	2.70	2.55	2.47	2.39	2.30	2.21	2.11	2.01
40	7.31	5.18	4.31	3.83	3.51	3.29	3.12	2.99	2.89	2.80	2.66	2.52	2.37	2.29	2.20	2.11	2.02	1.92	1.80
60	7.08	4.98	4.13	3.65	3.34	3.12	2.95	2.82	2.72	2.63	2.50	2.35	2.20	2.12	2.03	1.94	1.84	1.73	1.60
120	6.85	4.79	3.95	3.48	3.17	2.96	2.79	2.66	2.56	2.47	2.34	2.19	2.03	1.95	1.86	1.76	1.66	1.53	1.38
∞	6.63	4.61	3.78	3.32	3.02	2.80	2.64	2.51	2.41	2.32	2.18	2.04	1.88	1.79	1.70	1.59	1.47	1.32	1.00

SOURCE: E. S. Pearson and H. O. Hartley *Biometrika Tables for Statisticians*, vol. 1 (London: Cambridge University Press, 1962). Reprinted by permission of the publisher

Table 5—Cumulative Terms, Poisson Distribution

$$\text{Entries} = \sum_{n}^{\infty} \frac{\lambda^k}{k!} e^{-\lambda}$$

n	λ: 0.1	0.2	0.3	0.4	0.5	0.6	0.7	0.8	0.8	1.0
0	1.0000	1.0000	1.0000	1.0000	1.0000	1.0000	1.0000	1.0000	1.0000	1.0000
1	.0952	.1813	.2592	.3297	.3935	.4512	.5034	.5507	.5934	.6321
2	.0047	.0175	.0369	.0616	.0902	1219	1558	.1912	.2275	.2642
3	.0002	.0011	.0036	.0079	.0144	.0231	.0341	.0474	.0629	.0803
4	.0000	.0001	.0003	.0008	.0018	.0034	.0058	.0091	.0135	.0190
5	.0000	.0000	.0000	.0001	.0002	.0004	.0008	.0014	.0023	.0037
6	.0000	.0000	.0000	.0000	.0000	.0000	.0001	.0002	.0003	.0006
7	.0000	.0000	.0000	.0000	.0000	.0000	.0000	.0000	.0000	.0001

n	λ: 1.1	1.2	1.3	1.4	1.5	1.6	1.7	1.8	1.9	2.0
0	1.0000	1.0000	1.0000	1.0000	1.0000	1.0000	1.0000	1.0000	1.0000	1.0000
1	.6671	.6988	.7275	.7534	.7769	.7981	.8173	.8347	.8504	.8647
2	.3010	.3374	.3732	.4082	.4422	.4751	.5068	.5372	.5663	.5940
3	.0996	.1205	.1429	.1665	.1912	.2166	.2428	.2694	.2963	.3233
4	.0257	.0338	.0431	.0537	.0656	.0788	.0932	.1087	.1253	.1429
5	.0054	.0077	.0107	.0143	.0186	.0237	.0296	.0364	.0441	.0527
6	.0010	.0015	.0022	.0032	.0045	.0060	.0080	.0104	.0132	.0166
7	.0001	.0003	.0004	.0006	.0009	.0013	.0019	.0026	.0034	.0045
8	.0000	.0000	.0001	.0001	.0002	.0003	.0004	.0006	.0008	.0011
9	.0000	.0000	.0000	.0000	.0000	.0000	.0001	.0001	.0002	.0002

n	λ: 2.1	2.2	2.3	2.4	2.5	2.6	2.7	2.8	2.9	3.0
0	1.0000	1.0000	1.0000	1.0000	1.0000	1.0000	1.0000	1.0000	1.0000	1.0009
1	.8775	.8892	.8997	.9093	.9179	.9257	.9328	.9392	.9450	.9502
2	.6204	.6454	.6691	.6916	.7127	.7326	.7513	.7689	.7854	.8009
3	.3504	.3773	.4040	.4303	.4562	.4816	.5064	.5305	.5540	.5768
4	.1614	.1806	.2007	.2213	.2424	.2640	.2859	.3081	.3304	.3528
5	.0621	.0725	.0838	.0959	.1088	.1226	.1371	.1523	.1682	.1847
6	.0204	.0249	.0300	.0357	.0420	.0490	.0567	.0651	.0742	.0839
7	.0059	.0075	.0094	.0116	.0142	.0172	.0206	.0244	.0287	.0335
8	.0015	.0020	.0026	.0033	.0042	.0053	.0066	.0081	.0099	.0119
9	.0003	.0005	.0006	.0009	.0011	.0015	.0019	.0024	.0031	.0038
10	.0001	.0001	.0001	.0002	.0003	.0004	.0005	.0007	.0009	.0011
11	.0000	.0000	.0000	.0000	.0001	.0001	.0001	.0002	.0002	.0003
12	.0000	.0000	.0000	.0000	.0000	.0000	.0000	.0000	.0001	.0001

SOURCE: Reproduced and excerpted from *Standard Mathematical Tables*, 14th edition, ed. S. M. Selby (Cleveland, Ohio: The Chemical Rubber Company, 1965). © 1965. With permission of the publisher.

Table 6—Studentized Range, 90% Points

Entries r are determined from the equation

$$P(R < r) = .90$$

$n - J$ \ J	2	3	4	5	6	7	8	9	10	11	12	13	14	15	16	17	18	19	20
1	8.93	13.44	16.36	18.49	20 15	21.51	22.64	23.62	24.48	25.24	25.92	26.54	27.10	27.62	28.10	28.54	28.96	29.35	29.71
2	4.13	5.73	6.77	7.54	8.14	8.63	9.05	9.41	9.72	10.01	10.26	10.49	10.70	10.89	11.07	11.24	11.39	11.54	11.68
3	3.33	4.47	5.20	5.74	6.16	6.51	6.81	7.06	7.29	7.49	7.67	7.83	7.98	8.12	8.25	8.37	8.48	8.58	8.68
4	3.01	3.98	4.59	5.03	5.39	5.68	5.93	6.14	6.33	6.49	6.65	6.78	6.91	7.02	7.13	7.23	7.33	7.41	7.50
5	2.85	3.72	4.26	4.66	4.98	5.24	5.46	5.65	5.82	5.97	6.10	6.22	6.34	6.44	6.54	6.63	6.71	6.79	6.86
6	2.75	3.56	4.07	4.44	4.73	4.97	5.17	5.34	5.50	5.64	5.76	5.87	5.98	6.07	6.16	6.25	6.32	6.40	6.47
7	2.68	3.45	3.93	4.28	4.55	4.78	4.97	5.14	5.28	5.41	5.53	5.64	5.74	5.83	5.91	5.99	6.06	6.13	6.19
8	2.63	3.37	3.83	4.17	4.43	4.65	4.83	4.99	5.13	5.25	5.36	5.46	5.56	5.64	5.72	5.80	5.87	5.93	6.00
9	2.59	3.32	3.76	4.08	4.34	4.54	4.72	4.87	5.01	5.13	5.23	5.33	5.42	5.51	5.58	5.66	5.72	5.79	5.85
10	2.56	3.27	3.70	4.02	4.26	4.47	4.64	4.78	4.91	5.03	5.13	5.23	5.32	5.40	5.47	5.54	5.61	5.67	5.73
11	2.54	3.23	3.66	3.96	4.20	4.40	4.57	4.71	4.84	4.95	5.05	5.15	5.23	5.31	5.38	5.45	5.51	5.57	5.63
12	2.52	3.20	3.62	3.92	4.16	4.35	4.51	4.65	4.78	4.89	4.99	5.08	5.16	5.24	5.31	5.37	5.44	5.49	5.55
13	2.50	3.18	3.59	3.88	4.12	4.30	4.46	4.60	4.72	4.83	4.93	5.02	5.10	5.18	5.25	5.31	5.37	5.43	5.48
14	2.49	3.16	3.56	3.85	4.08	4.27	4.42	4.56	4.68	4.79	4.88	4.97	5.05	5.12	5.19	5.26	5.32	5.37	5.43
15	2.48	3.14	3.54	3.83	4.05	4.23	4.39	4.52	4.64	4.75	4.84	4.93	5.01	5.08	5.15	5.21	5.27	5.32	5.38
16	2.47	3.12	3.52	3 80	4.03	4.21	4.36	4.49	4.61	4.71	4.81	4.89	4.97	5.04	5.11	5.17	5.23	5.28	5.33
17	2.46	3.11	3.50	3.78	4.00	4.18	4.33	4.46	4.58	4.68	4.77	4.86	4.93	5.01	5.07	5.13	5.19	5.24	5.30
18	2.45	3.10	3.49	3.77	3.98	4.16	4.31	4.44	4.55	4.65	4.75	4.83	4.90	4.98	5.04	5.10	5.16	5.21	5.26
19	2.45	3.09	3.47	3.75	3.97	4.14	4.29	4.42	4.53	4.63	4.72	4.80	4.88	4.95	5.01	5.07	5.13	5.18	5.23
20	2.44	3.08	3.46	3.74	3.95	4.12	4.27	4.40	4.51	4.61	4.78	4.78	4.85	4.92	4.99	5.05	5.10	5.16	5.20
24	2.42	3.05	3.42	3.69	3.90	4.07	4.21	4.34	4.44	4.54	4.63	4.71	4.78	4.85	4.91	4.97	5.02	5.07	5.12
30	2.40	3.02	3.39	3.65	3.85	4.02	4.16	4.28	4.38	4.47	4.56	4.64	4.71	4.77	4.83	4.89	4.94	4.99	5.03
40	2.38	2.99	3.35	3.60	3.80	3.96	4.10	4.21	4.32	4.41	4.49	4.56	4.63	4.69	4.75	4.81	4.86	4.90	4.95
60	2.36	2.96	3.31	3.56	3.75	3.91	4.04	4.16	4.25	4.34	4.42	4.49	4.56	4.62	4.67	4.73	4.78	4.82	4.86
120	2.34	2.93	3.28	3.52	3.71	3.86	3.99	4.10	4.19	4.28	4.35	4.42	4.48	4.54	4.60	4.65	4.69	4.74	4.78
∞	2.33	2.90	3.24	3.48	3.66	3.81	3.93	4.04	4.13	4.21	4.28	4.35	4.41	4.47	4.52	4.57	4.61	4.65	4.69

SOURCE: H. L. Harter, "Percentage Points, Studentized Range," *The Annals of Mathematical Statistics* 31 (1960): 1122–1147. Reprinted by permission of the publisher.

Table 7—Sign Test

Entries k are the largest integers satisfying

$$P(k < S_n < n - k) \geq 1 - \alpha$$

n \ α	.01	.05	.10	.25	n \ α	.01	.05	.10	.25
1					46	13	15	16	18
2					47	14	16	17	19
3				0	48	14	16	17	19
4				0	49	15	17	18	19
5			0	0	50	15	17	18	20
6		0	0	1	51	15	18	19	20
7		0	0	1	52	16	18	19	21
8	0	0	1	1	53	16	18	20	21
9	0	1	1	2	54	17	19	20	22
10	0	1	1	2	55	17	19	20	22
11	0	1	2	3	56	17	20	21	23
12	1	2	2	3	57	18	20	21	23
13	1	2	3	3	58	18	21	22	24
14	1	2	3	4	59	19	21	22	24
15	2	3	3	4	60	19	21	23	25
16	2	3	4	5	61	20	22	23	25
17	2	4	4	5	62	20	22	24	25
18	3	4	5	6	63	20	23	24	26
19	3	4	5	6	64	21	23	24	26
20	3	5	5	6	65	21	24	25	27
21	4	5	6	7	66	22	24	25	27
22	4	5	6	7	67	22	25	26	28
23	4	6	7	8	68	22	25	26	28
24	5	6	7	8	69	23	25	27	29
25	5	7	7	9	70	23	26	27	29
26	6	7	8	9	71	24	26	28	30
27	6	7	8	10	72	24	27	28	30
28	6	8	9	10	73	25	27	28	31
29	7	8	9	10	74	25	28	29	31
30	7	9	10	11	75	25	28	29	32
31	7	9	10	11	76	26	28	30	32
32	8	9	10	12	77	26	29	30	32
33	8	10	11	12	78	27	29	31	33
34	9	10	11	13	79	27	30	31	33
35	9	11	12	13	80	28	30	32	34
36	9	11	12	14	81	28	31	32	34
37	10	12	13	14	82	28	31	33	35
38	10	12	13	14	83	29	32	33	35
39	11	12	13	15	84	29	32	33	36
40	11	13	14	15	85	30	32	34	36
41	11	13	14	16	86	30	33	34	37
42	12	14	15	16	87	31	33	35	37
43	12	14	15	17	88	31	34	35	38
44	13	15	16	17	89	31	34	36	38
45	13	15	16	18	90	32	35	36	39

SOURCE: W. J. Dixon and A. M. Mood, "Table of Critical Values for the Sign Test," *Journal of the American Statistical Association* 41 (1946): 557–566. Reprinted by permission of the publisher.

Table 8—Wilcoxon Matched Pairs Signed Ranks Test

Entries k are the largest integers satisfying

$$P(T \leq k) \leq \alpha$$

α	$n = 5$	$n = 6$	$n = 7$	$n = 8$	$n = 9$	$n = 10$
.10	1	2	4	6	8	11
.05		1	2	4	6	8
.02			0	2	3	5
.01				0	2	3
	$n = 11$	$n = 12$	$n = 13$	$n = 14$	$n = 15$	$n = 16$
.10	14	17	21	26	30	36
.05	11	14	17	21	25	30
.02	7	10	13	16	20	24
.01	5	7	10	13	16	19
	$n = 17$	$n = 18$	$n = 19$	$n = 20$	$n = 21$	$n = 22$
.10	41	47	54	60	68	75
.05	35	40	46	52	59	66
.02	28	33	38	43	49	56
.01	23	28	32	37	43	49
	$n = 23$	$n = 24$	$n = 25$	$n = 26$	$n = 27$	$n = 28$
.10	83	92	101	110	120	130
.05	73	81	90	98	107	117
.02	62	69	77	85	93	102
.01	55	61	68	76	84	92
	$n = 29$	$n = 30$	$n = 31$	$n = 32$	$n = 33$	$n = 34$
.10	141	152	163	175	188	201
.05	127	137	148	159	171	183
.02	111	120	130	141	151	162
.01	100	109	188	128	138	149
	$n = 35$	$n = 36$	$n = 37$	$n = 38$	$n = 39$	
.10	214	228	242	256	271	
.05	195	208	222	235	250	
.02	174	186	198	211	224	
.01	160	171	183	195	208	
	$n = 40$	$n = 41$	$n = 42$	$n = 43$	$n = 44$	$n = 45$
.10	287	303	319	336	353	371
.05	264	279	295	311	327	344
.02	238	252	267	281	297	313
.01	221	234	248	262	277	292
	$n = 46$	$n = 47$	$n = 48$	$n = 49$	$n = 50$	
.10	389	408	427	446	466	
.05	361	379	397	415	434	
.02	329	345	362	380	398	
.01	307	323	339	356	373	

SOURCE: F. Wilcoxon and R. A. Wilcox, *Some Rapid Approximate Statistical Procedures* (Pearl River, N.Y.: American Cyanamid Company, 1964). Reprinted by permission of the publisher.

Table 9—Spearman's Rank Correlation Coefficient

Entries ρr are determined from the equation

$$P(|\rho_s| \leq r) = 1 - \alpha$$

n	$\alpha = 0.10$	$\alpha = 0.05$	$\alpha = 0.02$	$\alpha = 0.01$
5	0.900	—	—	—
6	0.829	0.886	0.943	—
7	0.714	0.786	0.893	—
8	0.643	0.738	0.833	0.881
9	0.600	0.683	0.783	0.833
10	0.564	0.648	0.745	0.794
11	0.523	0.623	0.736	0.818
12	0.497	0.591	0.703	0.780
13	0.475	0.566	0.673	0.745
14	0.457	0.545	0.646	0.716
15	0.441	0.525	0.623	0.689
16	0.425	0.507	0.601	0.666
17	0.412	0.490	0.582	0.645
18	0.399	0.476	0.564	0.625
19	0.388	0.462	0.549	0.608
20	0.377	0.450	0.534	0.591
21	0.368	0.438	0.521	0.576
22	0.359	0.428	0.508	0.562
23	0.351	0.418	0.496	0.549
24	0.343	0.409	0.485	0.537
25	0.336	0.400	0.475	0.526
26	0.329	0.392	0.465	0.515
27	0.323	0.385	0.456	0.505
28	0.317	0.377	0.448	0.496
29	0.311	0.370	0.440	0.487
30	0.305	0.364	0.432	0.478

Note: The distribution of ρ_s is symmetric around zero under the no-trend hypothesis. Therefore, the above table can be used to compute the boundary of the one-sided tests.

SOURCE: E. G. Olds, "Critical Values of Spearman's Rank Correlation Coefficient," *The Annals of Mathematical Statistics* 9 (1938): 133–148. Reprinted by permission of the publisher.

Table 10—Random Numbers

Entries are five digit numbers drawn from an equiprobable distribution on all five digit numbers; placing a decimal point in front of each number will convert this to a sample of numbers drawn from a uniform distribution on [0, 1] (to five figure accuracy)

13284	16834	74151	92027	24670	36665	00770	22878	02179	51602	07270	76517	97275	45960
21224	00370	30420	03883	99648	89428	41583	17564	27395	63904	41548	49197	82277	24120
99052	47887	81085	64933	66279	80432	65793	83287	34142	13241	30590	97760	35848	91983
00199	50993	98603	38452	87890	94624	69721	57484	67501	77638	44331	11257	71131	11059
60578	06483	28733	37867	07936	98710	98539	27186	31237	80612	44488	97819	70401	95419
91240	18312	17741	01929	18163	69201	31211	54288	39296	37318	65724	90401	79017	62077
97458	14229	12063	59611	32249	90466	33216	19358	02591	54263	88449	01912	07436	50813
35249	38646	34475	72417	60514	69257	12489	51924	86871	92446	36607	11458	30440	52639
38980	46600	11759	11900	46741	27860	77940	39298	97838	95145	32378	68038	89351	37005
10750	52745	38749	87365	58959	53731	89295	59062	39404	13198	59960	70408	29812	83126
36247	27850	73958	20673	37800	63835	71051	84724	52492	22342	78071	17456	96104	18327
70994	66986	99744	72438	01174	42159	11392	20724	54322	36923	70009	23233	65438	59685
99638	94702	11463	18148	81386	80431	90628	52506	02016	85151	88598	47821	00265	82525
72055	15774	43857	99805	10419	76939	25993	03544	21560	83471	43989	90770	22965	44247
24038	65541	85788	55835	38835	59399	13790	35112	01324	39520	76210	22467	83275	32286
74976	14631	35908	28221	39470	91548	12854	30166	09073	75887	36782	00268	97121	57676
35553	71628	70189	26436	63407	91178	90348	55359	80392	41012	36270	77786	89578	21059
35676	12797	51434	82976	42010	26344	92920	92155	58807	54644	58581	95331	78629	73344
74815	67523	72985	23183	02446	63594	98924	20633	58842	85961	07648	70164	34994	67662
45246	88048	65173	50989	91060	89894	36063	32819	68559	99221	49475	50558	34698	71800
76509	47069	86378	41797	11910	49672	88575	97966	32466	10083	54728	81972	58975	30761
19689	90332	04315	21358	97248	11188	39062	63312	52496	07349	79178	33692	57352	72862
42751	35318	97513	61537	54955	08159	00337	80778	27507	95478	21252	12746	37554	97775
11946	22681	45045	13964	57517	54919	58045	44067	58716	58840	45557	96345	33271	53464
96518	48688	20996	11090	48396	57177	83867	86464	14342	21545	46717	72364	86954	55580
35726	58643	76869	84622	39098	36083	72505	92265	23107	60278	05822	46760	44294	07672
39737	42750	48968	70536	84864	64952	38404	94317	65402	13589	01055	79044	19308	83623
97025	66492	56177	04049	80312	48028	26408	43591	75528	65341	49044	95495	81256	53214
62814	08075	09788	56350	76787	51591	54509	49295	85830	59860	30883	89660	96142	18354
25578	22950	15227	83291	41737	79599	96191	71845	86899	70694	24290	01551	80092	82118
68763	69576	88991	49662	46704	63362	56625	00481	73323	91427	15264	06969	57048	54149
17900	00813	64361	60725	88974	61005	99709	30666	26451	11528	44323	34778	60342	60388
71944	60227	63551	71109	05624	43836	58254	26160	32116	63403	35404	57146	10909	07346
54684	93691	85132	64399	29182	44324	14491	55226	78793	34107	30374	48429	51376	09559
25946	27623	11258	65204	52832	50880	22273	05554	99521	73791	85744	29276	70326	60251
01353	39318	44961	44972	91766	90262	56073	06606	51826	18893	83448	31915	97764	75091
99083	88191	27662	99113	57174	35571	99884	13951	71057	53961	61448	74909	07322	80960
52021	45406	37945	75234	24327	86978	22644	87779	23753	99926	63898	54886	18051	96314
78755	47744	43776	83098	03225	14281	83637	55984	13300	52212	58781	14905	46502	04472
25282	69106	59180	16257	22810	43609	12224	25643	89884	31149	85423	32581	34374	70873
11959	94202	02743	86847	79725	51811	12998	76844	05320	54236	53891	70226	38632	84776
11644	13792	98190	01424	30078	28197	55583	05197	47714	68440	22016	79204	06862	94451
06307	97912	68110	59812	95448	43244	31262	88880	13040	16458	43813	89416	42482	33939
76285	75714	89585	99296	52640	46518	55486	90754	88932	19937	57119	23251	55619	23679
55322	07589	39600	60866	63007	20007	66819	84164	61131	81429	60676	42807	78286	29015
78017	90928	90220	92503	83375	26986	74399	30885	88567	29169	72816	53357	15428	86932
44768	43342	20696	26331	43140	69744	82928	24988	94237	46138	77426	39039	55599	12655
25100	19336	14605	86603	51680	97678	24261	02464	86563	74812	60069	71674	15478	47642
83612	46623	62876	85197	07824	91392	58317	37726	84628	42221	10268	20692	15699	29167
41347	81666	82961	60413	71020	83658	02415	33322	66036	98712	46795	16308	28413	05417

SOURCE: Reproduced and excerpted from *Standard Mathematical Tables*, 14th edition, ed. S. M. Selby (Cleveland, Ohio: The Chemical Rubber Company, 1965).

Table 11—Random Normal Numbers

Entries form a sample of numbers drawn from a $N(0, 1)$ distribution

0.464	0.137	2.455	−0.323	−0.068	0.296	−0.288	1.298	0.241	−0.957
0.060	−2.526	−0.531	−0.194	0.543	−1.558	0.187	−1.190	0.022	0.525
1.486	−0.354	−0.634	0.697	0.926	1.375	0.785	−0.963	−0.853	−1.865
1.022	−0.472	1.279	3.521	0.571	−1.851	0.194	1.192	−0.501	−0.273
1.394	−0.555	0.046	0.321	2.945	1.974	−0.258	0.412	0.439	−0.035
0.906	−0.513	−0.525	0.595	0.881	−0.934	1.579	0.161	−1.885	0.371
1.179	−1.055	0.007	0.769	0.971	0.712	1.090	−0.631	−0.255	−0.702
−1.501	−0.488	−0.162	−0.136	1.033	0.203	0.448	0.748	−0.423	−0.432
−0.690	0.756	−1.618	−0.345	−0.511	−2.051	−0.457	−0.218	0.857	−0.465
1.372	0.225	0.378	0.761	0.181	−0.736	0.960	−1.530	−0.260	0.120
−0.482	1.678	−0.057	−1.229	−0.486	0.856	−0.491	−1.983	−2.830	−0.238
−1.376	−0.150	1.356	−0.561	−0.256	−0.212	0.219	0.779	0.953	−0.869
−1.010	0.598	−0.918	1.598	0.065	0.415	−0.169	0.313	−0.973	−1.016
−0.005	−0.899	0.012	−0.725	1.147	−0.121	1.096	0.481	−1.691	0.417
1.393	−1.163	−0.911	1.231	−0.199	−0.246	1.239	−2.574	−0.558	0.056
−1.787	−0.261	1.237	1.046	−0.508	−1.630	−0.146	−0.392	−0.627	0.561
−0.105	−0.357	−1.384	0.360	−0.992	−0.116	−1.698	−2.832	−1.108	−2.357
−1.339	1.827	−0.959	0.424	0.969	−1.141	−1.041	0.362	−1.726	1.956
1.041	0.535	0.731	1.377	0.983	−1.330	1.620	−1.040	0.524	−0.281
0.279	−2.056	0.717	−0.873	−1.096	−1.396	1.047	0.089	−0.573	0.932
−1.805	−2.008	−1.633	0.542	0.250	−0.166	0.032	0.079	0.471	−1.029
−1.186	1.180	1.114	0.882	1.265	−0.202	0.151	−0.376	−0.310	0.479
0.658	−1.141	1.151	−1.210	−0.927	0.425	0.290	−0.902	0.610	2.709
−0.439	0.358	−1.939	0.891	−0.227	0.602	0.873	−0.437	−0.220	−0.057
−1.399	−0.230	0.385	−0.649	−0.577	0.237	−0.289	0.513	0.738	−0.300
0.199	0.208	−1.083	−0.219	−0.291	1.221	1.119	0.004	−2.015	−0.594
0.159	0.272	−0.313	0.084	−2.828	−0.439	−0.792	−1.275	−0.623	−1.047
2.273	0.606	0.606	−0.747	0.247	1.291	0.063	−1.793	−0.699	−1.347
0.041	−0.307	0.121	0.790	−0.584	0.541	0.484	−0.986	0.481	0.996
−1.132	−2.098	0.921	0.145	0.446	−1.661	1.045	−1.363	−0.586	−1.023
0.768	0.079	−1.473	0.034	−2.127	0.665	0.084	−0.880	−0.579	0.551
0.375	−1.658	−0.851	0.234	−0.656	0.340	−0.086	−0.158	−0.120	0.418
−0.513	−0.344	0.210	−0.735	1.041	0.008	0.427	−0.831	0.191	0.074
0.292	−0.521	1.266	−1.206	−0.899	0.110	−0.528	−0.813	0.071	0.524
1.026	2.990	−0.574	−0.491	−1.114	1.297	−1.433	−1.345	−3.001	0.479
−1.334	1.278	−0.568	−0.109	−0.515	−0.566	2.923	0.500	0.359	0.326
−0.287	−0.144	−0.254	0.574	−0.451	−1.181	−1.190	−0.318	−0.094	1.114
0.161	−0.886	−0.921	−0.509	1.410	−0.518	0.192	−0.432	1.501	1.068
−1.346	0.193	−1.202	0.394	−1.045	0.843	0.942	1.045	0.031	0.772
1.250	−0.199	−0.288	1.810	1.378	0.584	1.216	0.733	0.402	0.226
0.630	−0.537	0.782	0.060	0.499	−0.431	1.705	1.164	0.884	−0.298
0.375	−1.941	0.247	−0.491	−0.665	−0.135	−0.145	−0.498	0.457	1.064
−1.420	0.489	−1.711	−1.186	0.754	−0.732	−0.066	1.006	−0.798	0.162
−0.151	−0.243	−0.430	−0.762	0.298	1.049	1.810	2.885	−0.768	−0.129
−0.309	0.531	0.416	−1.541	1.456	2.040	−0.124	0.196	0.023	−1.204
0.424	−0.444	0.593	0.993	−0.106	0.116	0.484	−1.272	1.066	1.097
0.593	0.658	−1.127	−1.407	−1.579	−1.616	1.458	1.262	0.736	−0.916
0.862	−0.885	−0.142	−0.504	0.532	1.381	0.022	−0.281	−0.342	1.222
0.235	−0.628	−0.023	−0.463	−0.899	−0.394	−0.538	1.707	−0.188	−1.153
−0.853	0.402	0.777	0.833	0.410	−0.349	−1.094	0.580	1.395	1.298

Notation

The notation listed below is used in this text and is introduced and defined in the previous volume *Probability and Stochastic Processes* (Houghton Mifflin Co., 1969). The page numbers refer to the definitions in the previous volume.

Notation	*Meaning*	*Page*
X, Y, Z	random variables	44 ff
X, Y, Z	random vectors; i.e., $\mathbf{X} = (\mathsf{X}_1, \ldots, \mathsf{X}_n)$	81
x, y, z	real numbers	
x, y, z	numbers in some n-dimension space; i.e., $\mathbf{x} = (x_1, \ldots, x_n)$	
$P(dx)$, $P(dt)$	the probability distribution of some random variable	47
$P_\mathsf{X}(dx)$, $P_\mathsf{Z}(dz)$	the probability distribution of X, Z, respectively	47
$P(I)$ or $P(\mathsf{X} \in I)$	the probability that the outcome of X is in the interval I	47
$F(x) = P(\mathsf{X} < x)$	the cumulative distribution function of the random variable X	53 ff
$\int g(x)P(dx)$	the integral of $g(x)$ with respect to the distribution $P(dx)$ $= \lim_{\max_n \lVert I_n \rVert \to 0} \sum_n g(x_n)P(I_n),\ x_n \in I_n$	57 ff
$Eg(\mathsf{X})$	expectation of the random variable $g(\mathsf{X})$ $= \int g(x)P_\mathsf{X}(dx)$	57 ff

Notation	Meaning	Page
X is discrete	X only takes on the discrete values $\{x_n\}$, $\int Eg(X) = \sum_n g(x_n)P(X = x_n)$	50
X has a density $f(x)$	$P(X \in dx) = f(x)\,dx$ $Eg(X) = \int g(x)f(x)\,dx$	14 ff, 51
X has the density $f(x_1, \ldots, x_n)$	$P(X_1 \in dx_1, \ldots, X_n \in dx_n)$ $= f(x_1, \ldots, x_n)\,dx_1, \ldots, dx_n$ $Eg(\mathbf{X}) = \int g(\mathbf{x})f(\mathbf{x})\,d\mathbf{x}$	17, 77
$V(X)$, or $\sigma^2(X)$, or σ_X^2	the variance of $X = EX^2 - (EX)^2$	63
σ_X or $\sigma(X)$	standard deviation of $X = \sqrt{\sigma^2(X)}$	64
$\Gamma(X, Y)$	the covariance of X, Y $= E(XY) - (EX)(EY)$	74
$\rho(X, Y)$	the correlation of X, Y $= \Gamma(X, Y)/\sigma_X, \sigma_Y$	74
$\lVert\Gamma\rVert$	the covariance matrix of the random variables $X_1, \ldots, X_n$ $\Gamma_{ij} = \Gamma(X_i, X_j)$	218
	the means vector of the random variables $X_1, \ldots, X_n$ $\mu_i = EX_i$	218

Index

ABCDEFGHIJ–H–798765432